Quaderni di Acme

Università degli Studi di Milano
Facoltà di Lettere e Filosofia

Quaderni di Acme
86

Dipartimento di Scienze del Linguaggio
e Letterature Straniere Comparate
Sezione di Comparatistica

ANGLO-AMERICAN MODERNITY
AND THE MEDITERRANEAN

Milan, 29-30 September 2005

Edited by
Caroline Patey, Giovanni Cianci and Francesca Cuojati

Cisalpino
Istituto Editoriale Universitario
www.monduzzieditore.it/cisalpino

QUADERNI DI ACME – Comitato scientifico

Isabella Gualandri (dir.) – Livio Antonielli, Giorgio Bejor, Claudia Berra, Elisa Bianchi, Giovanni Cianci, Gianfranco Fiaccadori, Renato Pettoello

Pubblicazione realizzata con il contributo MIUR per i Progetti di Ricerca Scientifica di Rilevante Interesse Nazionale (2003-2005) conferito alle Università degli Studi di Milano, Genova e Trieste

In copertina: Nicolas de Staël, *Montagne Sainte-Victoire*, 1954 (olio su tela 89 × 130 cm).

Realizzazione editoriale: GRAFORAM – Milano
www.graforam.it

ISBN 88-323-6062-4
Copyright © 2006
CISALPINO. Istituto Editoriale Universitario – Monduzzi Editore S.p.A.
VIA B. EUSTACHI, 12 – 20129 MILANO
Tel. 02/20404031
cisalpino@monduzzieditore.it

CONTENTS

Contents   7

ACKNOWLEDGEMENTS

The essays collected in this volume were first presented as papers at the conference *Anglo-American Modernity and the Mediterranean* held at the University of Milan in September 2005. We wish to thank the colleagues of the Universities of Genoa and Trieste for their fruitful cooperation in the project. Special thanks are due to the Dipartimento di Letterature Straniere, Comparatistica e Studi Culturali for organizing and hosting an important seminar in Trieste in the month of October 2004. The editors are grateful to the staff of the Dipartimento di Scienze del Linguaggio e Letterature Straniere Comparate in Milan for their generous help with the organization of the conference; to John Young for lending his linguistic skills; to Marco Manunta and to Valentina Pontolillo D'Elia for their assistance with almost all aspects of the event, and for helping in the production of the final script. We owe a special debt of thanks to Valentina Pontolillo D'Elia for the prompt and efficient compilation of the index. The research and the publication of the proceedings collected here have had the financial support of the Ministero dell'Istruzione, dell'Università e della Ricerca (Cofinanziamento Programmi di Ricerca Scientifica di Rilevante Interesse Nazionale).

FOREWORD

Caroline Patey

Whether as a site of longing and learning or as the land of pillage, or indeed all of these at one and the same time, the Mediterranean is at the heart of European culture, having nurtured for millennia its economy, art and languages and fed, periodically, the thirst for *renovatio* that has punctuated the historical process since the Carolingian middle ages. The medieval historian Georges Duby stresses with energy the liveliness of such an early dissemination on the Northern map of what he calls 'Mediterranean aesthetics' (Duby 1987: 267-282): flourishing from Aachen to St. Gall as in Scotus's scholarship, or, not many centuries later, in the gigantic collective task of translating Arabic and Greek manuscripts, down to Aristotle and Plato. As to the sixteenth-century Renaissance, it enhances these already vital links, thanks to the continued circulation of artists, words, images and forms, in the wake of a "fully assimilated classical memory" (*Ibid.*, translation mine). In Duby's perception, the catastrophe of the fruitful exchange between Northern and Mediterranean Europe is to be ascribed to the prevailing of archaeology and its related antiquarian modes; if, in this sense, Winckelmann is the historian's *bête noire*, scholars of English literature are well acquainted with the connoisseur modes that frame, increasingly, the Grand Tour culture and the collecting mania of eighteenth-century Britain. Thus, in place of the animated transactions between the Mediterranean areas and the North, museums and taxonomies were born: to make the classical heritage visible, of course, but also to keep it as well at a safe distance; a language soon stiffened into the canonical discipline of the Canovas and the Thorvaldsens. No surprise, therefore, if so many nineteenth-century artists and intellectuals, precisely to escape the strictures of a Mediterranean turned academic, leave Northern Europe – temporarily – in search of the actual and real Mediterranean, anxious to experience living memory, living Latins and Greeks and the possible raptures of flesh and sun: the other side, in a word, of the urban, industrial and philological coin!

The essays that follow are all situated at this juncture of the tug-of-war between Northern Europe and the Mediterranean. They lie in the shadow of Nietzsche's *The Birth of the Tragedy* (1872), a text that had notoriously taken classical certainties by storm and invited even the unphilosophically minded to go beyond stereotypes and rigid dichotomies; and, in particular for the Anglo-Americans considered in this volume, they come on the morrow of John Ruskin's major works which, albeit in contradictory ways, had reactivated a vivid conversation with Mediterranean Italy: echoes ineluctable, Nietzsche's and Ruskin's, to be heard more or less softly in many of the modernist voices gathered in these pages (Jane Harrison, W.B. Yeats, Adrian Stokes). To which must be added the specific weight of British Imperialism, a major force in the shaping of both exotic images of the Mediterranean and subversive reactions to them (E.M. Forster, Wyndham Lewis, W.S. Blunt). The Mediterranean of the Modernists is in turmoil, aesthetically and politically: no longer, or not solely, a culture or a humanity to be consumed on location or at home, but a source of debate, ideas and forms; not "the Mediterranean submitted to being recreated in the image of British longings and aversions, hopes and fears" (Pemble 1987: 274), but a culture addressed in its bewildering diversity and linguistic wealth; not a promised land nor a necropolis, but a laboratory and a workshop.

Being curious and perhaps disliking trodden routes, the Modernists inscribe unvisited places on the map. Precious few people, in early-twenties Anglo-Saxondom, knew about Puglia and the Baroque marvels investigated by Sacheverell Sitwell and his brother; and though Robert Adam had visited – and drawn extensively – Split and Diocletian's palace in 1757, Dalmatia remained a virtually estranged territory at the beginning of the twentieth century: a fact that renders the artist Ivan Meštrović's British reputation and the critic Adrian Stokes's Adriatic musings most intriguing. And while Trieste is of course not new to modernist geography, the volume lifts the veil on fresh cultural and critical data concerning the part played by its poets and politics in the genesis of English-speaking Modernism. Attracted by the unknown or the less known, the Modernists have a knack for discovering it in the familiar: under Ezra Pound's formidable gaze, Roger Fry's more gentle if not less acute eyes, and Ford Madox Ford's adventurous pen, slumbering Provence reveals powerful thematic and formal potentialities, ready to be blended, as it were, in Virginia Woolf's prose (among others'). Only a few miles away from Cézanne's Provence, the French Riviera of the American expatriates Hemingway and Scott Fitzgerald holds up a mirror to the United States while offering a fragile refuge from them.

More important still is the modernist steering away from *idées reçues* and simplistic polar oppositions. In quite un-Ruskinian fashion, industrialization and nature, past and present, modernity and archaeology, are enmeshed in the fabric of Mediterranean images. It is no random attention that Fry lends to the industrial harbour of La Ciotat, as years before W.S. Blunt had kept an eye on the politics of Egypt rather than on its pyramids; and while Yeats imports Byzantine aesthetics into the politics of the Irish Free State, Ford blends in one single striking image the medieval tower in *provençale* Beaucaire and the Flatiron in New York (Ford 1995: 16). Unabashed and unafraid of dissonance, the modernist Mediterranean accommodates the ruins and the skyscrapers, Knossos and the clanking sounds of machines. In its huge and often harsh contrasts, the Mediterranean speaks to modernist dramatists and novelists the language of a liberated prose and perhaps even of a liberated temporality. To the poets, as Ezra Pound well knew, as indeed he taught to his friends and disciples, to Stokes and to Stevens, the Mediterranean offered a fresh sense of the objects, the crisp reality to be laid on the page, as clearly as the outlines of a limestone sculpture; it gave the language of Anglo-American modernity the time-old chisel of a long forgotten and most needed *lima amorosa*.

Bibliography

Duby, Georges, [1985] 1987, 'L'eredità', in Fernand Braudel (ed.), [1985] 1987, *Il Mediterraneo. Lo spazio, la storia, gli uomini, le tradizioni*, Bompiani, Milano, pp. 267-282.

Ford, Ford Madox, 1995, *The Good Soldier*, ed. by Martin Stannard, Norton, New York, London.

Pemble, John, 1987, *The Mediterranean Passion. Victorians and Edwardians in the South,* Clarendon Press, Oxford.

Part I

RECOLLECTIONS AND REVISITATIONS

W.B. YEATS, BYZANTIUM AND THE MEDITERRANEAN

J.B. Bullen

Yeats's poetry and prose is infused with the myths, the legends and the history of the Mediterranean, yet as a young man he never travelled there. His first journey was not made until 1907 when, at the age of forty-two, he visited Italy with Lady Gregory and her son and his second came in 1924 when he went to Sicily with his wife. On both occasions he visited Byzantine sites,[1] and of all the images that Yeats draws from Mediterranean sources, it is Byzantium that is best-remembered and best known. The two famous poems 'Sailing to Byzantium' (composed in 1926) and 'Byzantium' (composed in 1930) are alembics within Yeats's work, distilling his life-long preoccupations with human old age, the mutability of nature, and the eternity of art. As D.J. Gordon and I. Fletcher put it, these poems served to sum up for several generations an idea of Byzantium and its culture. "His crystallisation," they said "has become one of the meanings of Byzantium" (Gordon *et al.* 1961: 86). Yet Yeats was writing towards the end of a Romantic phase in the historiography of Byzantium. It was one that had begun with Ruskin's encomium for San Marco in Venice in mid century, and culminated in the Art & Crafts movement and the building of Westminster Cathedral at the end (see Bullen 2003: 106-185). But even while Yeats was writing, professional historians were annexing Byzantium, the discipline of Byzantine studies was coming into existence and romance was fading.

---

[1] It has often been suggested (notably by Jeffares 1984: 214-5) that the procession of martyrs in Apollinare Nuovo in Ravenna provided Yeats with the inspiration for the 'sages standing in God's holy fire'.

In this way, Yeats was tapping into long established myths about Byzantium and informing them with a powerful spirit of his own. The two poems, therefore, have dimensions that are both public and private. On the one hand they communicate ideas and emotions about Orientalism that were not uncommon in the second decade of the twentieth century, while on the other they express sentiments that were personal to him alone. In this way Byzantium was among the private myths which Yeats used, as Massimo Bacigalupo correctly claims, "to reflect upon the falling apart of the world" as he knew it (2000: 44).

Yeats's interest in Byzantium was not confined to the 1920s, however, and its presence can be felt throughout his writing life more, perhaps, as a shadow than a substance. As he conceived it, it is less a place with a specific geographic location than an idea or a congeries of feeling. It is a spot where he never set foot, but one to which he often mentally aspired. Perhaps for this reason Byzantium is often associated in his mind with movement, with travel, with voyaging or with sailing, and it is this journey that I wish to pursue here, tracing something of his trajectory towards and away from the ancient capital.[2]

One of the most important stages in that journey began on 14 November 1923 in Merrion Square, Dublin (Foster 1997: 245). Yeats and his young wife, Georgie Hyde were thinking about going to bed when the telephone rang. Yeats was 58 years old, a recently appointed senator in the recently formed government of the Irish Free State. In the somewhat chaotic condition of Irish politics, Yeats was concerned with cultural issues. He was preoccupied with the place of Ireland in European history, about the part played by Irish legend in Irish art and about the future of Irish culture. Two issues in particular dominated his thinking. The first was the status of the Irish National Theatre and the second was the health of the Irish National Gallery in Dublin.

His private life was also in a critical state. Six years previously he had married Georgie Hyde, then only twenty-five years old, and now in middle age Yeats, with high blood pressure and indifferent health, was feeling the weight of his years. He was in a reflective, introspective phase. He was engaged with writing *A Vision* in a mood, as he put it, "between spi-

---

[2] This path, or parts of it, have been previously explored by Melchiori (1960), Fletcher and Gordon (1961), McAlindon (1966-7), Levine (1983), and Jeffares (1984).

ritual excitement and sexual torture and the knowledge that they are some-how inseparable" (quoted in Jeffares 1984: 215). When the telephone sounded in Merrion Square it was to announce a significant event in his personal Byzantine voyage. It came from the editor of the *Irish Times*, ring-ing to tell Yeats that he had won the Nobel Prize for literature. He had not expected this. In the face of stiff competition from Thomas Mann and Thomas Hardy the Swedish committee had awarded him the considerable sum of £7,500 (or €300,000 at today's value) together with an invitation to Stockholm in December of that year to attend the honours ceremony. Yeats agreed to go, travelling with Georgie first to London and then to Harwich; on 6 December he boarded a Danish boat that sailed eastward to Copenhagen and then on to Stockholm. Since the weather was clement and the sea unusually calm, the couple hugely enjoyed the trip and, as Mrs Yeats told John Stallworthy, Yeats spent much time in conversation with the Danish mariners. (Stallworthy 1963: 96)

When the couple arrived in Stockholm Yeats was overwhelmed by his reception. He was treated like a god by the Nobel committee and as an equal by the Royal Family. The latter seemed to him to be a mixture of sophistication and culture, and to have helped create a Swedish society that was a modern meritocracy rather than an outworn aristocracy. "Swe-den," he wrote, "has achieved more than we have hoped for in our own country. I think most of all, perhaps, of that splendid spectacle of your Court, a family beloved and able that has gathered about it not the rank only but the intellect of its country" (Yeats 1977: 418). But the journey to Stockholm had for Yeats strong personal implications. The experience of winning such a prestigious international award created in him a sense of euphoria where he was extremely receptive to everything around him. He was moved by the dignity of the honours ceremony, by the regal pro-cessions, but above all he was amazed by his first experience of Swedish art and craft and in particular by his visit to the recently opened Stock-holm town hall. This building and all it contained, said Yeats, was "the greatest work of Swedish art, a masterwork of the Romantic movement…" (Yeats 1977, 406) The *stadshus* was certainly an impressive building as it stood proudly overlooking Lake Mälaren. It had been some ten years in the making and it is not difficult to see how it seemed to Yeats to resolve in one gesture many of the issues with which he had been grappling re-cently, back in Ireland. It was a collective enterprise, centrally funded and acting as a kind of showcase for Swedish arts and crafts using the vehicle of Swedish history and legend. A writer in the *Studio* spoke of how "during the process of building, workshops for modelling, art smith's work, coppersmith's work furniture and textiles ... were all under the supervision of the architect' (Anon. 1925: 205). The architect in question was Ragnar Östberg (1866-1945). He was the same age as Yeats and made a deep impression on the poet, who treated him with great generosity when he came to Dublin three years later in 1926.[3] For the town hall Östberg had chosen an eclectic style, but one which was dominantly neo-Byzantine, and once inside Yeats might well have felt that he had indeed come "to the Holy city of Byzantium". (Yeats 1973: 217)

Immediately behind the main entrance there is what appears to be an open Mediterranean courtyard comprising colonnades supported by Byzantine cushioned capitals, but all roofed over as protection against the Swedish climate [Fig. 1]. Inside the building cavernous rooms with large wall spaces are broken with round-arched fenestration where the architect made provision for the all-important frescoes and mosaics. It was these decorations that most impressed Yeats and particularly the collective nature of their production. The town hall, he claimed was "the organizing centre [...] for an art imaginative and amazing" (Yeats 1977: 406), a centre which combined decorative work with fine art to produce a living emblem of what he called elsewhere "unity of culture". It had, he wrote, been "decorated by many artists, working in harmony with one another and with the design of the building as a whole, and yet all in seeming perfect freedom" (Yeats 1977: 407). The Royal Family had been hugely supportive and one of its members, Prince Eugene, Duke of Nerike, a painter, worked for two years alongside the other artists. On completion of the project he was photographed (as Yeats himself pointed out: Yeats 1977: 395), not with other members of the family, but in the company of his fellow artists.

The highlight of Yeats's visit to the town hall, however, was the so-called "Golden Room", a long chamber decorated from end to end with mosaics. These were designed by a young artist, Einar Forseth (1892-1988). He had worked historically with subjects chosen from Swedish myth and legend to create what appeared to Yeats to be a secular cathe-

---

[3] When Östberg came to Dublin in 1926 Yeats entertained him lavishly. See Letter to Olivia Shakespear, Dec 7 1926. (Yeats 1954: 719-200).

Figure 1 – Ragnar Östberg, Stockholm Town Hall, 1923. Inner courtyard.

dral of benevolent nationalism. "All that multitude and unity," said Yeats, "could hardly have been possible, had not love of Stockholm and belief in its future so filled men of different minds, classes, and occupations that they almost attained the supreme miracle, the dream that has haunted all religions, and loved one another" (Yeats 1977: 407). Moving from the ancient Swedish past to the modern Swedish present, it was as if a young man had broken away from the "sensual music" of contemporary art and had managed to create a series of "monuments of unageing intellect" which reached its climax on the north wall. Here there is a grand Deisis in the Byzantine manner, with one major difference. Instead of the traditional image of God the Father, Forseth has represented a regal female [Pl. 1]. She is Stockholm herself, the Queen of the Mälaren. She wears Byzantine sandals, sits in a Byzantine throne and holds in her hands the Byzantine sceptre and crown. In her lap lies what appears to be an ancient Byzantine church, but on closer inspection turns out to be the modern Stockholm town hall [Fig. 2]. From left and right she receives homage, but not from the saints of traditional mosaic. Instead, Forseth

Figure 2 – Einar Forseth, *Queen of the Mälaren,* 1923. Mosaic, Golden Hall Stockholm. Detail.

has depicted emblems of countries from the New World in the West and the old one in the East. The West is represented by America, with its skyscrapers and Statue of Liberty, and by France with the Eiffel Tower; the East is represented by the Turkish flag and by an outline of Hagia Sophia in Constantinople. The old and the new come together in a striking form of romantic archaising, carrying the mind, as Yeats put it, "backward to Byzantium" (Yeats 1977: 406).

Some lines written for "Sailing to Byzantium" but later removed suggest that this moment in the town hall was a critical one in the composition of the Byzantine poems. They run:

> *This Danish merchant on a relic swears*
> *That he will be*
> *All that afflicts me, but this merchant swears*
> *To bear me eastward to Byzantium*
>
> *But now these pleasant dark skinned mariners*
> ~~*Carry Bear*~~
> *Carry me toward that great Byzantium* (Stallworthy 1963: 95).

It was the Danish seamen who carried Yeats to the new Byzantium of Ragnar Östberg and it was the mosaics of Einar Forseth that "carried" him back to ideas stored in his mind about the old Byzantium. But what did "Byzantium" mean to Yeats as he stood in Stockholm town hall in 1923? To answer that we have to travel backward into Yeats's own past, back to his time as a young man of twenty-two when, in 1887, he came to live in Bedford Park, London, and struck up a friendship with someone he called "the most many-sided man of our times [...] a man of genius" (Yeats 1970-75: 183). This was William Morris (1910a: 7) and in the 1880s Morris had become passionately interested in Byzantium.

For reasons probably connected with his interest in the "Eastern Question" and the threat of a Russo-Turkish war in the mid-1870s, Morris's attention had been drawn to Constantinople and Hagia Sophia. In 1878 he asked a Greek friend, Aglaia Coronio, to send him photographs of the building and told his daughter that he had been "reading a lot about the Byzantine Empire" (Morris 1984-96: 463). In the previous year he described Byzantium as "rich and fruitful" (1910a: 7), singling out Hagia Sophia as "lovely and stately" (Morris 1910b: 78) and the "most beautiful" European building, after which "the earth began to blossom with beautiful buildings." (Morris 1910e: 208). One of those buildings was, of course, San Marco (which Yeats was to see in 1907), and in November 1879 Morris corresponded with Ruskin to protest against the damage that was being done to the building through its restoration. But Morris's understanding of Byzantine art and culture was much broader than Ruskin's, and in the 1880s the city of Constantinople, and Hagia Sophia in particular, became central to Morris's thinking about the relationship between art and society. For Morris, Gothic art was the true art of the modern period, and Byzantine art was its earliest manifestation. It was, he said, 'new born' out of the decadence of Greece and Rome. (Morris 1910c: 185)

Morris's first extended account of Byzantium came in his 1882 lecture, "History of Pattern Designing". In this, he illustrated the rise and fall of modern organic art by reference to three buildings, the palace of Diocletian at Spalato, St Peter's in Rome, and Hagia Sophia. The first, he says, is encumbered by outworn classicism, the second by "pedantry and hopelessness", but in Hagia Sophia organic art "has utterly thrown aside all pedantic encumbrances" and is "the most beautiful ... of the buildings raised in Europe before the nineteenth century" (Morris 1910c: 207-

8). Hagia Sophia continued to be a measure for Morris of the greatness of Byzantine art. In 1889 when Yeats was closest to Morris he several times gave a lecture entitled "Gothic Architecture".[4] There can be no doubt that Yeats absorbed this most elaborate account of the importance of Byzantine art and culture in which Morris pointed to the "freedom" of the human spirit in breaking the deadly grip of classicism. "The first expression of this freedom", he wrote, "is called Byzantine Art, and there is nothing to object to in the name. For centuries Byzantium was the centre of it and its first great work [Hagia Sophia] remains its greatest work. The style leaps into completion in this most lovely building". (Morris 1936: 273-4)

Morris romanticized Byzantium, stressing its democratic spirit, social unity, and equality. "Who built Hagia Sophia?" he asks. The answer is: "men like you and me, handicraftsmen who left no names behind them" (Morris1910a: 6-7). The essential characteristic of Byzantine art, he said in another lecture from 1881 was that it was "the work of collective rather than individual genius" (Morris 1910: 159) and just as it united architect and craftsman, it drew together East and West in a richly synthesizing process. Byzantium constituted "a kind of knot to all the many thrums of the first days of modern Europe" (Morris 1910e: 229). It gathered together the arts of India, Mesopotamia, Syria, Persia, Asia Minor and Egypt, which it "mingled" with the older arts of Greece, and it joined, too, Eastern love of freedom, mystery and intricate design with Western respect for discipline, structure and fact. "It is the living child and fruitful mother of art, past and future'. (Morris 1910e: 208)

Morris's dithyrambic, historically panoramic account of Byzantine architecture was most persuasively expressed in "Gothic Architecture". Its structure owes much to Ruskin's famous chapter "Nature of Gothic" in his *Stones of Venice*, but Ruskin's version of Byzantium began and ended in Venice; Morris, however, taking the hint from later archaeological work, gave Byzantine architecture a far wider significance in the history of European culture. With Ruskin's brilliant aerial journey across Europe in mind, Morris creates a similar cultural map that extends to Scandinavia in one direction and the Near East in the other but with Byzantium as

---

[4] Morris delivered this lecture five times between 1889 and 1890. In 1893 it was given again to the Arts and Crafts Exhibition Society and published by the Kelmscott Press later the same year.

its centre. In the early days architecture, he wrote, "met with traditions drawn from many sources. In Syria, the borderland of so many races and customs, the East mingled with the West, and Byzantine art was born. Its characteristics are simplicity of structure and outline of mass; amazing delicacy of ornament combined with abhorrence of vagueness: it is bright and clear in colour, pure in line, hating barrenness as much as vagueness; redundant, but not florid, the very opposite of Roman architecture in spirit, though it took so many of its forms and revivified them. Nothing more beautiful than its best works has ever been produced by man, but in spite of its stately loveliness and quietude, it was the mother of fierce vigour in the days to come, for from its first days in St Sophia, Gothic architecture has still one thousand years of life before it." Byzantine art, says Morris, extends not only backwards in time, but also across Europe in space. "East and West," he says, "it overran the world wherever men built with history behind them. In the East it mingled with the traditions of the native populations, especially with Persia of the Sassanian period [...] In the West it settled itself in the parts of Italy that Justinian had conquered, notably Ravenna, and thence came to Venice. From Italy, or perhaps even from Byzantium itself, it was carried into Germany and pre-Norman England, touching even Ireland and Scandinavia". (Morris 1936: 274)

Which carries the mind back to Stockholm and back, too, to Yeats's conversations with Ragnar Östberg, who must have told him that he, too, when he visited England in 1899 had fallen under the spell of William Morris and the British Arts and Crafts movement. In the decoration of Stockholm town hall Yeats saw many of Morris's ideas fulfilled. He admired the execution of the mosaics and frescos as an expression of artistic community, the choice of themes expressive of communal history and the freedom that had been given to individual artists under the collective direction of the architect. The parallels in Yeats's mind between Sweden's egalitarianism, the guild systems of Byzantium and his aspirations for the new Irish Free State are very evident. "I think that in early Byzantium,' he wrote in A Vision, "maybe never before or since in recorded history, religious, aesthetic and practical life were one, that architect and artificers spoke to the multitude and the few alike." (1962: 279)

By the time he left Stockholm the political and social elements of Yeats's Byzantine myth were in place but he was yet to clarify the personal, private meanings that he attributed to this period. In this way the attractive austerity of Byzantine art was to become clearer to him only when he 'sailed' to Byzantium for the second time in his life. He knew that Einar Forseth had travelled to Sicily in 1921 in search of ideas for the mosaic decoration of the town hall. And he probably knew that in the figure of San Pietro in a side apse of the cathedral in Cefalù Forseth had found a model for his Queen of the Mälaren [Fig. 3]. So in November 1924 he made his own pilgrimage to Cefalù, Palermo, and Monreale in search of an emblem of the spiritual life. Before leaving, however, he prepared himself by spending some of the Nobel prize-money on books that would tell him more about Byzantine art. Among these were Edward

Figure 3 – San Pietro, c. 1180. Mosaic. Cefalù Cathedral, Sicily.

Gibbon's *Decline and Fall of the Roman Empire*, W.G. Holmes's *The Age of Justinian and Theodora* (1905-7), and a recent translation of Josef Strzygowski's *Origin of Christian Church Art* (1923). He also bought a copy of the *Encyclopaedia Britannica* that contained articles by W.R. Lethaby on this period. Lethaby's, *The Church of Sancta Sophia, Constantinople* (1894) had been an epoch-making study of a Byzantine building, and in an article on Byzantium in the *Encyclopaedia Britannica* Yeats would have read that Byzantine art involved "the elimination of non-essentials" and "an intensity of expression such as may nowhere else be found". In Byzantine art all was "solemn, epical cosmic". It was an art of "the spirit not of the body", "hieratic, impersonal (Lethaby 1910-11: 908). He also bought O.M. Dalton's *Byzantine Art and Archaeology* (1911) where Dalton stressed the incorporeality of Byzantine art, "elect and spiritual" (Dalton 1911: 33); it is an art, he said, that "frankly renounces nature" (Dalton 1911: 35). This struck a sympathetic chord in Yeats. "Consumed with desire", as he wrote in "Sailing to Byzantium", he was, alas, "an aged man", "a paltry thing", and "a tattered coat upon a stick": thus the necessity to distance himself from "the young in one another's arms" and to seek solace with the "singing masters" of his soul in Byzantine art. (Yeats 1973: 217)

Back in 1907 in Ravenna and as a much younger man he had been struck by the hieratic nature of Byzantine art. In the following year Yeats put words into the mouth of Owen Aherne in "The Tables of the Law", saying how "the Byzantine style [...] moves me because these tall emaciated angels and saints seem to have less relation to the world about us than to an abstract pattern of flowing lines, that suggest an imagination absorbed in the contemplation of Eternity" (Yeats 1992: 150 n.). Now, seventeen years later, old age had given a spur to Yeats's desire to find an existential alternative to youthfulness and organic plenitude; to identify a place that endorsed and valorized his maturity. So late in 1924, Yeats and his wife set out for Sicily, where his experience of Byzantine art did not disappoint him. He began a collection of sixty-two black-and-white photographs[5] of what in *A Vision* he called the "supernatural splendour" of the mosaics and where he made his famous pronouncement that if he could "be given a month of Antiquity and leave to spend it where [he] could [he] would spend it in Byzantium, a little before Justinian opened

---

[5] Russell E. Murphy documents these photographs in Murphy (2004: 114-129).

St Sophia and closed the Academy of Plato" (Yeats 1962: 279). Like Morris before him, Yeats never saw Hagia Sophia, but he was intensely moved by the Cappella Palatina in Palermo and an art form where the spirit took precedence over flesh. 'Nobody,' he said, 'can stray into that little Byzantine chapel at Palermo, which suggested the chapel of the Grail to Wagner, without for an instant renouncing the body and all its works' (Yeats 1975: 478). In Palermo the personal and the historical came together. Stimulated by his journey to Stockholm, where the Arts and Crafts ideals of the Byzantine revival provided him with a social and political model for the "Unity of Culture", on the shores of the Mediterranean he rediscovered an art which in its hieratic detachment offered an image of the body transcended by the spirit. The unity of image in the Byzantine world offered Yeats a model for the new world of the Irish Free State, whose own history was closely linked to that of Byzantium. As early as 1881 Morris had written that Byzantine art was felt "from Bokhara to Galway, from Iceland to Madras" and that "all the world glittered with its brightness and quivered with its vigour" (Morris 1881: 158). Fifty years later, in 1931, after completing both his Byzantine poems Yeats said: "I have been writing about the state of my soul [...] When Irishmen were illuminating the Book of Kells and making jewelled crosiers [...] Byzantium was the centre of European civilization [...] so I symbolize the search for the spiritual life by a journey to that city."[6]

Bibliography

Anon., 1925, "The Stockholm City Hall", *The Studio*, 90, pp. 205-207.

Bacigalupo, Massimo, 2000, "Yeats and the 'Quarrel over Ruskin'," in Toni Cerutti (ed.), *Ruskin and the Twentieth Century: the Modernity of Ruskinism*, Mercurio, Vercelli, pp. 37-46.

Bullen, J.B., 2003, *Byzantium Rediscovered*, Phaidon, London.

Dalton, Ormonde Maddock, 1911, *Byzantine Art and Archaeology*, Clarendon Press, Oxford.

Fletcher, Ian - Donald Gordon, 1961, *Images of a Poet,* Manchester University Press, Manchester.

[6] Yeats's words come from a BBC broadcast in September 1931 and are quoted in Jeffares (1984: 213).

Foster, Robert Fitzroy, 1997, *W.B. Yeats: a Life*, Oxford University Press, Oxford.

Jeffares, Norman A., 1984, *A New Commentary on the Poems of W. B. Yeats,* Macmillan, London and Basingstoke.

Lethaby, William Richard, 1910-11, "Byzantine Art," in *Encyclopaedia Britannica*, 11th ed., 29 vols, vol. 4, Cambridge University Press, Cambridge.

Levine, Herbert J., 1983, "Yeats's Ruskinian Byzantium", *Yeats Annual 2,* pp. 25-34.

McAlindon, Thomas, 1966-7, "The Idea of Byzantium in William Morris and W. B. Yeats", *Modern Philology*, 64, pp. 307-319.

Melchiori, Giorgio, 1960, "The Dome of Many-Coloured Glass", in *The Whole Mystery of Art*, Greenwood Press, Westport, Connecticut, pp. 200-233.

Morris, William, [1877] 1910a, "The Lesser Arts", *The Collected Works of William Morris*, ed. by May Morris, Longmans Green & Co., London, 24 vols, Vol. 22, pp, 3-27.

——, [1880] 1910b, "The Beauty of Life", *The Collected Works of William Morris*, ed. by May Morris, Longmans Green & Co., London, 24 vols, Vol. 22, pp. 51-80.

——, [1881] 1910c, "Some Hints on Pattern Designing", *The Collected Works of William Morris*, ed. by May Morris, Longmans Green & Co., London, 24 vols, Vol. 22, pp. 175-205.

——, [1881] 1910d, "Art and Beauty of the Earth", *The Collected Works of William Morris*, ed. by May Morris, Longmans Green & Co., London, 24 vols, Vol. 22, pp. 155-174.

——, [1882] 1910e, "The History of Pattern Designing", *The Collected Works of William Morris*, ed. by May Morris, Longmans Green & Co., London, 24 vols, Vol. 22, pp. 206-234.

——, [1889] 1936, "Gothic Architecture", in *William Morris: Artist, Writer, Socialist*, ed. by May Morris, Basil Blackwell, Oxford, 2 vols, Vol. 1, pp. 266-286.

——, 1984-1996, *The Collected Letters of William Morris*, ed. by Norman Kelvin, 4 vols, Princeton University Press, Princeton.

Murphy, Russell Elliott, 2004, *The Meaning of Byzantium in the Poetry and Prose of W.B. Yeats*, Lampeter University Press, Lampeter.

Stallworthy, John, 1963, *Between the Lines: Yeats's Poetry in the Making*, Clarendon Press, Oxford.

Yeats, William Butler, 1954, *The Letters of W. B. Yeats*, ed. by Allan Wade, Rupert Hart-Davis, London.

——, 1962, *A Vision*, 2nd ed., Macmillan, London.

——, 1970-1975, "An Exhibition of William Morris", in *Uncollected Prose by W.B. Yeats,* ed. by John P. Frayne, Macmillan, London, 2 vols, Vol. 1, pp. 182-186.

——, 1970-1975, "The Censorship and St. Thomas Aquinas," in *Uncollected Prose by W.B. Yeats,* ed. by John P. Frayne, 2 vols, Vol. 2, Macmillan, London

——, 1973, *The Collected Poems of W.B. Yeats*, Macmillan, London.

——, 1977, "The Bounty of Sweden" and "The Irish Dramatic Movement" in *Autobiographies*, Macmillan, London.

——, 1992, *The Secret Rose: Stories: A Variorum Edition*, ed. by Warwick Gould *et al.*, 2nd edition, Macmillan, London.

"FORTH ON THE GODLY SEA":
THE MEDITERRANEAN IN POUND, YEATS AND STEVENS

Massimo Bacigalupo

> *They that go down to the sea in ships, that do business in great waters;*
> *These see the works of the LORD, and his wonders in the deep.*
>
> Psalms, 107:23-30, KJV

How does the Modernist imagination respond to the Mediterranean? British and American writers in the first decades of the twentieth century presented diverse views of the Mediterranean heritage, often in contrast to the culture and religion of their home countries. D.H. Lawrence sought in his "Etruscan places" and in Sardinia an older and primeval world. Ezra Pound lived permanently between Liguria and Venice, soaking up the sea air and stripping away layers of habit to return to a timeless world of natural cycles. He portrayed himself as a Ulysses travelling over and beyond the Mediterranean and expressed in his poetry the sensual delight to be had in the South. This is reminiscent of Keats's wish of a hundred years earlier, as voiced in "Ode to a Nightingale":

> O for a beaker full of the warm South,
>   Full of the true, the blushful Hippocrene...

The major Modernists were heirs of the great Romantics not only in their revolt against convention and in the power of their imagination but also in their confrontation with other cultures and other times. Like Byron and Shelley before them, Lawrence, Pound and Joyce made Italy their second home, and at least in the case of Pound this led to conclusions as dramatic as any of Byron's and Shelley's adventures. Indeed the U.S. Army prison camp in which Pound was detained in the summer of 1945 is not far from the Viareggio beach where Shelley's body was cremated in August 1822.

Byron's legendary swims in the Gulf of Spezia are celebrated in a draft, circa 1930, of what was to become Pound's renowned "Seven Lakes" canto 49 (2002: 92). When W.B. Yeats visited his friend Pound in Rapallo in 1928 and decided to move there permanently, he wrote to Lady Gregory:

> This is an indescribably lovely place – some little Greek town one imagines – there is a passage in Keats describing just such a town. Here I shall put off the bitterness of my Irish quarrels, and write my most amiable verses. (1955: 738)

The reference to Keats found its place in the stately opening of *A Vision* (1937):

> Mountains that shelter the bay from all but the south wind, bare brown branches of low vines and of tall trees blurring their outline as though with a soft mist; houses mirrored in an almost motionless sea; a verandahed gable a couple of miles away bringing to mind some Chinese painting. Rapallo's thin line of broken mother of pearl along the water's edge. The little town described in the *Ode on a Grecian Urn*. (1962: 3)

Yeats's metaphoric reference to Keats has been mistaken by a number of readers to mean that Rapallo *is* the "little town by river or seashore" of "Ode on a Grecian Urn", though of course Keats only came to Italy after the Ode was composed, and then to Rome. Harold W. Thompson, writing in the *Hamilton Alumni Review* for March 1936, said: "[Pound] has certainly moved from London to Paris to Keats' Rapallo – now his and the great Irish poet's" (84). Though here "Keats" could be a misprint for "Yeats", Pound himself was intrigued by the reference and wrote another Hamilton professor in April 1936:

> Has Thomp/ any grounds for his "Keats' Rapallo".
> Shelley was at Spezia; Sh/ and Byr/ at Pisa/ all of 'em must have gone down by Via Aurelia and therefore passed here.
> Nietzsche was here, tho' I didn't know it till I had been here a long time
> [...]

The magic casements are here/ will try to get post card of some of 'em/
though I have always suspected a different and less perfect edifice suggested 'em/

and that is prob/ wrong as I suppose
Keats wrote the line before he ever got out of 'Ighgate. (McWhirter
2001: 118)

Pound seems interested in proving that the Romantics have all been
through his beloved Rapallo, since he believes they must have travelled
south or north along the coastal road. This may be the same thought in
the back of some lines in Canto 76:

Sigismundo by the Aurelia to Genova
    by la vecchia sotto S. Pantaleone (1995: 472)

San Pantaleo is a little church near the house where Pound lived in
1943-45. The Via Aurelia goes though a tunnel beneath the church, but
the older road still goes around the hill to the west of the tunnel. So Pound
is saying that if his favourite Renaissance hero Sigismondo travelled to
Genoa, he must have passed along the Vecchia Aurelia rather than through
the modern tunnel!

All of this is unhistorical. Travellers in Byron's day, not to mention
Sigismondo's, rarely attempted the dangerous coastal road, but travelled
by boat from Pisa to Genoa (which is precisely what Byron did, by way
of Lerici and Sestri Levante, in September 1822). With Nietzsche Pound
is on safer ground, and he reports some local gossip about the German
visitor (as docs Yeats in a letter of April 1930). He goes on to note fancifully that he has found in Rapallo the "magic casements opening on the
foam" that were "charmed" by Keats's nightingale, and this shows Pound's
local passion for detail, and the emotional intensity with which he charges
his chosen landscapes. He adds reasonably that so far as he knows Keats
wrote the Ode before coming to Italy. But he remains interested, as the
letter shows, in establishing precedent, and perhaps implicitly in seeing
himself as a follower in the footsteps of Byron and Shelley.

Of course Pound had more remote fore-runners as well. Canto 1 opens:

And then went down to the ship,
Set keel to breakers, forth on the godly sea, and
We set up mast and sail on that swart ship... (Pound 1995: 3)

The alliteration is Anglo-Saxon. (Pound had scored an early success in a free rendering of "The Seafarer".) But the "godly sea" is mostly Homer's Mediterranean. In the early cantos there are many seascapes, joining literary sources and personal memory. For example, the following from Canto 2 is partly derived from the meeting of Odysseus and Tyro in *Odyssey* 11:

> And   by   the   beach-run,   Tyro,
>   Twisted   arms   of   the   sea-god,
> Lithe   sinews   of   water,   gripping   her,   cross-hold,
> And   the   blue-gray   glass   of   the   wave   tents   them,
> Glare   azure   of   water,
>   cord-welter,   close   cover.
> Quiet   sun-tawny   sand-stretch,
> The   gulls   broad   out   their   wings,
>   nipping   between   the   splay   feathers;
> Snipe   come   for   their   bath,
>   bend   out   their   wing-joints,
> Spread   wet   wings   to   the   sun-film (Pound 1995: 6-7)

The extra spacing between words is Pound's, in the first printing of the poem (as "The Eighth Canto") in *The Dial* for May 1922 – the year of *The Waste Land*. By distancing the word-clusters, many of them hyphenated, Pound is trying to communicate in blocks (what he was later to call "ideograms"), while at the same time evoking (as in canto 1) Anglo-Saxon compounds or kennings. In Homer, Tyro's marine seduction by Poseidon is described as follows:

> ...she was in love with a river, godlike Enipeus [...]
> and she used to haunt Enipeus' beautiful waters;
> taking his likeness, the god who circles the earth and shakes it
> lay with her where the swirling river finds its outlet,
> and a sea-blue wave curved into a hill of water reared up
> about the two, to hide the god and the mortal woman;
> and he broke her virgin zone and drifted a sleep upon her...
>   (Lattimore 1967: 174)

For this Pound presents a series of discrete images, as if he were half-seeing the lovers embracing in the surf, and in the first sentence the only finite verb describes the wave covering the two. Here it is worth noting how two of Homer's lines (in Lattimore's translation) are compressed into the one line: "And the blue-gray glass of the wave tents them". When "The Eighth Canto" was reprinted as Canto 2 in *Draft of XVI Cantos* (1925), lines 5-6 were run together and a curious change (misprint?) occurred (preserved in all subsequent reprints):

Glare azure of water, cold-welter, close cover. (1995: 6)

"Cold-welter" is obviously meaningless, whereas "cord-welter" is part of the images of bodies and sinews entwining like cords. But since the whole passage is rather hypnotic, perhaps Pound did not notice or did not care to correct the misprint. By normalizing the spacing an effect is lost: "cord-welter" isolated as it originally appears is something to consider, an image, whereas run together it passes unnoticed.

Pound wrote "The Eighth Canto" in Paris shortly after revising *The Waste Land*, and sent it to Ford Madox Ford for comment. Ford questioned an earlier line, "And the wave runs in the beach-groove", as well as "cord-welter" and other compounds:

Do <u>waves</u>, whirling or billowy things run in the valleys between hillocks of beach – <u>beach-grooves</u>? I don't think they do: they are converted into surf or foam as soon as they strike the pebbles & become wash or undertow in receding & then run.
<u>Per se</u> that does not matter – but the weakening of the attention does. It is the same with your compound words like "spray-whited" & "cord-welter". – But as to these I am not so certain: my dislike for them may be my merely personal distaste for Anglo-Saxon locutions which always affect me with nausea & yr. purpose in using them may be the purely aesthetic one of roughening up yr. surface [...]
<u>Pulling</u> seas would not matter if you did not have oarsmen in the next line [...]
Of course, I think that, in essence, you're a mediaeval gargoyle, Idaho or no! And it's not a bad thing to be. (Lindberg-Seyersted 1982: 64)

The "pulling sea" comment refers to the half-line "seas pulling to eastward", in a passage (rpt. in Pound 2005: 179) that was cancelled in the final version. In his response, Pound defended the compounds on account of rhythm ("MUST bang up the big-bazoo a bit, I mean rhythm must strengthen here if [the reader] is to be kept going"). As for Ford's comment on "pulling seas", Pound began referring to a "receding wave" as if this were a justification:

Surely one speaks of "receding wave". It may be a technical looseness of
phrase, but it is certainly "english". "Wash" is impos. Homophone with
laundry [...]
Re/ pulling. Surely, you must have been at sea in storm and know how
the bloody wave pulls the whole boat. Boat makes a heave at wave, cuts
in a bit, then gets dragged off course.
Gorm, I've spewed eleven times onto the broad gray buttocks of the
swankin Atlantic.
It is anything but a "run", it's a pull. (Lindberg-Seyersted 1982: 66)

Of course it was Ford, not Pound, who had used the good word "receding". In any case, the "pulling sea" passage was discarded when the
revised "Eighth Canto" took on its final collocation as Canto 2, so that
The Cantos begin with *two* Mediterranean voyages: Odysseus' in Canto 1
and Dionysus' in Canto 2. But Ford's good word, "receding", remained
in Pound's memory and was to perform its function of *mot juste* in the
Pisan Cantos. Here is another scene of lovemaking, but it is a "connubium terrae", man (the poet-priest) embracing the earth:

man, earth : two halves of the tally
but I will come out of this knowing no one
neither they me
            connubium terrae   ἔφατα πόσις ἐμὸς
                                        ΧΘΟΝΙΟΣ, mysterium

fluid ΧΘΟΝΙΟΣ o'erflowed me
    lay in the fluid ΧΘΟΝΙΟΣ;
                        that lie
under the air's solidity
    drunk with 'ΙΧΩΡ of ΧΘΟΝΙΟΣ
        fluid ΧΘΟΝΙΟΣ, strong as the undertow
    of the wave receding
but that a man should live in that further terror, and live
    the loneliness of death came upon me
            (at 3 P.M. for an instant) (1995: 546-47)

Here Pound in fact celebrated his Mediterranean "mysteries", embracing the earth (ΧΘΟΝΙΟΣ), which is also the sea, hence the finally compelling image of the undertow and the "wave receding". The lines are
staggered so as to suggest waves, and the indeterminacy is used to effect.
It is not explained what the "further terror" is, though the next line may

provide an indication. It is all very well to see death as a holy embrace of the body by the earth. However, there is a rather different way of considering it. Pound's mystique is even open to everyday realities, and the text performs brilliantly in recording states of mind, and mixing the "big-bazoo" with the banal (also in diction): "3 P.M.".

There is a historical dimension to Pound's Mediterranean landscapes. In the scene quoted above there is a reference to the Renaissance:

By Ferrara was buried naked, fu Nicolò
      e di qua di là del Po
wind: ἐμὸν τὸν ἄνδρα
lie into earth to the breast bone, to the left shoulder
      Kipling suspected it
to the height of ten inches or over (*Ibid.*: 546)

This is a characteristic example of Poundian association: Nicolò d'Este is the prince who fathers many children "di qua e di là del Po", the Greeks have poems about invoking one's husband, and Kipling was not averse to such knowledge of the earth as invoked here. Except for the latter, the references are all Mediterranean. And the basic myth is that of the potent king who brings back fertility to the Waste Land.

An Edwardian disciple of Browning and James, Pound started out on a quest for psychological subtleties among the characters of history and literature, but finally espoused a mythical perspective (or "method", to use Eliot's term) and reduced the world to impersonal forces of good and evil: art, sexuality, usury. The "Mediterranean sanity" (as he called it in 1928 – Pound 1954: 154) was a touchstone for all that was life-giving in his universe.

In an essay of 1939, "European Paideuma", inspired by the anthropologist Leo Frobenius, he attempts a rudimentary geographical division between northern and southern Mediterranean:

Chinese ethics from the 40th parallel; Athens 38th; Rome 42nd. [...] How far South of the 38th parallel can we go without great alertness and unsleeping suspicion of every belief, every idea, every ceremonial gesture or every form-characteristic. (Bacigalupo 2001: 228)

This paper was written for a German audience in August 1939, so there may be a bit of opportunism in its bow to the "nordic". (Pound in his Italian years became more and more alienated from Paris and London). In any case, this would seem to underline that it is the European Mediterranean that is his model, while he is suspicious of the Southern Mediterranean to which we owe the Bible. Pound's system is even more alarming than Yeats's. He thought in mythical terms, and was given to gross generalizations but also contradictions. So the African cultures described by Frobenius, whatever their "parallel", provide the *Cantos* with some of their paradisal moments. Also, China spreads over thirty parallels from North to South, as against the Mediterranean's scarce fifteen. In the later cantos Pound's Ulysses comes to rest in an Ithaca in remote Yünnan, a Tibetan culture where everything is simple, beautiful and in accord with nature. Like early or ideal America.

The "wave receding" of Canto 83 has another precedent, besides Pound's shop-talk with Ford. During his first extended stay in Rapallo, on 2 March 1929, Yeats sent Olivia Shakespear (Pound's mother-in-law, by the way) the following quatrain:

> Though the great song return no more
> There's keen delight in what we have –
> A rattle of pebbles on the shore
> Under the receding wave. (1955: 759)

He later collected this, under the title "The Nineteenth Century and After", in *Words for Music* (1932), then in *The Winding Stair* (1933), a collection which has at least one reference to Pound's cantos in the notes. So Pound in Pisa may have remembered his old master's comment on the passing of the "great song". After all, the early version of "The Eighth Canto" which he had discussed with Ford had begun in a similar vein:

> The weeping Muse
>   Mourns Homer,
> Mourns the day of long song,
> Mourns for the breath of the singers,
> Winds stretching out, seas pulling to eastward (Pound 2005: 179)

These are in fact the "pulling" seas that became "receding" in the Ford correspondence. Yeats's quatrain, though suggested by Browning and Morris, may well sum up the difference between, say, "Sailing to Byzantium" and *The Cantos*, which are very much about the rattle of the pebbles. Yeats was impressed enough with the very Mediterranean (Ligurian-Venetian) Canto 17 to recommend it to an acquaintance and to include it in the *Oxford Anthology of Modern Verse*. During his stay in Rapallo he went on to draft (early 1930) his second Byzantine poem, "Byzantium", which does remind us of the impressionism and mytho-history of *The Cantos*. In *A Vision* Yeats compared his friend's attempt at reawakening the "long song" mourned by Homer to the unfinished painting in Balzac's *Chef-d'oeuvre inconnu*,

> where everything rounds or thrusts itself without edges, without con-tours – conventions of the intellect – from a splash of tints and shades [...] a work as characteristic of the art of our times as the paintings of Cézanne [...] as *Ulysses* and its dream association of words and images, a poem in which there is nothing that can be taken out and reasoned over, nothing that is not a part of the poem itself. (Yeats 1962: 4)

This seems a better definition of "Byzantium" than of *The Cantos*:

The unpurged images of day recede;
The Emperor's drunken soldiery are abed;
Night resonance recedes, night walkers' song,
After great cathedral gong;
A starlit or a moonlit dome disdains
All that man is,
All mere complexities,
The fury and the mire of human veins. (Yeats 1965: 280)

The ostensible content points the way to a purification, as we leave this world for the superhuman "dome", but the poem is full of the car-nality of light and shade: drunkenness, lechery (the "night-walkers"), blood. Pound's mystical "connubium terrae" is more abstract, more like a litany, than Yeats's descent into the heart of darkness. Likewise, the em-brace of Poseidon and Tyro in canto 2 is certainly less shocking than that other Mediterranean embrace evoked in "Leda and the Swan", that terri-ble sonnet. Still, "Byzantium" resonates with repetition as with the gong that opens the poem. Written high above the Mediterranean, in Portofino Vetta where Yeats was recovering from Maltese fever that had brought him to death's door, it is a vision of another purgatorial world, a "dream association of words and images" (Yeats's very Yeatsian reading of *Ulysses*). Indeed, the ending speaks of

> Those images that yet
> Fresh images beget,
> That dolphin-torn, that gong-tormented sea. (*Ibid.*: 281)

The Mediterranean is there, and seems rather choppy: a churning of images like the ones that emerge from it in "The Eighth Canto". (There are also the compound words: a coincidence, probably.)

We know that to Yeats Byzantium was an image of human perfection in art and thought: "religious, aesthetic and practical life were one [...] architect and artificers [...] spoke to the multitude and the few alike" (Yeats 1962: 279). The same utopia, which seems to reflect John Ruskin's ideas, we find again in the mid 1930s in Pound's usura Cantos:

> With usury has no man a good house
> made of stone, no paradise on his church wall
> With usury the stone cutter is kept from his stone
> the weaver is kept from his loom by usura (Pound 1995: 250)

Yeats's strenuous ideal becomes in Pound the more down-to-earth "Mediterranean sanity", and its loss is not due to the procession of the gyres but to actual economic malpractice. It is interesting that Pound describes his paradise as lost, in the negative, so that instead of peace of mind it suggests fury and war. In "Byzantium" Yeats tells us that the two, the ideal and the drunken violence, are always present together. Thus Yeats was more the spectator (see "Meditations in Time of Civil War"), Pound more the active agent. Though this led to the latter's aberrations, it also brought about his most notable poetry. The Mediterranean dream, if I may call it that, is celebrated vividly in the Pisan Cantos, which (unlike earlier Cantos) justify themselves as poetry:

> as by Terracina rose from the sea Zephyr behind her
>   and from her manner of walking
>   as had Anchises
> till the shrine be again white with marble
> till the stone eyes look again seaward (Pound 1995: 455)

Even if this be not "the great song" but "the rattle of the pebbles", it is persuasive music which has behind it the force of a life's convictions, as Yeats's work does more evenly.

In "Sailing to Byzantium", the first great Byzantine poem, Yeats had travelled from Ireland and the mire of human life to an ancient Mediterranean holy world of meditation and song, from which to consider the present with detachment. At the end he had identified with an artificial bird "of hammered gold and gold enamelling", a bird that would sing from "a golden bough" of "what is past, or passing, or to come". All history can be comprehended, though Yeats regrets renouncing the world of flesh and blood so movingly portrayed in the first stanza. Wallace Stevens, who was Yeats's junior by only fourteen years, responded to the Irishman's four great stanzas with a poem called "Farewell to Florida", which he placed at the beginning of *Ideas of Order* (1936) just as Yeats had placed "Sailing to Byzantium" at the beginning of *The Tower* (1928). Stevens never visited Europe or the Mediterranean: in most of his poems he used Florida (see Pinkerton 1971) as his locus amoenus of sun, sea and the natural life to set against the strenuous money-making and cold North (he lived in the rather unromantic and dour Hartford, Connecticut). "Farewell to Florida" travels in the opposite direction of "Sailing to Byzantium", from Florida to New England:

> My North is leafless and lies in a wintry slime
> Both of men and clouds, a slime of men in crowds... (Stevens 1954: 118)

The poem is ambiguous. The speaker appears to take pride in his choice of sailing North, as if it were in any case necessary, but everything he says about Florida shows that his heart is divided to say the least:

> I hated the weathery yawl from which the pools
> Disclosed the sea floor and the wilderness
> Of waving weeds. I hated the vivid blooms
> Curled over the shadowless hut, the rust and bones,
> The trees like bones and the leaves half sand, half sun. (*Ibid.*)

Did he really hate the blooms over his hut? We would do well to doubt it. Perhaps the poem is a parody of the tourist returning to his wintry clime and making the best of it. In some ways more abstract and hard to pin down than "Sailing to Byzantium", on the other hand it may be perfectly understandable as the sardonic reflections of the insurance executive returning from Key West to Hartford. Stevens turns "Sailing to Byzantium" on its head but the two poems come to the same; in fact

Stevens is much more partial to his Mediterranean (Florida) than Yeats to Byzantium. Besides, Florida is not a never-never land but a perfectly realistic tourist destination. So the American poet manages to be down-to-earth and at least as rhetorical as his great Irish contemporary in expressing his ambivalence. After all, he does need the North to measure his attraction to the South. As in the usura Canto, but less venomously, he appreciates the Florida blooms all the more even as he renounces them – for the time being.

There are also direct references to the Mediterranean and its contexts in Stevens's *Collected Poems*. He cherished the idea of Southern Europe and its culture of leisure, of good life (and good food) which – he is at pains to tell us – is not averse to spiritual enlargements. As in "Farewell to Florida", in "Landscape with Boat" (1940) he presents a man who has chosen austerity "by rejecting what he saw and denying what he heard", and goes on to criticize this asceticism:

> Had he been better able to suppose:
> He might sit on a sofa on a balcony
> Above the Mediterranean, emerald
> Becoming emeralds. He might watch the palms
> Flap green ears in the heat. He might observe
> A yellow wine and follow a steamer's track
> And say, "The thing I hum appears to be
> The rhythm of this celestial pantomime". (*Ibid.*: 243)

Through disengagement, by sitting on a balcony over a beach like one of Dufy's or Matisse's, and sipping "a yellow wine", one might find that accord with reality (a favourite Stevensian phrase) that the self-denier searches for in vain. Whatever the spectator hums nonchalantly is really the secret "rhythm" of the sensual world, here called "celestial pantomime" because it is or becomes an allegory, a show, but in any case is heavenly insofar as it is earthly. Stevens was always preoccupied with countering the demands of Christianity and its otherworldliness, and this rather tropical Mediterranean serves this function, admittedly as a hypothesis, a might-have-been, an opportunity missed by the ascetic, though realized by the poet.

The same hypothetical strain occurs in "Esthétique du Mal" (1944), a complex sequence which sets the scene in Naples and contrasts the aesthete's satisfactions with the drama of an eruption of Vesuvius. This was in the news at the time, besides the military occupation of Naples by the Allies.

> He was at Naples writing letters home
> And, between letters, reading paragraphs
> On the sublime. Vesuvius had groaned
> For a month. It was pleasant to be sitting there,
> While the sultriest fulgurations, flickering,
> Cast corners in the glass. (Stevens 1954: 313-14)

This is how the poem begins, commenting apparently on theories of the sublime that speak of grandeur and violence (the stormy sea, Niagara Falls, a volcanic explosion) witnessed from afar, and of the "troubled pleasure" (to cite Wordsworth) this apparently affords us. In section 13 of this impressive and forbidding poem, Stevens again portrays the good life:

> And it may be
> That in his Mediterranean cloister a man,
> Reclining, eased of desire, establishes
> The visible, a zone of blue and orange
> Versicolorings, establishes a time
> To watch the fire-feinting sea and calls it good,
> The ultimate good ... (Stevens 1954: 324)

Here the Mediterranean man (be he a visiting Northener or a native) is both an ascetic and a sensualist, and he performs the function of the God of Genesis, pronouncing the things that he establishes "good". As a painter would, Stevens presents us with words that are unexpected and unusual, like "versicolorings", and "fire-feinting". The latter means that the sea feigns or resembles fire, perhaps also because it reflects the eruption. By speaking directly of a "Mediterranean cloister" Stevens, as is his wont, gives us something to think about, a conciliation of opposites, the opposites that he is attracted to: solitude and *joie de vivre*.

The "reclining" figure of Stevens' hero may be compared with the Lotus Eaters that Ezra Pound presents in a more diffuse scene of the early Cantos on the background of a sea whose noise he tries to suggest with an onomatopoeia reminiscent of Futurism:

> And the blue water dusky beneath them,
>   pouring there into the cataract,
> With noise of sea over shingle,

striking with:
hah hah ahah thmm, thunb, ah
woh woh araha thumm, bhaaa.
And from the floating bodies, the incense
    blue-pale, purple above them.
Shelf of the lotophagoi,
Aerial, cut in the aether.
                Reclining,
With the silver spilla,
The ball as of melted amber, coiled, caught up and turned.
Lotophagoi of the suave nails, quiet, scornful,
Voce-profondo:
    "Feared neither death nor pain for this beauty;
If harm, harm to ourselves". (Pound 1995: 93)

In a Rapallo letter of 1927 Pound explained to his father that the *lo-tophagoi*, "main subject of the Canto", were "lotus eaters, or respectable dope smokers" (1950: 210). Thus he defends their right to abstain from action and flow with the current against the background of the eternal sea, in this atmosphere which is very much that of the aesthetic Twenties, though of course Pound has a more challenging project for himself as the hero Odysseus. The impressionistic Pound seeks to draw us into his Mediterranean visionary drowsiness, whereas Stevens remains alert and presents a crisp reality of discrete and arresting objects ("Mediterranean cloister", "fire-feinting sea") even when the point is mere "reclining" enjoyment.

The contrast with Odysseus and the yielding to drowsiness can already be found in the classic Victorian treatment of the subject, Tennyson's "The Lotos-Eaters", which like Pound's revisionary vignette has a narrative introduction and a chorus expressing the desire to join the happy addicts:

Only to hear and see the far-off sparkling brine,
Only to hear were sweet, stretch'd out beneath the pine.
(Tennyson 1910: 56)

Pound's writing is more taught and energetic, but it is likewise an evocation of a sensual mood, and typically insists on an abstraction, "Beauty", which Tennyson does not define but evokes by his lush descriptions.

James Joyce had also pointed the way to his early champion Pound by devoting chapter 5 of *Ulysses* to "Lotus-Eaters". Here there are also images of indulgence, though Leopold Bloom thinks more accurately (given the opium theme) of the Orient:

> The far east. Lovely spot it must be: the garden of the world, big lazy leaves to float about on, cactuses, flowery meads, snaky lianas they call them. [...] Those Cinghalese lobbing around in the sun, in *dolce far niente*. Not doing a hand's turn all day. Sleep six months out of twelve. Too hot to quarrel. Influence of the climate. Lethargy. Flowers of idleness. (Joyce 1969: 73)

While both Stevens and Pound (in his more uncouth way) directly tackle the question of the morality of the aesthetic and sensual life, Joyce the novelist is content with recording Bloom's attraction to *dolce far niente*. Later in the chapter the same desire for rest is shifted to a conventional Mediterranean background in the Holy Land, which Bloom has seen in a painting:

> Nice kind of evening feeling. No more wandering about. Just loll there: quiet dusk: let everything rip. Forget. Tell about places you have been, strange customs. The other one, jar on her head, was getting the supper: fruit, olives, lovely cool water out of the well stonecold [...] She listens with big dark soft eyes. Tell her: more and more: all. Then a sigh: silence. Long long long rest. (Joyce 1969: 80)

Bloom's version of the satisfied traveller is much more prosaic than Stevens's, whose character is probably wearing a coat and tie even while reclining in his "Mediterranean cloister". But the sentiment is analogous, as is the maker's alighting on the telling word. The lines vibrate with perception, though Stevens's are not immediately available, but remain tantalizing and a little remote.

Stevens's ascetic aesthete is not really a Lotus Eater, though he is partial as we saw to "yellow wine". But the Lotus Eater type seems to have touched in different ways all of these writers. I will not pursue Mediterranean themes in James Joyce, but I do want to mention that, like Stevens, Ezra Pound finds in the Mediterranean a strong model or utopia. See the passage about "the shrine white with marble" quoted above from the Pisan cantos. In the 1930s Pound left his Lotus Eaters behind and wrote the so-called "fertility Cantos" (39, 47), which contrast the contemporary Usura

Cantos (45, 521) by presenting a Mediterranean community celebrating its sexual rites in the summer night. The Yankees Pound and Stevens can never be content to relax on their balcony, but must offer it as a model. They can't put away the "missionary spirit", as Pound once called it (1950: 338). Around 1930 Pound described as follows the quiet revelations he said were to be had in Brancusi's Paris studio:

> When all France (after the war) was teething and tittering, and busy, oh BUSY, there was ONE temple of QUIET. There was one refuge of the eternal calm that is no longer in the Christian religion. There was one place where you cd. take your mind and have it sluiced clean, like you can get it in the Gulf of Tigullio on a June day on a bathing raft with the sun on the lantern sails. I mean there was this white wide intellectual sunlight at 5 francs taxi ride from one's door. In the Impasse Ronsin. (Pound 1980: 307)

The sea and the sun are cleansing, so the ideal remains ascetic, unworldly like Brancusi's bird. The image of Pound imbibing moral vigour while "reclining" on his bathing raft in Rapallo, and the association with Brancusi, illuminate the sense of purpose that the Modernists brought to the Mediterranean.

I would still like to mention a few patterns in Stevens that can be associated with the Mediterranean. One is the myth of Rome as a city of man and god, evoked in the magnificent elegy "To an Old Philosopher in Rome", about the dying George Santayana, who had literally retired to a "Mediterranean cloister". Rome is the "head of the world" – Stevens wrote in an essay (1951: 148) – so it is appropriate for a "central mind" to find refuge there and reconsider his creations, the sense of his (and Stevens's) life. In his late poetry Stevens also resurrected Ulysses as a type of the seeker and Penelope as an image of the waiting *anima* that prepares and creates its revelations (in the poem "The World as Meditation"). However, the actuality of the South is more directly present in a late poem like "St. Armorer's Church from the Outside", which celebrates a "chapel of air" built by the poet, of which the Mediterranean provides an actual example:

> It is like a new account of everything old,
> Matisse at Vence and a great deal more than that,
> A new-colored sun, say... (Stevens 1954: 529)

Matisse's *Chapelle du Rosaire* at St-Paul-de-Vence is only an example after all. The simile is the same, though reversed, as Pound's equation of Brancusi and summer bathing. Stevens's list of analogies goes on with the image of a trip South, a variation on Goethe's well-known "land where the lemon trees flower":

> The first car out of a tunnel en voyage
> Into lands of ruddy-ruby fruits, achieved
> Not merely desired, for sale… (*Ibid.*: 530)

We know that these lines were suggested by the postcards that a friend of Stevens, Barbara Church, sent him as she travelled into Italy by car. Stevens contemplated the Mediterranean vicariously but no less intensely for that.

The last occurrence of the word in the *Collected Poems* is wholly metaphorical ("Things of August", 1949):

> … He that in this intelligence
> Mistakes it with a world of objects,
> Which, being green and blue, appease him,
> By chance, or happy chance, or happiness,
> According to his thought, in the Mediterranean
> Of the quiet of the middle of the night,
> With the broken statues standing on the shore. (Stevens 1954: 491)

Meditation is seen to become a world of objects, and these can provide a permanent satisfaction or even happiness, just as the imaginary Mediterranean of meditation (note the extended list of complements in the penultimate line: quiet… middle… night…) becomes finally a real place to be visited or at least contemplated, "with the broken statues standing on the shore". One may object that this too is only a painting by De Chirico, yet the process is clear: a place wholly of the mind, a metaphor, becomes a fully realized image, and an orotund one at that (note the massive alliteration): a complete resolution. We would be hard put to find in English poetry a vignette of the Mediterranean more evocative than this bold one-line sketch of Stevens, as essential as a drawing by Picasso.

The above readings suggest that to Modernists like Yeats, Stevens and Pound the Mediterranean presented models that they used in their ideological constructs, for they perceived themselves as torch-bearers and deployed a purposefulness and single-mindedness that distinguishes the period and its artistic attainments. Though their immediate antecedents were the Victorians, they were closer in their ambition to the major inventors of the Romantic age, whose paths they often crossed, both in spirit and in actuality. Taking his clue from the Bible's "men who go down to the sea in ships" Pound opened his *Cantos* describing their common enterprise:

> And then went down to the ship,
> Set keel to breakers, forth on the godly sea ...

Bibliography

Bacigalupo, Massimo, 2001, "Ezra Pound's 'European Paideuma' ", *Paideuma*, 30.1-2, pp. 225-245.

Joyce, James, [1922] 1969, *Ulysses*, Penguin, London.

Lattimore, Richmond, 1967, *The Odyssey of Homer: A Modern Translation*, Harper, New York.

Lindberg-Seyersted, Brita (ed.), 1982, *Pound/Ford: The Correspondence Between Ezra Pound and Ford Madox Ford and Their Writings About Each Other*, Faber, London.

McWhirter, Cameron, 2001, "'Dear Poet-General and Walloper': The Correspondence of Ezra Pound and Harold W. Thompson 1936-1939", *Paideuma*, 30.3, pp. 109-144.

Pinkerton, Jan, 1971, "Wallace Stevens in the Tropics", *Yale Review*, 60.2, pp. 215-27.

Pound, Ezra, 1950, *The Letters of Ezra Pound*, Harcourt Brace, New York.

——, 1954, *Selected Essays*, ed. by T.S. Eliot, Faber, London.

——, 1980, *Ezra Pound and the Visual Arts*, ed. by Harriet Zinnes, New Directions, New York.

——, 1995, *The Cantos*, 13th printing, New Directions, New York.

——, 2002, *Canti postumi*, ed. by Massimo Bacigalupo, Mondadori, Milano.

——, 2005, *Early Writings: Poetry and Prose*, ed. by Ira B. Nadel, Penguin, London.

Stevens, Wallace, 1951, *The Necessary Angel: Essays on Reality and the Imagination*, Knopf, New York.

——, 1954, *Collected Poems*, Knopf, New York.

Tennyson, Lord Alfred, 1910, *Poems of Alfred Lord Tennyson 1830-1865*, intr. T. Herbert Warren, Oxford University Press, London.

Yeats, William Butler, 1933, *The Winding Stair and Other Poems*, Macmillan, London.
——, 1955, *The Letters of W.B. Yeats*, ed. by Allan Wade, Macmillan, New York.
——, 1962, *A Vision*, Macmillan, London.
——, 1965, *Collected Poems*, Macmillan, London.

NEGOTIATING WITH GAUGUIN'S 'SOLAR MYTH': ART, ECONOMY AND IDEOLOGY IN FORD MADOX FORD'S PROVENCE

Laura Colombino

Two contemporary readings

A few months ago, I happened on "The Life of Paul Gauguin", a review which Geoff Dyer wrote in 1995 for *Modern Painters* and republished in his 1999 collection *Anglo-English Attitudes*. I realised with amazement that, after a century, the story of Gauguin going off to Tahiti could still linger undiminished in late twentieth-century imagination. Two points in Dyer's piece caught my attention. Firstly, the image of Gauguin's journey "to the ends of the earth", which "gives such literal expression to the idea of someone going to the edge, pushing himself as far as possible" and, through this "massive gamble" claiming his "'right to dare all'". Secondly, Gauguin's search for the primitivist myth, "something", in his own words quoted by Dyer, "indescribably ancient, august, religious in the rhythm of their gestures, in their extraordinary immobility" (1999: 123, 127). An experience which is best expressed by D.H. Lawrence in a letter recounting a similar response to the Pacific coast of Mexico, and which certainly evokes Gauguin's art:

> It's very like the South Seas Isles in quality [...] a queer bay with tropical huts and natives very like islanders, soft, dark, some almost black, and handsome. That Pacific blue-black in the eyes and hair, fathomless, timeless. They don't know the meaning of time. And they *can't* care. All the walls and nooks of our time-enclosure are down for them. Their eternity is vast, they can't care at all. Their blue-black eyes... I have learnt

something from them. The vastness of Pacific time, unhistoried, undivided. (Dyer 1999: 128)

Insofar as coming across Dyer's review was for me a somehow reassuring experience, involving the excitement produced by a passion for the modernist myth of originality[1] which I shared – but which, as a living philosophy, I had thought of as long over and done with in contemporary culture, it was for me just as disquieting to bump into another recent, but this time much more disenchanted, reading of Gauguin's myth by J.G. Ballard:

> I often think that the most radical thing one can do is to *deliberately* choose the bourgeois life – *get* that house in the suburbs, the job with the insurance company or the bank, wear a blue suit and a white shirt and a tie and have one's hair cut short, buy the right fabrics and furnishings, and pick one's friends according to the degree to which they fit into all the bourgeois standards. Actually *go for* the complete bourgeois life – do it without ever smiling; do it without ever winking. In a way, that may be the late 20th century's equivalent of Gauguin going off to Tahiti – it's possible! (1984: 9)

There is no way out of our globalised and 'bourgeoisified' world, Ballard contends, no trajectory in space and time which can make a difference between here and there, before and after. "Travel is the last fantasy the 20th century left us", he writes in *Millennium People*, "the delusion that going somewhere helps you reinvent yourself". "All the upgrades in existence" he adds, "lead to the same airports and resort hotels, the same pina colada bullshit". "There's nowhere to go. The planet is full. You might as well stay at home" (2004: 54-55). Ballard's philosophy is that if you cannot beat the market system, your only option is to join it, even to the extreme of compliance. But the chance of a rebellion to the capitalist system is still open for him, because there is an extant place where defiance might be cultivated and where truth escapes the massive fiction we live in: that place is inside our heads.

---

[1] Of course one should also investigate the extent to which Gauguin's encounter with the South Seas was affected by his cultural biases and imperial adventurism. But this will not be discussed here, inasmuch as my attention will be focused only on others' interpretations of his myth.

Commenting on the fate of the modernist myth of originality, Hal Foster contends that in our global economy the postulation of a pure outside is almost impossible. This does not imply a premature totalization of our world system, but recognises the need to conceive both identity and otherness "as immanent relations rather than transcendental events". In other words, as implying a constant work of negotiation between opposing categories: "Only recently", he argues, "have postcolonial artists and critics pushed practice and theory from binary structures of otherness to relational models of difference, from discrete space-times to mixed border zones" (1999: 177). And here I finally come to Ford, for, as I will argue, his Provence, especially as it is described in his novels *The Rash Act* (1933) and *Henry for Hugh* (1934) but also in *Provence. From Minstrels to the Machine* (1935), is precisely this sort of threshold zone, somehow at the crossroads between Dyer and Ballard's divergent perspectives, in that it shares on the one hand the search for full humanity and Gauguin's myth of primitive authenticity and on the other hand the unifying and standardising ideology of the capitalist world.

Utopia versus contemporary history and mass society

In *Provence* Ford contends that the French region is essentially a frame of mind rather than a geographically well-defined place. This is not just because you might find some of the qualities it stands for also in other parts of the world – from "New York" to "London" (1938: 80) – or because, according to a typically impressionist displacement of perception, one may better focus Provence at a great distance while, for example, "writing in the garret of a gloomy, fog-filled, undignifiedly old London House" (1938: 79). What makes it a mental place is, first and foremost, the fact that its virtues – traditional food and means of production,[2] the crafts, the continuum between technique and art, the peaceful *joie de vivre* – have to be carefully sought on the miscellaneous ground of the region, by tracing a network of virtuous gastronomic, aesthetic and economic

---

[2] On Ford's proposal for the return to self-sufficient farming as opposed to modern mass production see Neilson (2002).

itineraries, which the traveller is asked to mentally abstract from the wider
map of a Provence already tainted by mass culture and tourism. And this
by tracking, as the case may be, the hidden strongholds of the native, authentic *cuisine* with *foie gras* (besieged by the evil Northern butter and
pork-fat invaders!) or by leaving out of this ideal map "the parasitic
bathing towns of the *Cote d'Azur*", which, though "historically and geographically" located in Provence, are alien to its true nature: "To say that
these little cities of rather mechanical and monotonous pleasure are not
true Provence would be as unjust to them as it would be unjust to Provence
to include them" (94).

Intermittently in *Provence*, more consistently in *The Rash Act* and *Henry
for Hugh*, this region is represented as a "Solar Myth" (1934: 21) with its
Mediterranean landscape closely evoking the South Seas. The tropical version of this ideal had been "a widespread imaginative possession of all in
the trenches who were cold, tired and terrified" (Fussell 1980: 5), prompting Henry Major Tomlinson to sail off to the South Seas in 1923 (an event
he records the following year in his travel book *Tidemarks*) and Edward
Marsh to juxtapose, in his *Rupert Brooke: A Memoir* (1918), the image of
the poet swimming in the warm Tahiti waters and that of his death on a
hospital ship off Gallipoli (Fussell 1980: 7-8). Along the same lines, after the war the Mediterranean too was drawn into this new heliophily.
This solar myth included, and sometimes mixed, aesthetic practices (such
as the hedonistic fashion of nude sunbathing) and therapeutic concerns
(the sun being prescribed at the time as a remedy for both tuberculosis
and children's vitamin-deficiency).[3] In a similar way, the *provençal* sun-
shine which literally fills Ford's two novels of the thirties is associated
with artistic and pictorial features as well as (but perhaps more loosely)
with the hero's convalescence.

Ford, however, senses that history has partly breached and contaminated this prevalent myth of the 1920s and 1930s. Indeed this ideal, as
it emerges from *Provence*, is a thing of bits and pieces scattered over the
map of Southern France and beyond, picked here and there to weave a
web of itineraries which make up a guided tour as well as a safety net –
alas how fragile! – for a world precipitating towards the apocalypse of the

[3] In 1923 the Swiss physician Auguste Rollier devoted his influential treatise *He-
liotherapy* to this subject. See Fussell (1980: 138).

imminent Second World War.[4] It is no accident that, as a guidebook, *Provence* is first and foremost an attempt at directing the tourists to the production of correct impressions – both those which leave an imprint on their minds and those which they themselves are bound to produce on the locals. And by 'correct' I mean fostering mutual comprehension and respect. For example, advising on the amount of gratuities a waiter should be given, Ford suggests the following:

> And above all, give your tip with the air that you are the one that accepts favours. Remember that he has as much right as you to be there – and more. Remember that he is as honest a man as you, with as engaging a family to support and that, the human cosmogony being what it is, he has as much right to his tip as you to the emoluments that by force or guile you extract from the universe – and then more. And remember above all that, whilst you are a mere transitory nuisance, he is carrying on for his town the great work of civilization since for you, and how many others, he can and, if decently treated, will infinitely soothe your way through life and his town. (1938: 38)

Inspired by Ford the writer-"ambassador", the tourist will turn into a sort of amateur diplomat "influenc[ing] the policy of his home country towards the country he is visiting" (1938: 59) at a time of harsh international confrontations. Impressions, which have always been a malleable instrument for Ford, now come to represent the flimsy, hypersensitive, almost touchy surface between domestic and foreign.

Another significant aspect of *Provence* is that, as in a truly modern tourist guide where, in Roland Barthes's words, "the number of bathrooms and forks indicating good restaurants vies with that of 'artistic curiosities'" (2000: 76) – cafés feature as equal to ateliers and recipes compete with paintings by Cézanne. Yet Ford's cultural mixture is not con-

---

[4] Indeed, the solar myth Ford shares with other contemporary writers may be included in the more general "desperate" "search for a mythology that could somehow straighten society out in such troubled times": Ezra Pound and William Carlos William propounded the ideal of machine rationality and efficiency; the Bauhaus advocated "rational order" "for socially useful goals"; surrealism, constructivism and socialist realism mythologized the proletariat; Picasso plundered the universe of the African primitive (Harvey 1989: 34, 31).

ceived in the anything-goes logic of Postmodern culture but more in tune
with the historical materialism Walter Benjamin describes in his 1937
essay entitled "Edward Fuchs, Collector and Historian" (2002: 261). In
spite of the apparent differences between the two, for Ford as for Benjamin artistic production at all levels belongs to a cultural continuum
where art is part and parcel of social and economic life:

> The point is that in Provence the arts live, if hidden from Missouri then
> in the hearts of people. And you cannot call it either a proletariat art
> nor one induced from above, since it is the product of peasant propri-
> etors – and not of peasant proprietors only. The sons of not too rich news-
> paper proprietors paint pictures; those of millionaire tanners write epics;
> naval officers paint water-colours from Cap Sète to Annam…and proud
> of it!…That's the point.
>   I do not say that the production of masterpieces is enormous; but
> the presence of the celluloid doll before the three-thousand-year de-
> scended *saintons* is a proof of how intimately the native arts enter into
> the real life of the people even to-day. They are a part of life as unno-
> ticed as the daily bread, the prayers, the games of boules, the furniture
> and the Sunday bull-fights. (Ford 1938: 237)

High and low are not regarded as separate, symptomatic of different
social strata and ideologies. They belong to the same homogeneous cultural reality, the small-proprietor model implying no harsh severance between spiritless manual work and algid intellectual products, technique
and art. But also, and most interestingly, no class distinction between
workers and capitalists with all the social tensions such separation entails.
The philosophy of the small agricultural producer somehow counteracts
the idea, so paramount in 1930s Marxist theories, of history and society
as a dialectic reality of groups in conflict, with deriving waves of thought
overtaking each other. Ford's *provençal* myth, therefore, becomes the an-
tidote both to this class model propounded by the contemporary philosophy of evolutionary historicism and to the reality of a decade which, in
Frederic Jameson's words, "projected a tangible model of the antagonism
of the various classes toward each other, both within the individual na-
tion-states and on the international scene as well – a model as stark as the
Popular Front or the Spanish Civil War" (1971: xvii).

From eternity to impotent suspension

Now, let us focus on *The Rash Act* and *Henry for Hugh*. If Henry's *provençal* Villa Niké is perceived as an appendix to Gauguin's[5] South Seas it is because it represents a spatio-temporal dimension where history, which is rapidly moving towards the inevitable war, seems, at best, suspended and, at worst, slowed down. But there is a major difference between the Tahitian original myth and its *provençal* offspring. The solar hopes of ripeness and full humanity exemplified, for example, by Lawrence's short story *Sun* (1926) are partly negated, for Ford, by the imminent catastrophe. This is why Henry's *provençal* reign involves not so much Lawrence's idea of eternity as that of impotent suspension. Let me just remind you that the story recounted in these novels is that of Henry, a bankrupt aspiring suicide, who adopts the new identity – forced on him by converging circumstances – of the rich, crossed-in-love suicidal Hugh. But in the other's shoes, he finds his experiences and encounters unexpectedly and uncannily analogous to those of his previous life. And this includes the fact that, if Hugh was unable to reach the South Seas, Henry himself, for the repetition automatism by which his life seems to be governed, is faced with the same impossibility.

It is almost superfluous to recall how this resembles the well-known dream experience of 'moving immobility': in spite of all our frenetic activity we are stuck in the same place. What this Sisyphus-like condemnation to stasis through the infinite repetition of the same act adds up to, metaphorically, is Ford's perception of the 1930s as an age characterised essentially by economic and ethical impotence. Henry Martin incarnates a period whose "chief characteristic is want of courage – physical and moral", incapable, as he is, to resolve "either to spend money, commit suicide or" do "anything else" (Ford 1965: 210) which may represent a final solution to his problems. No Gauguin-like massive gamble is any longer conceivable for him, no decision, however momentous, is "enough to carry him through to the end of his destiny" (Dyer 1999: 122).

This idea of impotent suspension is also evident in the image of the women characters revolving around Henry. They may well remind us, in some respects, of the statuesque stillness and the dark but often radiant

[5] The painter is mentioned twice in *Henry for Hugh* (71, 119).

complexion of Gauguin's natives – take the instance of Eudoxie whose "mahogany" "skin" is like a "transparent surface contain[ing] luminous blood" (1982: 307) – as well as, in general, the exoticism of Gauguin's Tahiti, but they deliberately share none of the profound significations of his paintings. Consider, in this respect, how the image of Becquerel, as it emerges from the following highly pictorial description, is, in a way, the very reverse of Gauguin's *Te Tamari No Atua* or *Nativity* (1896):

> Jeanne Becquerel was sleeping, uncovered, on her bed that was bluish because of the dim light of the dawn and the moon. Her left hand was across her chest and her right was extended above her head. She was moaning a little in her sleep. She lay flat in bed, her hair spread over the pillow above her – like a South Sea island fan of perfumed fibre [...]. Nevertheless, until that moment Jeanne Becquerel had seemed to him as being entirely "static," as the phrase is. And even at that moment he stood looking down at her asleep and could not believe that anything went on inside her – as if she had been of marble – warm and soft but still marble that neither grew nor felt. She seemed to him most typical in the hours after lunch, as if she were really fitted to be a denizon [*sic*] of a land where it was always afternoon. (1934: 105)

The details of the exact position of each hand and the representation of the sleeping and dreaming (as "moaning" suggests) figure, lying static in the dim light evoke the foreground of the canvas (with the reference to the South Seas working as proof positive of this association with Gauguin). Yet, the references to the absence of thought and oneiric content, and, for that matter, to the insensitiveness of that marble body unable to feel and, particularly, "grow" (and here the allusion to conception and pregnancy is a perfectly legitimate inference) appear as references, in reverse, to the painting's background theme of the nativity; a reversal which deprives Ford's image of the dimensions of mystery and sacredness present in the canvas and, at the same time, points to the issue of sterility or impotence which, as we have seen, is so paramount in Ford's vision of the 1930s.

Now, let me return to Dyer's analysis: Gauguin's women, he contends, "gaze out of the paintings as if they were staring into them, like ideal all-comprehending spectators who", in a "self-circling movement", "see them exactly as Gauguin intended" – namely, in the painter's own words, as "Animal figures rigid as statues" with their "dreaming eyes" like "the blurred surface of some unfathomable enigma" (1999: 126-127). But in

Ford's novels, the exotic timelessness of Henry's women does not entail the stratification of meaning pointed out by Dyer. In their all-knowing gaze, Henry grasps not so much the enigma of comprehending women spectators whose perception reflects his own, as the shallower "secret" of the lesbianism of these "beautiful, always gay, young women" (1934: 250). The homosexuality of Henry's harem girls cannot but unman his male and, in a broad sense, 'colonial'[6] gaze, revealing his sexual impotence, and, in so doing, somehow suggesting, by analogy, his moral, political and economic powerlessness as well.

So, the *virtuous circle* of reciprocity described by Dyer, whereby the woman sees herself seeing herself through the mediation of the painter's eye – the two subjects being in perfect agreement and both "all-knowing" – has given way to a bewildering *short circuit* of gazes which fails to convey any shared knowledge whatsoever. But then again, Provence itself is represented in these novels as an entrapping catoptric theatre, where we are captured in the baffling kaleidoscope of Henry's memories and experiences.

Myth as delusion

Interestingly, through this prismatic vision, Ford also questions the modernist myth of the uniqueness and purity of avant-garde vision, testifying to the broad cultural – rather than merely aesthetic – concerns already seen in operation in *Provence*. The way he conveys this idea is by presenting art and mass society as being, in Provence's 'border zone', like reflections of a purely optical device:

> To them he must have been like a God… He seemed still to stand astride his fabulous grip that was a bursting purse of Fortunatus. He stood there chucking largesse to everyone who came in sight – eight thousand francs to a weeping *poule*; a Pheidias Venus to the British Museum; a fifty pound

[6] That is, to the extent this *provençal* microcosm of which Henry is, in a way, the king, resonates with allusions to imperial domains and vaguely exotic women: from the oriental "Cochin-Saigon", "the colony" which Jeanne Becquerel "had once decorated" (1934: 118), through the representation of Villa Niké as a sort of harem, to the insistent associations with, precisely, the South Seas.

note to the waiter; two Simone Martinis, an El Greco, and a Van Gogh
to the National Gallery; twenty thousand-pound notes to him, Henry
Martin; the three remaining best second folios to the British Museum
Reading Room; an infinite number of magnums of Perrier Jouet to Gloria Malström's husband, Mr. Bumblecumpspumpje Bumblepuppy; —
the scissors of Cleopatra, the tiara of the Empress Eugénie, two Rolls
Royce's, a Monckton de luxe, a private travelling orchestra, and a yacht
to Gloria Malmström – and fabulous collections that were to bear her
name in the great museums of the world [...] He stood astride his grip
and threw miraculous objects to whoever would catch them [...]
Chucked them, he would have said... (1934: 22)

Let us consider how, in this passage, Ford makes the two polarities of
art and consumerism react: firstly, he reproduces them in an endless series of items; then he piles them up and packs them into the spring of the
écriture, so that, when pressure is carried to the limit, the text may release
them in a sort of Catherine wheel. Perceived at a great distance, the overall effect radiating from the text is that of a prism refracting light in all
directions and, in so doing, suitably rendering, on the iconic level, the
idea of the Solar Myth incarnated by Hugh. But if the text is examined
at very close range, there emerge in relief the innumerable folds of the
page where an indefatigable tension is at work: that between the "permanent things" – namely the collector's items intended, according to a
Jamesian motif, for the eternity of the museum – and the transitory, "fugi-
tive" (1982: 144) consumer pleasures which seem to taint that aesthetic
ideal.

But the most significant aspect of the negotiation between utopia and
consumerism lies in that illusive visuality Ford detects not just within
the myth's confines but also in the world out there, where advertising and
cinematography are starting to shape nothing less than – what an outstanding anticipation on Ford's part! – our modern society of the spectacle. Take the following passage from *Henry for Hugh* where a film director suggests how to end the promotional advertisement for Henry's Monckton cars:

"We've been explaining to Mademoiselle," Mr. Crape interrupted gently, "that we propose to end the picture with a series of close-ups showing Monckton cars in some of the remotest parts of the globe..."
    "Speeding," Mr. Old-Smith interrupted him in turn, "from Green-
land's icy mountains at dawn to sunset over India's coral strands [...]with

of course oodles of heathens whose untutored minds…"

"You would," Henry Martin said, astonishingly to himself, "use Miss Becquerel – my other secretary for – [*sic*] Far Eastern scenes – amongst the palm groves here…"

Mr. Old-Smith said that Mahdamahsell here had told them that already. It appeared that the other lady had gorgeous robes and gadgets and would dance Siamese fashion better than a native… (1934: 236).

The references to "some of the remotest parts of the world" and the "Far Eastern scenes" cannot but recall, by analogy, the South Seas for which Hugh and Henry vainly try to sail. What the return of the idea of primitivism in an advertisement ("oodles of heathens whose untutored mind") seems to add up to is that the only primitive experience we are allowed to attain in the contemporary world is the simulated, mystifying one of cinematographic fiction, where a movie "*vedette*" (172) can "dance Siamese fashion better than a native". What is cultivated beyond the confines of Provence, Ford seems to suggest, is a symbolic reality as unreal as that in which Henry seeks refuge.

One should also consider, in this respect, the footage representing the initial phases of car production, which involves the exploitation both of raw materials and human beings:

He proposed to make the film run as you might say from pig-iron to paradises […] The first picture would show assembled all the raw metal that went into the making of the Monckton car […] Steel, rubber, nickel, alloys, fine woods, and so on [   ] He trusted that this exposition was not overtiring for Mr. Smith […] Then, to give an idea of the vast distances these products came from, they would have close-ups of the natives tapping rubber trees in the forests of Para, Spaniards getting out pig-iron, natives toiling in copper mines at the Cape or in Montana; the molten pig-iron pouring into the sand-troughs […] Say at Middlesborough […] A fine effect you could get of that […] All that would give an idea of the vast number of workers that were benefited by the manufacture of automobiles […] Then you would show the thousands of workers pouring out of the Smithville works: some close-ups again of the more decorative parts of the works themselves, showing the consideration that the family had always shown for the mental development of their employees […] the famous reading-room, the debating hall, the schools, swimming pools, sport grounds…and then the directorial offices to symbolise the brains that made all these things possible […] (225-226)

The paternalistic concern for the "mental development" of the labour force evokes the economic phenomenon of 'welfare capitalism' and Fordism in the United States. In the 1920s, many large American corporations, impelled by legitimacy problems, began to build houses, schools and libraries for their employees on a wide scale, and provide medical and legal services. Rational capitalists, starting with Henry Ford (whose thriving car industry, just like Hugh's, managed, almost alone, to survive in the years of the Depression), had realised that this was the only way to create favourable opinions, together with the fact that a better educated work force would be a more efficient one. As Harvey reminds us, this patronising attitude of contemporary American industry which F. M. Ford so intelligently grasps was meant to shape an ideal worker-consumer:

> Questions of sexuality, the family, forms of moral coercion, of consumerism, and of state action were, in Gramsci's view, all bound up with the search to forge a particular kind of worker 'suited to the new type of work and productive process.' Yet, even two decades after Ford's opening gambit, Gramsci judged that 'this elaboration is still only in its initial phase and therefore (apparently) idyllic.' […] In 1916 Ford sent an army of social workers into the homes of his 'privileged' (and largely immigrant) workers to ensure that the 'new man' of mass production had the right kind of moral probity, family life, and capacity for prudent (i.e. non-alcoholic) and 'rational' consumption to live up to corporate needs and expectations. The experiment did not last long, but its very existence was a prescient signal of the deep social, psychological, and political problems that Fordism was to pose. (Harvey 1989: 126)

Now, what strikes me as central to F.M. Ford's description of the Monckton car's assembly-line (a mode of production first used by H. Ford) is that the image of the "natives" as having "benefited" from the manufacturing process is as patently artificial as that which wanted them to remain as idyllic "heathens" among "coral strands" and "palm groves": in that case the mystification was aesthetic, while here it is economic. In other words, what we are faced with in these two passages is the reversal of both primitivist and productivist modernisms. The former was the fantasy according to which, in Foster's words, "the other, usually assumed to be of color, has special access to primary psychic and social processes from which the white subject is somehow blocked". The latter is the imaginary representation of the proletariat "as primitive": "both negatively (the mass as primal horde) and positively (the proletariat as tribal collective)" (Foster 1999: 175, 276). What Ford suggests, therefore, is the fictional quality of such enchanted aesthetic, social and productivist relationships, or in other words, the fact that, with the appropriation (falsification) of the pictorial ideal of the origin by the industries of the visual, Gauguin's myth is turning into what Barthes dubs as 'mythology'. By the term he defines an ideological mystification which does not deny reality but purifies it through the language of the image so that the world may appear harmonious and reassuring, deprived of the social conflict which, instead, as we have seen, was so paramount in the thirties (prompting Ford to turn to the model of the small agricultural producer).

But what does all this add up to in relation to Gauguin 'going to the edge'? Ford senses that cinema and advertising have changed the relation between spatial perception and bodily motion. Film viewing now implies the possibility for the modern gaze to wonder through space and consume it as a vast commodity: without moving from his standpoint in Provence, Henry can travel to the end of the earth through his exotic women, half Gauguin-like, half cinema "*vedette*[*s*]", who, when observed, give Henry the impression he is "looking at a film that only just held his sympathy" (1934: 172). Cinematography has turned the voyage into mental and spectatorial tourism, mobility into immobility.

Conclusion

I think we can maintain that, in the thirties, Ford is poised between myth and history. On the one hand, he nourishes hopes of social renewal and regeneration – which he sees as available through a glimpse at the Southern French model with its socially functional art and its old, but constantly revitalised, collective artistic practice. On the other hand, he realises that the modern world does not so readily lend itself to such restructuring. Partly because, outside the confines of Provence, it offers no models of social harmony but tends, instead, to promote irresponsibly "an increasing occultation of the real class structure" (Jameson 1971: xvii). Partly because the mystifying techniques of the new media, and particularly advertising, seem to sunder the myth of a really primitive place which may function as a regenerating source, as something quite other we may turn to. Modern man appears as hopelessly entrapped in a condition of suspension. Therefore, as Ford the utopian has an ambassadorial mission to fulfil in spreading the *provençal* frame of mind, so (only more so) Ford the historian of the scopic regime of his time realises that, in the incipient society of the spectacle – with its dream world of artificial stimuli and experience – the idea, in Ballard's words, "that going somewhere helps you reinvent yourself" can turn out to be nothing but a "delusion".

Bibliography

Barthes, Roland, 2000, "The *Blue Guide*", in *Mythologies*, trans. Annette Lavers, Vintage, London, pp. 74-77.

Ballard, James Graham, 1984, "J.G. Ballard, interviewed by Andrea Juno and [Vivian] Vale", in Andrea Juno and Vivian Vale (eds), *J.G. Ballard*, Re/Search Publishing, San Francisco, pp. 6-35.

——, 2004, *Millennium People*, Harper, London.

Benjamin, Walter, 2002, "Edward Fuchs, Collector and Historian", in Howard Eiland - Michael William Jennings (eds), *Walter Benjamin. Selected Writings. Volume 3: 1935-1938*, trans. Edmund Jephcott, Howard Eiland, and Others, The MIT Press, Cambridge (Mass.) and London, pp. 260-302.

Dyer, Geoff, 1999, "The Life of Paul Gauguin", in *Anglo-English Attitudes. Essays, Reviews, Misadventures 1984-99*, Abacus, London, pp. 119-128.

Ford, Ford Madox, 1934, *Henry for Hugh*, Lippincott, Philadelphia.

——, 1938, *Provence from Minstrels to the Machine*, George Allen & Unwin, London.

——, 1965, *Letters of Ford Madox Ford*, ed. by Richard Milton Ludwig, Princeton University Press, Princeton.

——, 1982, *The Rash Act*, Carcanet, Manchester.

Foster, Hal, 1999, *The Return of the Real*, The MIT Press, Cambridge (Mass.) and London.

Fussell, Paul, 1980, *Abroad. British Literary Traveling Between the Wars*, Oxford University Press, New York and Oxford.

Harvey, David, 1989, *The Condition of Postmodernity. An Enquiry into the Origins of Cultural Change*, Basil Blackwell, Oxford and Cambridge (Mass.).

Jameson, Fredric, 1971, *Marxism and Form. Twentieth-Century Dialectical Theories of Literature*, Princeton University Press, Princeton.

Neilson, Brett, 2002, "Ford's Biopolitics: *Great Trade Route* and the Philosophy of the Kitchen Garden", in Vita Fortunati - Elena Lamberti (eds), *Ford Madox Ford and 'The Republic of Letters'*, CLUEB, Bologna, pp. 118-127.

EDITH WHARTON'S *MOTOR-FLIGHT THROUGH FRANCE*

Gabriella Ferruggia

To date, there are two published records about motoring by American writer Edith Wharton (1862-1937), *A Motor-Flight through France* (1908),[1] and *In Morocco* (1920). Both were serialized prior to their appearance in book form, the former as a series of essays published in the *Atlantic Monthly* in seven instalments, from December 1906 to April 1908, and the latter being featured in *Scribner's Magazine* and the *Yale Review* in 1919. *MFF* is considered the best of all of Wharton's travel books. It recounts three different trips, lasting one to three weeks, taken from Paris through, respectively, the North and Centre, the South-West and the South, and the North-East of France, in May 1906, March-April 1907, and May 1907. The Whartons were accompanied in turn by Edith's brother Henry Edward Jones, and Henry James. The book is primarily concerned with the art, architecture, archeology, and the history of France.

In her autobiography *A Backward Glance*, published in 1934, at seventy-two, Wharton draws attention to her "travel fever" (Wharton 1990: 853), and her "incurable passion for the road" (*Ibid.*: 808). The roots of this passion she traces back to her parents' adventurous crossings of the Atlantic, in the time before steamships, well before the American Civil War, and to her early exposure to the bliss of travel in Europe: "[B]y the time I was four", as one of her most renowned statements goes, "I was playing in the Roman Forum" (*Ibid.*: 806). She also recalls how, aged five, in Paris, still unable to read, she had found "the necessary formula" for

---

[1] Hereafter *MFF*.

story-telling, "weaving stories" while holding in her hands the mysterious, "thick black type" of Washington Irving's *Alhambra*, a "relic" (*Ibid.*:
809) of the family's pilgrimage to Spain. Wharton makes clear how travel
and writing had always been inextricably linked for her.

Marriage, in 1885, also brought long spells of travel: "[I]t was then",
as Wharton states, "that I felt really alive" (*Ibid.*: 853). The four month
cruise in the Mediterranean on board the chartered English steam yacht
Vanadis, which ensued three years later, is described as a "mad" scheme
(*Ibid.*: 858), to secure which Wharton, then twenty-six, had been ready to
give "everything" she owned (*Ibid.*: 857). She never published the record
of that adventure, not only failing to mention in her memoirs that such a
record existed, but even denying that she ever kept "even the briefest of
diaries" before 1918 (*Ibid.*: 780).[2] All the same, in her memoirs she terms
her voyage as "the greatest step forward in my making" (*Ibid.*: 859). A further, and more definite advance in her career were the explorations of Italy,
which took place in the following decade. They resulted in her first novel,
The Valley of Decision (1902), set in eighteenth-century Italy; the first two
of her travel books, *Italian Villas and Their Gardens* (1904); and *Italian
Backgrounds* (1905). By then, "the pendulum woman" (*Ibid.*: 916), according to James's famous definition, had become a successful writer.

"The day of the motor" (*Ibid.*: 886) heralded a new era for Wharton.
She recalls how she first experimented with what she considered at the
time mere "toys of the rich" (*Ibid.*) in Caprarola, in the *Campagna Romana*,
in 1903, thanks to her friend George Meyer, the American ambassador
in Rome. Enthusiastically aware of the possibilities offered by this new
means of transport, Wharton vowed to buy herself a motor as soon as she
"could make money enough" (*Ibid.*: 887), a feat which she soon accomplished. The following year she bought her first automobile, a Panhard-
Levassor, a French make, with the royalties from *The Valley of Decision*.
The car was the first of a series which were christened, in turn, with names
such as "Alfred", after Alfred de Musset, or "George", after George Sand.

---

[2] In chapter V of her autobiography Wharton writes: "Our Mediterranean cruise
took place in 1888; but, owing to my not having kept a diary, I find it impossible to
disentangle the chronology of our travels in Italy" (Wharton 1990: 861). The cruise's
record was published for the first time by French scholar Claudine Lesage in 1992. Lesage
had found the typescript in the Hyères Municipal Library by chance, when doing research on Joseph Conrad, in 1989 (Lesage 2004: 18-24).

In the same year, the Whartons hired a chauffeur, Charles Cook, an American whom they retained for seventeen years. Edith motored extensively, in particular during the years preceding World War I, with and without her husband, often with a companion chosen from her closely-knit group of friends. She was on the road in the United States and in England, Scotland, France, Italy, Spain, Luxembourg, Germany, Sicily, Northern Algeria, Tunisia, and French Morocco. *In Morocco*, her last record of a motor-trip, is also her last travel book.

The concluding chapters of Wharton's autobiography ring a note of mourning for travel writing as a genre. She remarks that, given the ravages of mass tourism, which has turned journeys once magic into commonplace, ventures in personal reminiscence have become perilous. "Though such recollections constitute the traveller's joy", ponders Wharton, "they may easily become the reader's weariness" (*Ibid.*:1031). No such lack of faith is traceable in *MFF*. When the book appeared in print, in October 1908, Wharton had already published, besides *The Decoration of Houses* (1897) and the above mentioned novel and essays on Italy, four collections of stories, and her second, extremely successful novel, *The House of Mirth* (1905), which by early 1906 had reached the top of the bestseller list (Benstock 1994:150). Moreover, the translation of *The House of Mirth* into French, with an introduction by Paul Bourget, in the same period, had secured her reputation in French literary circles.[3] Though she was then still a temporary resident of France, these years are regarded as the "decisive time" in view of her final decision to live as an expatriate (Wright 1996: 5). The novella *Madame de Treymes* (1907), set in Paris, was well received. Wharton, in her memoirs, credits those "busy happy Parisian years" for grounding her new "sense of mastery" (Wharton 1990:1002) more firmly.

The issue of professionalism had been a crucial one for Wharton for some time when she set down to writing the articles on her motor-runs through France for the *Atlantic Monthly*, and it continued to affect her in the years immediately following. A preoccupation with amateurism, in connection with travel writing in particular, is evident in her correspondence with Richard Watson Gilder, the editor of *Century*, since 1903.

---

[3] *Chez les heureux du monde*, translated by Charles du Bos, had appeared in the *Revue de Paris* in December 1907. It was published in book form by Plon, Paris, in October 1908 (Schriber 1980: 62).

When renegotiating the fee for the six essays which would later be collected in the volume *Italian Villas and Their Gardens*, Wharton declares that she has met unexpected expenses "in order to write with some sort of system & comprehensiveness on a subject which, hitherto, has been treated in English only in the most amateurish fashion" (Wharton 1988: 83). Later, recollecting, in her autobiography, the publication by *Century* of the first of these articles, Wharton points out how the editors had been disturbed by the "serious, technical" approach of her writing, in relation to a subject traditionally "associated with moonlight and nightingales" (Wharton 1990: 889). She adds that, on the same occasion, she had curtly refused to meet the request that she embellish her text with "a few anecdotes, and a touch of human interest" (*Ibid.*: 888). Dilettantism is associated with female writing in her novel *The Reef*, published a few years later (1912). Here, Wharton's irony touches on the "vague yearnings" of a little circle of English and American expatriates in Paris. In particular, she satirizes the literary efforts of a Mrs Farlow, from Massachusetts, who has had her "Inner Glimpses of French Life" appear in "a leading New England Journal", and those of "a clergyman's widow from Torquay", England, who has authored an *English Ladies' Guide to Foreign Galleries* (Wharton 1996: 80).[4]

Both Wharton's correspondence and autobiography reveal that she believed *MFF* to be paramount in validating her position as a professional writer. In February 1908, while proof-reading her book, she thanks William Crary Brownell, literary consultant for Charles Scribner's Sons, whom she addresses as "cher maître", for his "sympathetic and stimulating comments". Brownell's appreciation counteracts William Morton Fullerton's disquieting suggestion, previously offered to Wharton, that "'The idea is charming, if you'd only put something into the articles".[5] Brownell corroborates the writer's opinion that she has, on the contrary, achieved something of value, if only "for those who have eyes to see" (Wharton 1988: 128). Soon after publication, in the fall of the same year, Wharton states, in a letter to the ailing Charles Eliot Norton, that she

---

[4] For *The Reef*, see Wright 1997: 51.

[5] Wharton refers to Fullerton in the same letter as "a past master on questions of Burgundian topography and history" (Wharton 1988:128). He had published *Terres Françaises: Bourgogne, Franche-Comté, Narbonnaise*, A. Colin, Paris, in 1905. The two had met for the first time in Paris in January 1907.

had barely restrained herself, a few days before, from asking him "to cast a glance at [her] little 'Motor-flight book'". The "dear note" in praise of the volume, which "has come of its own accord" from him has, as Wharton points out, given her "a quite peculiar pleasure" (*Ibid.*: 163-164, *passim*). In November, Wharton writes to her friend Sarah, Norton's daughter, of a recent motor-trip in Surrey, one of the many taken in that period, with James, and of James's sudden decision to have his younger friend visit George Meredith, then eighty and an invalid, at Box Hill. Wharton records her surprise at being welcomed by the illustrious novelist "with outstretched hands & the exclamation: 'My dear child, I have read every word you've written, & I've always wanted to see you! I'm flying through France in your motor at this moment'" (*Ibid.*: 165). This sketchy episode is recounted at length, kindled with strong touches of emotion and nostalgia, by Wharton in her memoirs. Here, the importance of *MFF* is attested by the venerable portrait of Meredith, smiling at the unannounced Wharton, and with his "beautiful strong hand", lifting up the book which "laid open at his elbow". To this she adds that, upon reading the title, "the blood rushed over [her] like fire!" (Wharton 1990: 973).

Wharton's correspondence, with her keen attention to royalties, cheques, estimates of fees and records of expenses, the quality of paper used by her publishers, the price and added value of illustrations, shows that she was well aware of the "selling capacity", as she phrases it (Wharton 1988, 82), of her magazine articles and books. The market for travel writing was well saturated by the beginning of the twentieth century. Critic Mary Suzanne Schriber has directed attention to the fact that, in 1857, an American reviewer complained that "the subject of European travel is so hackneyed that only a first-rate ability is competent to make a new book on the subject interesting". Fifty years later, one publisher was even more explicit, as he saw fit to introduce Adelaide S. Hall's *Two Women Abroad* with the comment that: "[To] publish a book describing the scenes and incidents of a six months' tour in foreign lands seems almost absurd" (Schriber 1995: xxvi).[6] With the new century, there was no lack of guidebooks, hand-books variously specialized, and travel narratives, leading both tourists and simple readers through Europe, and through France in par-

---

[6] Schriber lists 1,605 books of foreign travel by men and 643 by women, published by Americans before 1900.

ticular. The January 1908 issue of the *Atlantic Monthly*, which carries the fourth of Wharton's articles, entitled "A Second Motor-Flight Through France", for example, also advertises the six volumes included in its Library of Travel,[7] all illustrated. James, not the least of Wharton's competitors among her compatriots, opened the first chapter of his *A Little Tour in France*, published in the United States in 1884, with the following remark: "I am ashamed to begin with saying that Touraine is the garden of France; that remark has long ago lost its bloom" (James 1993: 19). In her recollections, Wharton's irony focuses on the "artless" American tourists of her parents' generation, travelling in Europe in search of "scenery, ruins and historic sites; places about which some sentimental legend hung". She depicts these "timid" sight-seers as being "gently led" in their discoveries by "Scott, Byron, Hans Andersen, Bulwer, Washington Irving or Hawthorne" (Wharton 1990: 832). As for the volumes of travel and art criticism appearing in the 'seventies and 'eighties, Wharton dismisses them as being "agreeable", and "of the cultured dilettante type". Among them she includes the work of two of her friends and contemporaries, the "deliciously desultory volumes" of "Vernon Lee" (Violet Paget) and the "delicate 'Sensations d'Italie'" of Paul Bourget (*Ibid.*: 889).

These remarks make clear that Wharton's primary concerns, when writing *MFF*, were with competence, with modernity, and with the market. Her title, which she retained after publication of her first article on the *Atlantic Monthly*, links the automobile to the metaphor of flight, and foregrounds maximum speed. It is *per se* provokingly modern. Wharton recollects, in her autobiography, how her long-time friend Walter Berry, for many years the counsellor of the French Embassy in Washington, had accompanied a French military mission to Dayton, Ohio, as legal adviser in 1905. He had investigated for three weeks the "queer, new 'flying machine'" invented by "two unknown craftsmen", the Wright brothers. He had seen it "'levitate' a few inches above earth", and returned, as Wharton points out, "awed by the 'strange futures beautiful and new'" folded up between those clumsy wings" (*Ibid.*: 1020).[8] "Mo-

---

[7] The books listed in the issue are Nathaniel Hawthorne's *Our Old Home*, Willam Elliot Griffis's *The American in Holland*, Henry James's *A Little Tour in France*, John Hay's *Castilian Days*, William Dean Howells's *Italian Journeys*, and Charles Dudley Warner's *In the Levant*.

[8] Wharton describes the immense excitement she experienced when watching, by

tor-flight" must have captivated Wharton's contemporaries' imagination more firmly than the terms "motor-journey" or "motor-run" which surface at times in her narrative, or the more tranquil "motor-jaunt" which she uses in her autobiography. Her title is also more powerful than those of other English and American travellers of the first two decades of the century, who emphasize the thoroughness, the randomness, or the leisure of their journeys through France rather than their speed, while ignoring their means of transport – *Along French Byways* (1900), *France from Sea to Sea* (1925), *Zigzags in France* (1925), *A Dawdle in France* (1926).[9] Motoring was, as Wharton shows, the new, distinctive feature which could infuse new life into a genre which had long become obsolete. One of the most successful, widely read records, this one exactly contemporary to Wharton's, by the French author Octave Mirbeau, in similar fashion takes its title, *La 628-E8* (1907),[10] from his "merveilleuse automobile" (Mirbeau 1907: v). His book is enthusiastically dedicated to "cher Monsieur Charron" (*Ibid.*), the man who has not simply built, but "animé, d'une vie merveilleuse", this extraordinary object, "cette bête magique" (*Ibid.*: ix).

In Wharton's terse title, gender is erased. In choosing not to pose as a "Lady" traveller, the writer avoids association with the "Women on Wheels" stereotype (Clarke 2003) which was beginning to spread along with the increasing mobility of women. At the same time, she rejects the nineteenth-century convention which had female authors frequently emphasize in their book's title or subtitle, as Schriber has argued, that the scenes they described had been observed "THROUGH A WOMAN'S EYE" (Schriber 1995: xxvii). Moreover, the volume is not restricted to a specific geographical area or to a special field of interest. By eliminating all reference to the usual sights, cathedrals, cloisters and *châteaux*, it sug-

---

chance, an aeroplane high up over the Place de la Concorde, in a letter to Sarah Norton dated October 20 [1909]: "And it was", she explains, "the Comte de Lambert in a Wright biplane, who had just flown across from Juvisy – and it was the *first time* that an aeroplane has ever crossed a great city!" (Wharton 1988: 192).

[9] The first two are accounts of journeys by rail, the third is a motor tour, the fourth a bicycle tour.

[10] The first edition of *La 628-E8* was printed in 3,000 copies. The luxury edition featured illustrations by Pierre Bonnard. *MFF* is reported to have been printed in 7,500 copies (Schriber 1991: xlvi, n4).

gests, as its scope, the whole territory of France.[11] Wharton's title sounds all the more assertive if we compare it to the diminutive *A Little Tour in France* chosen by James for his own account. James stressed even further the unpretentiousness of his "notes" or "sketches" when he declared, in his preface to the first English edition (1900), that they had been meant to *"accompany* a series of drawings" (James 1993: 3, my emphasis), and that the text and illustrations, by Joseph Pennell, were simple "things of the play of eye and hand and fancy" (*Ibid.*: 4). In similar fashion, he addressed his readers as "the author of these little pages" (*Ibid.*: 18).

Wharton's attitude is not as restrained. In her autobiography, she recalls the publication of the first of her Italian Villas articles on *Century*, a reworking of the episode mentioned in her letters. Wharton's text had been deemed by the editors unfit ("too dry") for the "fairy-tale" style of the pictures accompanying it, the work of the popular magazine illustrator Maxfield Parrish. Warton's outrage that her writing be considered as a mere commentary to Parrish's work speaks clearly thirty years later: "But I added", writes Wharton, "that, if, on reflection, my articles were thought unsuitable to the illustrations [...] I was quite willing to annul my contract" (Wharton 1990: 888). As for *MFF*, it was published with 48 black-and-white photographs, a further indication that Wharton keenly wished to emphasize the new, modern character of her volume. She was much alarmed when she saw the drawings of Ernest Peixotto, the same artist who had supplied the illustrations to *Italian Backgrounds* a few years before, appearing in the proofs. She now reacted to the quaintness of those "little dabby sketches of cows and marchandes de quatre saisons" and insisted that they be eliminated. Her articles, she complained in an unpublished letter to Brownell, were not "exactly the kind [...] to be used as a margin for humorous sketches" (Wright 1997: 78). As for the form of her volume, Wharton conceived it as fragmentary, "a series of desultory essays" (*Ibid.*), three separate sections unrelated to each other. This seeming lack of organization is, undoubtedly, linked to the initial publication of the travel pieces as independent magazine articles. On the other hand, their randomness is consequent with the multiplicity of choices, and freedom from boundaries and fixed routes, granted by the automobile.

[11] See Waddington 1908; Rose 1910.

The fact that some of the book's traits, such as its title, its structure or illustrations, represent, as we believe, a deliberate choice in favor of modernity on Wharton's part, was easily lost on her contemporary reviewers. One, writing for the *New York Times Saturday Review* in October 1908, contended that the book lacked "automobile lore", and that the machine, on the whole, cut "the smallest figure". For that reason, he concluded, "its insertion so prominently in the title becomes an intrusion". Another, on the *Nation* in November of the same year, regretted that "in the place of Peixotto's charming drawings" which had embellished *Italian Backgrounds*, the new volume came "weighed down, both literally and metaphorically, with many stupid, obvious photographs on heavy glazed paper". The same critic went as far as indicting what he termed a total neglect of form. He lamented the excess of material, the "crude arrangement in three unrelated, unbalanced narratives", and the abrupt ending, "as of a letter of which the last page has been lost". To him, the book, "too crowded with detail, too hasty in movement", built upon the briefest impressions of a writer "whisked past" the very places she wanted to see, was far from satisfactory to readers, unless they were already acquainted with the sites. In short, it was useless as a guide, and more suggestive of "the potboiler". Curiously enough, the keenest appreciation of *MFF*'s modernity came from the reviewer writing for the English *Spectator*, in October 1908. He praised the "treasures of novelty" to be found in the volume, the perceptiveness of Wharton's eye, the attraction, for the ordinary reader, of many of the places described, as exotic, he pointed out, as "the interior of Africa", the large number of "interesting illustrations", and, last but not least, "the new and almost magic power of the automobile" (Tuttleton 1992: 163-165, *passim*).

That *MFF* lacks "automobile lore" is out of the question. No "anecdoctes", no "touch[es] of human interest" (Wharton 1990: 888) are found in Wharton's text, like the ones used for example in her memoirs, when describing her first motor-ride, with George Meyer "luckily" at the wheel. "In a thin spring dress", writes Wharton, "a sailor hat balanced on my chignon, and a two-inch veil over my nose, I climbed proudly to my perch, and off we tore across the Campagna" (*Ibid.*: 887). None of the stock in trade hazards related to automobile travel, such as breakdowns or punctured tyres, are presented. Wharton never describes her car, never mentions its make, nor one of the many nicknames which she was in the habit of using. She omits her *chauffeur* from her narrative. However, Cook, the "faithful and skilful driver", "born path-finder", is affectionately recalled

in her autobiography (*Ibid.*: 964), while James refers to him as "my idol"
(Powers 1990: 76). She also fails to mention the names of her travelling
companions.

Wharton's frequent hints, in *MFF*, about the misleading influence of
the *Baedeker* and the *Guide Continental*, and her repudiation of "maps and
guidebooks" (Wharton 1991:38),[12] reveal that the aim of her book is not
to open new roads for "the throng" (143). Rather, she has chosen to lead
the discerning reader along untrodden paths. The voyage suggested is no
ordinary one, but a pilgrimage, a "sacred quest" for the initiated (Schriber
1987: 263), and its true object "our France" (74), a territory intimately
possessed by the writer. Wharton does not, accordingly, trace clear itineraries, single out new "stars", point the way to "modern conveniences"
(56), report on the state of the inns or the quality of food. Truly, the motor is not portrayed as the main character of her narrative, nor is it used
as a pretext for anecdoctes. Rather, its pauses and sudden accelerations
structure her whole text. Heedless of "the strain of [...] time limit[s]"
(158), Wharton finds inspiration in the spokes of the wheel, which "radiate into so many different directions" (110), launching into a vortex of
multiple possibilities. Constantly "pressed on" (119), she dashes, in a
"comet flight" (111), in "pursuit of scenery" (59) and adventures into the
unknown. She treasures "the deeper pang" (68) of emotion over objective
description; the "glimpse" (92) over complete information; the repeated
experience over the novel sensation: "One returns", as she warns her readers when in sight of Reims, "to renew and deepen the relation" (177).

Motion creates a new perception. By extolling the "mere bird's-eye
view" (58), the "sweeping glance" (58), the "eclectic eye" (54), Wharton
signals throughout her book the empowerment of vision granted to the
"imaginative" (103) sight-seer by the automobile. Modern travel, as she
makes clear in the following assessment of motor versus railway, is less
consistent with accumulation of detail than it is with an enhancement of
intuition:

Had we visited by rail the principal places named in this itinerary, necessity would have detained us longer in each, and we should have had
a fuller store of specific impressions; but we should have missed what

---

[12] Quotations refer to Wharton: 1991, and will be cited parenthetically in the text
hereafter.

is, in one way, the truest initiation of travel, the sense of continuity, of relation between districts, of familiarity with the unnamed, unhistorical region stretching between successive centres of human history, and exerting, in deep unnoticed ways, so persistent an influence on the turn that history takes. (36-37)

Her rapid descents on towns often taken "unawares, stealing on [them] by back ways and unchronicled paths" (1), the sudden twists and sharp changes of the road, lead to abrupt revelations where nocturnal lights transform ordinary locations into unsettling landscapes:

[F]or in the darkness we took a wrong turn, coming out on a high suburb of the west bank, with the city outspread below in a wide network of lights against the holy hill of Fourvière. Lyon passes, I believe, for the most prosaic of great French towns; but no one can so think of it who descends on it thus through the night, seeing its majestic bridges link quay to quay, and the double sweep of the river reflecting the million lights of its banks. (147)

The road, now calmly heading onward, now "throw[ing] its loops" (121) dizzily about the sides of a mountain, bewitches the writer's imagination. Wharton is subjugated, on her way to Beauvais, by "[t]he same wonderful white road, flinging itself along coils and arrow-flights across the same spacious landscape" (15) which had led her on the previous day to Amiens.

The motor allows a new freedom, enhances the pleasure of the sudden whim, offers access to "quiet by-roads and unknown villages" (25). Wharton, a modern, privileged, omnivorous traveller, can not only exert control over time and space, but also exclude offending elements from her field of vision at will. As she announces in *MFF*'s opening, the automobile, freeing the approach to each town from the disorderly growth created by the railway, has "restored the romance of travel" (1). A similar assertion serves as introduction to Part III. The several roads radiating from Paris allow one, Wharton considers, to depart "without touching even the fringe of what, were it like other cities, would be called its slums" (172).

Compared to Wharton's, James's record of his six weeks railway travel in France, interspersed as it is with his resort to omnibuses, his hiring of coaches, "sorry carriole[s]" and "lumbering white mare[s]" (James 1993:

82), seems to speak from an age more distant than the twenty-odd years which separate the two books. James is frequently detained, exasperated, "provoked", by the trains, which, in Touraine for example, "serve as little as possible for excursions. If they convey you one way at the right hour", he complains, "it is on the condition of bringing you back at the wrong; they either allow you far too little time to examine the castle or the ruin, or they leave you planted in front of it for periods that outlast curiosity" (*Ibid.*: 85-86). James deplores those very restrictions and time limits from which Wharton boldly proclaims herself free: "The train whirled me away", laments the older writer, "and these are my only impressions" (*Ibid.*: 142). Whereas Wharton magnifies her run to Poitiers at night as "a ghostly flight through a moon-washed landscape, with here and there a church tower looming in the dimness, or a heap of ruined walls rising mysteri-ously above the white bend of a river (87), James withdraws, renouncing description of the same town, approached in the dark: "I gazed into the gloom from such an aperture [the window of the train] before we got into the station, for I remembered the impression received on another occasion; but I saw nothing save the universal night, spotted here and there with an ugly railway lamp" (James 1993: 134). Upon arrival to Vaucluse his gaze, unlike Wharton's when entering a new city, is subject to the disfigure-ment inflicted by "a pair of hideous mills for the manufacture of paper and of wool" (*Ibid.*: 246) on the surrounding scenery. James lists his many griev-ances on railway travel in France towards the end of his book. In short, this "last tribute" (*Ibid.*: 256) of his centres on the insufferable despotism of the train, a despotism which Wharton will rapturously leave behind.

In *A Backward Glance*, Wharton points out how James, having tried in turn motoring during his first sojourn with her at The Mount, in 1904, was keenly appreciative of the "immense enlargement of life" (Wharton 1990: 916) he experienced. James's letters bear witness, in many instances, to his awed admiration of his friend's mobility. He refers, in one instance, in mock distress, at the incoming approach of "the Bird o'-freedom – the whirr and wind of whose great pinions is already cold on [his] foredoomed brow" (Edel 1984: 620). In another instance, he pictures "The Firebird" arriving at Lamb House all "wound up and going", ready to move on (*Ibid.*: 622). An invitation by Wharton in October 1907 is hailed as "your silver-sounding toot that invites me to the Car – the wondrous cushioned *general* Car of your so wondrously india-rubber-tyred & deep-cushioned fortune" (Powers 1990: 75). In about the same period (May 1909), James pays homage to his friend by acknowledging that his story "The Velvet Glove" had been inspired by his runs through Paris in her motor: "the whole thing *reeks* with you, and with Cook, & with *our* Paris (Cook's and yours and mine)" (*Ibid.*: 112).

This motoring camaraderie between Wharton and James reached its climax in the trip they took to Nohant, the site of George Sand's house, in the Berry region, in March 1907. This was the first visit for James, the second for Wharton, who had been there with both her husband and her brother Henry in May of the previous year. James yearned to see Nohant, and uttered his wish in two successive letters, in July and November 1906. He replied to Wharton's announcement of her own journey there, writing that the news had come to him as a "Parthian shot", inflicting a "bleeding wound" in his side. He lamented that he had through all his " (motorless) years so abjectly failed to enlighten [his] eyes withal" (*Ibid.*: 65), and begged to be taken along on another occasion. Of still weightier import is (in the same letter) his entreaty that Wharton should publish a description of the site. The "javelin", proclaims James, will "stick there [in my side] and poison my blood, till you *write* – I mean till you PRINT, till you 'do' the place … You can't *not*. I yearn and languish" (*Ibid.*). James spurs his friend on, exalting her at the same time: "So you owe me the *récit*", he insists, "There has been, you know, no *récit* (of the impression of the place) of any sort of authority or value but George's own" (*Ibid.*: 66). His connecting Wharton's writing to Sand's *Histoire de ma vie* was no trifling bestowal of authority on his part.

James is not included in the description of Nohant which Wharton did publish in *MFF*,[13] nor in the whole of Part II, which relates his journey on the "Vehicle of Passion" (*Ibid.*: 67). His name, like the names of the other travelling companions, is sacrificed in favor of the portrayal of an anonymous little group, always alluded to as "we". All the same, the middle part of the book contains what are perhaps its most poignant accounts, namely the second trip to Sand's house, and the breathtaking descent on the Mediterranean.

---

[13] In her autobiography, Wharton offers an irreverent portrait of James at Nohant. "James stood there a long time, gazing and brooding beneath the row of closed shutters. 'And in which of those rooms, I wonder, did George herself sleep?' I heard him suddenly mutter. 'Though in which, indeed – with a twinkle – 'in which indeed my dear, did she *not?*" (Wharton 1990: 1012).

Taken together, the two sections dedicated to Sand ("The Loire and the Indre" and "Paris to Poitiers") are, in a text whose narrator is strictly "faceless, gender-neutral" (Batcos 1999: 183), the closest, if indirect, attempt at self-revelation on the author's part. In her first "pilgrimage", Wharton is not simply retracing the steps of, and metaphorically "setting out on the same road " (40) as, the "great personages – Liszt, Sainte-Beuve, Gautier, Dumas fils, Flaubert" (39-40) who had hastened to pay homage to the great woman. While contemplating the grounds, the walled graveyard and the exterior of the *château*, Wharton relives through her imagination the scenes described in *Histoire de ma vie.* The somber, dignified, façade of the house, "familiarly related to the high-road and the farm" (41), immediately counteracts Wharton's expectations of an almost palpable social disorder. She is left to ponder on the many inconsistencies of Sand's personality, and on the riddle whereby the "timid Madame Dudevant" (46) was turned into the great writer.

The second trip allows for penetration into the interior, and for a closer probing of "the real meaning of the place" (80). While Sand's oeuvre remains unheeded, save a passing reference to *Indiana*, the decor of her dining-room, the furniture of her *salon*, the ancestral portraits hanging on the walls, and the little theatre contrived out of a store-room by "the mistress of revels" (82) are meticulously assessed. Musing on the marionettes still grouped on shelves at the back of the stage, dressed up by those same "indefatigable fingers which drove the quill upstairs" (83), Wharton confronts "the strange unfathomable" (83) mystery of her predecessor's life by addressing, more than the artist, the "extraordinary woman" (41), "the woman who had made her pseudonym illustrious" (80), "the great woman" (81).

In a book more concerned with the art and history of France than with its natural landscape, the descent on the Mediterranean, included, as already noted, in Part II, is one notable exception. The "everincreasing loveliness" (129) of the land from Aix to Saint-Maximin, gives way to the sudden vision of the sea, in the radiancy and "full glory" (131) of Spring, at Hyères. "One's first feelings", considers Wharton, "is that no work of man, no accumulated appeal of history, can contend a moment against this joy of the eye so prodigally poured out" (131). The stretch of coast from Toulon to Saint-Tropez, with its "peculiar nobility, [and] Virgilian breadth of composition" (132), with its green plain "so divinely [...] open[ing] to the sea, between mountain lines of such Attic purity" (132), besets the mind with "classical allusions, analogies to to the golden age" (132).

Whilst, as we have seen, undefinable as far as gender is concerned, the status of the narrator as writer is very conspicuous. A crowd of authors past and present, French, English, and German, is summoned up in the book's pages, and images connected to writing abound. A sudden change of scenery is described as "the passing from a black-and-white page to one illuminated" (74); the incoming day promises to "turn a brighter page for us" (74); the buildings of Poitiers elucidate their history "like the consecutive pages of a richly illuminated chronicle" (89); and upon arriving in Amiens after dark, "one's mind presents a blank page for the town to write its name on" (7). Stories are invented, "vignettes" supplied, as when the narrator suggests of an old inn in Dourdan that it was such as "Manon and des Grieux" might have dined in "on their way to Paris" (32). More important still, the whole venture recorded in the book, the exploration of fields in which Wharton's opinion was at risk of being be judged amateurish, and her knowledge "unequipped for technical verdicts" (179), is justified as providing a training ground for the higher calling of writing. Even the pleasure, the narrator contends, of accumulating impressions of a "mixed and nebulous order" (179) can "enrich the aesthetic consciousness, prepare it for fresh and perhaps more definite impressions, enlarge its sense of the underlying relation between art and life" (180). In a text where, as already noted, the first person plural "we" is dominant, Wharton switches to the "I" in one, significant instance. Having described Etampes, on the road from Rouen to Fontainebleau, as the average French provincial town, "rather featureless and disappointing", she proclaims it, with a swift change, to be so typical, "that its one straight grey street and squat old church will hereafter always serve for the *ville de province* background in my staging of French fiction" (33). In *MFF*, Wharton minimizes both her identity as a woman and her private, personal sphere. At the same time, she magnifies her role as a professional writer, seeking authority, as has been noted, within a "public, masculinized" (Nowlin 1998: 445) space.

Bibliography

Anon. "Advertiser", 1908, *Atlantic Monthly*, Vol. CI, No. 1, pp. 27-28.

Batcos, Stephanie, 1999, "A 'Fairy Tale Every Minute'", in Clare Colquitt - Susan Goodman - Candace Waid (eds), *A Forward Glance. New Essays on Edith Wharton*, University of Delaware Press, Newark, NJ, pp.172-187.

Benstock, Shari, 1994, *No Gifts from Chance. A Biography of Edith Wharton*, Penguin, London.

Clarke, Deborah, 2003, "Women on Wheels: 'A threat at yesterday's order of things'", *Arizona Quarterly*, Vol. LIX, No. 4, pp. 103-133.

Edel, Leon (ed.), 1984, *Henry James. Letters*, Belknap Press/Harvard University Press, Cambridge, Mass.

James, Henry, [1884] 1993, *A Little Tour in France*, in Richard Howard (ed.), *Henry James. Collected Travel Writings: The Continent*, The Library of America/Literary Classics of the United States, New York, pp. 1-277.

Johnson, Clifton, 1900, *Along French Byways*, The MacMillan Co., London.

Lesage, Claudine, 2004, in Edith Wharton, 2004, pp. 19-24.

Lewis, R.W.B. - Nancy Lewis, 1988, "Introduction", in Edith Wharton, 1988, pp. 3-23.

Lucas, E.V., 1925, *Zigzags in France and Various Essays*, Methuen & Co., London.

Mirbeau, Octave, 1907, *La 628-E8*, Eugène Fasquelle Editeur, Paris.

Nowlin, Michael E., 1998, "Edith Wharton as Critic, Traveller, and War Hero", *Studies in the Novel*, Vol. XXX, No. 3, pp. 444-451.

Powers, Lyall H. (ed.), 1990, *Henry James and Edith Wharton. Letters: 1900-1915*, Charles Scribner's Sons, New York.

Riggs, Arthur Stanley, 1925, *France from Sea to Sea*, Robert M. McBride & Co., New York.

Rose, Elise Whitlock, 1910, *Cathedrals and Cloisters of the Isle de France*, G.P. Putnam's Sons, New York, London.

Schriber, Mary Suzanne, 1980, "Edith Wharton and the French Critics, 1906-1937", *American Literary Realism*, Vol. XIII, No. 1, pp. 62-72.

——, 1987, "Edith Wharton and Travel Writing as Self-Discovery", *American Literature*, Vol. LIX, No. 2, pp. 257-267.

——, 1991, "Introduction", in Edith Wharton, 1991, pp. xvii-l.

——, 1995, "Introduction", in Schriber (ed.), *Telling Travels. Selected Writings by Nineteenth-Century American Women Abroad*, Northern Illinois University Press, DeKalb, Ill., pp. xi-xxxi.

Sheldon-Williams, Inglis, 1926, *A Dawdle in France*, A. & C. Black, London.

Tuttleton, James *et al.* (eds), 1992, *Edith Wharton. The Contemporary Reviews*, Cambridge University Press, Cambridge.

Waddington, Mary King, 1908, *Château and Country Life in France*, Scribner, New York.

Wharton, Edith, [1908] 1991, *A Motor-Flight through France*, Northern Illinois University Press., DeKalb, Ill.

———, [1912], 1996, *The Reef*, in *Three European Novels*, Penguin, London, pp. 53-298.

———, [1920], 1997, *In Morocco*, Kessinger Publishing, Whitefish, Mt.

———, [1934], 1990, *A Backward Glance*, in Cynthia Griffin Wolf (ed.), *Edith Wharton. Novellas and Other Writings*, The Library of America/Literary Classics of the United States, New York, pp. 767-1068.

———, 1988, *The Letters of Edith Wharton*, ed. by R.W.B. Lewis - Nancy Lewis, Charles Scribner's Sons, New York.

———, 2004, *The Cruise of the Vanadis*, Bloomsbury, London.

Wright, Sarah Bird, 1996, "Introduction", in Sarah Bird Wright (ed.), *Edith Wharton Abroad. Selected Travel Writings, 1888-1920*, St. Martin's Griffin, New York, pp. 1-37.

———, 1997, *Edith Wharton's Travel Writing. The Making of a Connoisseur*, St. Martin's Press, New York.

IN THE MOOD FOR PROVENCE, IN THE HEART OF THE MODERN: BLOOMSBURY AND SOUTHERN FRANCE

Caroline Patey

Geography of form, forms of geography

While not covering extensive surfaces, the Mediterranean geography of frenchified or francophile Bloomsburyans has a definite personality of its own, in no way to be confused with the dissipated seductions of the neighbouring *Côte d'Azur*: theirs is a region that has none of the "Babylonian beauty of Montecarlo" (Sutton 1972: 621) and next to no taste for the glittering style and gambling modes associated with Nice and the Riviera (Silver 1988: 134). Rather, the area destined to play such an important part in the definition and formulation of modernist aesthetics originates, as it were, from Aix-en-Provence, extending westward to Arles and St. Rémy and heading South to embrace the seaside towns of Cassis and Martigues. True to the memory of Cézanne, whose ghost hovers all over, this Provence is defined, first of all, by its austerity and geometrical volumes; in Fry's words, Aix is

> a delightful town: all seventeenth and eighteenth century but much rougher and austerer and bigger than in the North [...] huge façades generally rather flat, and then in the big streets, immense plane trees in front and everywhere the life of a very old-fashioned country town. (Sutton 1972: 464)

As to the countryside, it conveys an economy and a lack of emphasis reminiscent indeed of Cézanne's pictures such as *La route du Château Noir*, in which "the colour is geometric"(Fry 1927: 77); Fry's cycling expedi-

tions lead him to a sort of reversal of the stereotyped lavishness of light
and intensity of tones:

> The colour is amazing and the secret of it is that there are no bright
> colours. I find I use almost entirely black, yellow, ochre, venetian red,
> raw umber, burnt umber, Indian red and terra verte.
> Terra verte pure is too bright for the sky and has generally to be toned
> with black and red [...] It's just the purity and beauty of the greys that
> makes it seem more coloured than England. (Sutton 1972: 463)

And though no ascetic and an amateur of local food and wine, Fry
lingers on the beloved motives of construction, abstraction, gravity and
geometry, used in the exploration of Cézanne's art and equally in the description of local landscapes. Long before the time came to share a *mas*
with his friends in St. Rémy, this stretch of Provence felt to Fry as the
object of a proper *nostos* and the place of homecoming: "By what quirk
of the laws of heredity I do not know, but I always find myself at home
here; I always feel as if I had come back from exile" (Sutton 1972: 467).
A sense of belonging that was not lost to his first biographer, Virginia
Woolf:

> He loved that corner of Provence with a passion that seemed to spring
> from some ancestral memory. He did his best to believe that there was
> southern blood in his veins. There was the name Mariabella and his
> mother's southern darkness to prove it. Even if the family annals were
> against him [...] 'both Margery and I always feel', he wrote, ' that we
> were born here'. (Woolf 2003: 282)

As years go by, Fry's letters to his friends Marie and Charles Mauron
are charged with an ever increasing sense of ease and intimacy, matched
only by the growing estrangement from his native culture:

> I have offended English snobbism. My painting is not sufficiently ac-
> centuated; there's nothing fashionable in it. There's no formula one can
> recognize at once. Finally there's nothing to excite the idle gaze of those
> in search of distraction [...] The misunderstanding between me and the
> English public is complete now. (Sutton 1972: 481)

Eager to share with friends his true homeland and the impressions
and ideas fostered in it and by it, Roger Fry enthuses endlessly about the

unique quality of the place in letters full of exhortations to join in the *provençal* bliss:

> In fact, I feel that it's ridiculous not to spend most of one's life in this kind of place [...] the thought now of Paris seems to me a chilling disquieting one and London sounds to me like a wild and improbable adventure into cimmerian darkness and savagery. (Sutton 1972: 462, 469)

Thus was born in Cassis Bloomsbury-sur-Mediterranée (Pl. 2), harbour to many an artist before and after the first world war (Caws 1994: 6): indeed, Fry had come in 1915 in the steps of Derain and Matisse, but not unaware either of the strong *provençal* identity of the small town, inscribed on the literary maps by Frédéric Mistral's fishermen's story, *Calendau* (1867). From 1925 onwards, the Bells, and the Woolfs and the Grants were to intertwine more or less fleetingly their destiny with the other inhabitants of this tract of coast; and if famous names ring sonorously, those of Larionov, Gontcharova, Roland Penrose or Maynard Keynes, to Virginia Woolf, however, the fashionable and cosmopolitan face of Cassis may not have been its most important feature. Rather, she recalls an experience of silence, of losing touch with language while sitting eight hours a day looking out of the window (Caws 2000: 157); and records visions of a landscape that defies prettiness and tends, on the contrary, towards the stern and the geometric:

> I am waiting to see what form of Cassis will finally cast up in my mind. There are the rocks. We used to go out after breakfast and sit on the rocks with the sun on us [..] It was stony, steep and very hot [...] The ragged red tulips were out in the fields; all the fields were little angular shelves cut out of the hill and ruled and ribbed with vines; and all red and rosy and purple here and there with the spray of some fruit tree in bud. Here and there was an angular white or yellow or blue washed house, with all its shutters tightly closed, and flat paths round it, and once rows of stocks; an incomparable cleanness and definiteness everywhere. (Woolf 1959: 73)

In Virginia Woolf's and Roger Fry's diaries and correspondence, the text and images unfolded by Provence, far from any hedonistic impression of light, colour and sun, appear to be the fabric of a larger aesthetic dimension and a site of form rather than of pleasure; there is a rhythmic quality attached to Woolf's repetitions (rocks, angular, stone) and she, too, is sensitive to the strangeness and dissonance of the surroundings. To her, as to Roger Fry, the perception of La Ciotat in particular, affords intense chromatic sensations but also a distinct plunge into the complexities and aporias of modernity, all perhaps redolent of Georges Braque who lived and worked there; in sheer Fauvist manner, Virginia notes that "At La Ciotat great orange ships rose up out of the blue water of the little bay"(Woolf 1959: 73), while Fry, besides the uncommon colour scheme, captures the richly dis-harmonic quality of the place:

> … what a queer rather mad landscape this is. There is a *calanque* that is pure Chinese something like this[1], with the rocks all sculptured into the most fantastic shapes you ever saw and such extraordinary colour, deep rich orange-brown, going to violet in the rocks and the pines a very rich yellow-green on it. But perhaps *à la longue* one will get most out of the port, which is quite different from S. Tropez, much less picturesque – very up-to-date machinery everywhere, huge cranes and immense workshops and great transatlantics all muddled up with small sailing boats and rather jolly old houses. (Sutton 1972: 546)

If part of the Bloomsbury diaspora chose the coast and Cassis, be it Hotel Cendrillon or Villa La Bergère (Gardes 2002: 18), the trust and taste of others went to inland St. Rémy which, thanks to the Maurons' hospitality and multiple intellectual interests, soon transformed into a lively literary laboratory. An anchor to Fry, St. Rémy also became a safe harbour for E.M. Forster, whose *Passage to India* was translated into French by Charles Mauron in 1925 while *Aspects of the Novel* was dedicated to both Charles and Marie, in a tribute to their role in the construction of British modernity; a modernity, incidentally, not limited to the generation of the Frys and Forsters but open to and fed by younger voices such as Julian Bell's – Vanessa's promising son and Fry's *protégé* – and his Cambridge friend Edward Playfair: *habitués* of the *mas*, close friends to its owners, engaged in passionate conversations with them, occasional interpreters of *Mireille* in Provençal and keen to learn the language Mauron loved[2].

---

[1] The letter is addressed to Vanessa Bell and includes a sketch of the 'pure Chinese' bay.

[2] The friendship between Charles Mauron and Julian Bell would require a page of its own; suffice it to say here that it was intimate and deeply felt, so much so that Julian's family invited Mauron to write a preface to the posthumous collection of his writ-

To the tightly knit community of St. Rémy, death, the civil war in Spain
and WWII were to bring a violent end: but the achievements and debates of this Mediterranean *cénacle* still radiate in the words and images
of its members, associates and sympathizers, in Woolf's narrative aesthetics as in Fry's visions and designs.

> It was not only the landscape that he loved; it was the pagan, classless
> society, where salads were held in common, where every peasant was an
> individual and the old man who trimmed his olive trees was a more ci-
> vilized human being than the citizens of Paris, Berlin or London. The
> Mas was always at the back of his mind, a centre of sanity and civiliza-
> tion, when the telephone rang at Bernard street and the loud speaker
> brayed next door. (Woolf 2003: 283)

There is a good deal of idealization, of course, in the utopia of class-
lessness and collective salads. But points are made: the pulsating heart of
France beats in St. Rémy, certainly not in 'dreary Paris'; and St. Rémy is
the Mediterranean end of a fabric that is being actively woven, between
London and this South, thanks to Fry: Fry who invites Woolf to join in
his affection for Mauron – my blind professor, as he call him – and in turn
unveils some of Virginia's beauties for his French friends:

> I hope to get you the translation of a very beautiful book by Virginia
> Woolf that will interest you perhaps more than Forster. There's noth-
> ing more curious in our modern literature – a search towards another
> kind of novel – a novel poem. The style is very unexpected. (Sutton
> 1972: 562)

In spite of Virginia not speaking French – a limitation that annoyed
her greatly - the conversation between these three takes pace, soon to include cross-channel translations, discussions and exchange on aesthetics,
soon to leave its mark on many a page, soon to bear fruits.

ings. "Writing to Quentin in November of 1938, about Julian's writings and death,
Mauron succinctly summed up the loss of Julian: 'The flavour of the words, and the explosion of life, and friendship and regret'. No one phrase could better describe the intimate, deep and lasting relation between Mauron and the Bloomsbury friends in England who had found their home, so often, with him". (Caws 2000: 287)

Novels born in Cassis?

It is important, in the light of such a tight *provençal* network, to remind that the Woolfs' stay at the charmingly named Hotel Cendrillon
predated by a few weeks only the composition of *To the Lighthouse*, started
in June 1925 – "I am under the impression of the moment, which is the
complex one of coming back home from the South of France to this wide
dim peaceful privacy..."(Woolf 1959: 71): a hint, perhaps, at a possible
confluence of the two lighthouses of Cassis and St. Ives; and a suggestion,
supported by Fry's perception, that the coherence, rhythm and gravity of
Cézanne's still-lifes and landscapes have a parallel in the novels of Virginia Woolf and her prose still-lifes, landscapes and seascapes (Caws 2000:
285). Indeed, dinner at the Ramsays, started with Mrs Ramsay "ladling
out soup" (Woolf 1992a: 113) and ending with the elaborate vision of a
fruit basket reads as a visual/verbal journey from still-life to still-life, and
seems to establish a continuity between Fry's study of Cézanne's soup-
tureen (*Nature morte à la soupière*, 1877) or *Compotier, verre et pommes* (1880)
and Woolf's narrative. Colours, textures and objects may be different,
but the mood is undoubtedly similar:

> The tablecloth is of dark, rich vermilion, modified by bluish reflection
> when the light falls on it from the window. The citron yellow of the ap-
> ples and the vivid black of the bottle stand out forcibly on the roses
> and pale greens which compose the background. (Fry 1927: 52)
> Now eight candles were stood down the table, and after the first stoop
> the flames stood upright and drew with them into visibility the long
> table entire, and in the middle a yellow and purple dish of fruit [...]
> Thus brought up suddenly into the light it seemed possessed of great
> size and depth, was like a world in which one could take one's staff and
> climb up hills... (Woolf 1992a: 131)

Yet again in tune with Fry's disquisition on the modes and techniques
of still-life, and regardless of Mr. Ramsay's books being about "Subject
and object and the nature of reality"(Woolf 1992a: 33), Woolf's novel
hinges on subject-less moments, "the purest self-revelation of the artist",
according to Fry (Fry 1927: 42), as if indeed Virginia were labouring for
narrative form in Cézanne's *atelier*, experiencing her own "bewildering
labyrinth" (Fry 1927: 39) on the "odd road to be walking, this of painting. Out and out one went, further and further, until at last one seemed

to be on a narrow plank, perfectly alone, over the sea" (Woolf 1992a: 232). Besides their temporal contiguity and their common interest for the process of composition, Fry's study of the French artist and Woolf's *Lighthouse* exhibit other moments of resonance; among these, Fry's analysis of the *Compotier*, concluded by a note on 'The Portrait of Mme. Cézanne' – "the rugged massiveness of the volumes"(Fry 1927:55); similarly, Mrs. Ramsay's verbal and pictorial portrait blends questions "of masses, of lights and shadows" (Woolf 1992a: 73) and the many commonplace and insignificant still-lifes, so dear to Cézanne (and Fry), because they are "dramas deprived of all dramatic incidents" (Fry 1927: 42). It should also be reminded that such intimacy between the arts and narrative aesthetics – a cliché today, not so then – was being debated in the very heart of Woolf's world, since, in that same fated 1927, The Hogarth Press published Mauron's *The Nature of Beauty in Art and Literature*, prefaced and translated, needless to say, by Roger Fry:

> We have made the thought the equivalent of the volume; we must now take the written text as equivalent to the surface which delimits the volume [...] surface and text are both skins through which we make contact with the reality they cover. (Mauron 1927: 82)

In Mauron's formula, "words are swollen with reality"(Mauron 1927: 45); and, more important still, their reality is agitated by a constant movement that seems to answer Woolf's anxieties about the quivering of things and how to represent the vibrating and pulsating quality of experience. Mauron's pages, one cannot help imagine, must have encountered an ear avid to hear them either as a rationale or the confirmation of an intuition:

> The mind weaves between the diverse impressions – and without distinguishing the sensual from the emotive – more or less tenuous threads which can cause each word to construct as an inner reality a kind of fragile system – often without any logic other than that of associations or potential resemblances – a system which would suggest to the chemist those enormous molecules which constitute living matter, whose groupings are so insecure and so complicated that the body undergoes incessant small changes, tightening here and loosening there, always analogous but never identical with itself and never defined. (Mauron 1927: 45)

Still-lifes, colour, vibrations, cancelled barriers between subject and form, volumes, planes: freewheeling across media and genres, the multi-

farious threads of a complex aesthetic fabric are being woven silently from
St. Rémy to Rodmell and Tavistock Square and back.

Similarly enmeshed in complexity is the all too famous image of the
moths, originated in Cassis and destined to an intense but altogether enigmatic textuality, from *The Waves* – first entitled to the insects in question
– to the short posthumous story relating their death.

> I sit with moths flying madly in circle round me and the lamp [...] One
> night some creature tapped [so] loudly on the pane [...] it wasn't man
> or bird, but a huge moth, half-a-foot literally across. (Dunn 2000: 150)

Virginia's strong emotional reaction to her sister's narration has been
made much of: " I could see nothing else but you and the moths..."(Dunn
2000: 150); in the stride of the emotions they provoke, the moths acquire
a textual life of great density, whether in their referential function or soon
as metaphors of time itself: "They have been crippled days, like moths
with shrivelled wings unable to fly" (Woolf 1992b: 41). Such thematic
and aesthetic prominence had been announced by various entries in the
diaries:

> Slowly ideas began trickling in; and then suddenly I rhapsodized [...]
> and told over the story of the moths which I think I will write very
> quickly, perhaps in between chapters of that long impending book on
> fiction – Now the moths will I think fill out the skeleton which I dashed
> in here: the play-poem idea; the idea of some continuous stream, not
> solely of human thought, but of the ship, the night etc.. all flowing
> together; intersected by the arrival of the bright moths. (Bell 1982:
> 139)

Greater intensity still during the genesis of *The Waves* – "The current
of the moths flying strongly this way" (Bell 1982: 229) – while the last
moth to be consigned to posterity shows increased complexity in the web
of its associations: "It was as if someone had taken a tiny bead of pure life
and decking it as high as possible with down and feathers, had set it danc-
ing and zig-zagging to show us the true nature of life" (Woolf 1942: 6).
Driving their way into so many narrow and intimate corridors, including those of the writer's and other human beings' own mind, the moths
have colonized the brain and its convoluted mass of nervous substance
and become in a synecdoche of all physical and intellectual existence. As

a kernel image, therefore, the humble moths tunnel their way through Woolf's work, wrapping somehow form and subject of narrative in one single little word; as to their almost incantatory repetition in *The Waves*, it should perhaps be related to the idea that the novel is written "to a rhythm" and a music rather than a plot (Lee 1996: 640).

Finally, and less referentially, the signifier 'moth' surely acquires a deeper meaning in the light of the intimate concern with motherhood that pervades the novel, albeit in contrasting and mysterious ways: "We are in the passive and exhausted frame of mind when we only wish to re-join the body of our mother from whom we have been severed"(Woolf 1992b: 194), argues Neville in almost Kleinian tones. And Gillian Beer reminds us that:

> The first version of *The Waves* includes, very near the opening, the im-age of the maternal sea: "many mothers, & before them many mothers, & again many mothers, have groaned and fallen... Like one wave, suc-ceeding each other. Wave after wave... And all the waves have been the prostrate forms of mothers. (Beer 1992: xxvi)

In the wake of such overwhelming attention for the underlying or ex-plicit themes of maternity and creation, the moth lends itself to being read, also, as a truncated mother, a mother deprived of its last syllable and of its completion; a mother that haunts Woolf's narrative, silent and often unnamed, and gives it its unceasing and quivering tension. Whether homely summery creatures, enigmatic messengers of the past and the else-where or volatile signifiers, the moths carry with them the mental and visual geography of Cassis, as the first notes regarding *The Waves* in Woolf's diary make clear: "France, near the sea, at night, a garden near the win-dow"(Bell 1982: 139).

Into the whirlpool of language

Nothing linguistic is foreign to the circle of St. Rémy. Much of Provence, it ought to be recorded, is passionately bilingual and involved in the revivalist fervour triggered by Frédéric Mistral and other *Félibristes*. It is an admiring Fry who marvels at the local eagerness for words in dif-ferent tongues:

He [an old peasant…] is the great authority on the Provençal language. He has translated Homer into Provençal and is now doing Dante. He has a passion for words and every now and then began a long account of the history and meaning of some unusual Provençal word. (Sutton 1972: 461)

Not surprisingly, translation is the buzzword and a major concern, from and to Provençal, from and to English, with a loving stress on the voyage of words rather than on their belonging to any national idiom; a stress nurtured, poetically and theoretically, by the great shadow of Stéphane Mallarmé, the riddler of words and a constant companion in aesthetics to the St. Rémy group. Fry is never quite without Mallarmé during the Twenties: busy with the translation of the *Poems*, a task always interrupted and always returned to; busy with the thought of the French man's poetics: "Almost every guest invited to dine with him about 1920 would find him manuscript in hand, seeking the right words with which to fill the gap in his translation…"(Woolf 2003: 239). Fry's translation – 29 poems out of 64 – was published after his death, edited and revised by Charles Mauron and Julian Bell; what is perhaps more relevant is that Mallarmé the obscure and the cryptic invites to a new critical approach to writing, namely to treat words as "painters treat paint" (Woolf 2003: 240), because "everything must come out of the *matière* of [the] prose and not out of the ideas and emotions [the artist] describes"(Woolf 2003: 240). Woolf has no second thoughts about it: "Mallarmé stood with Cézanne among his [Fry's] patron saints"(Woolf 2003: 239). The two saints are reunited in the early introduction to the Mallarmé volume:

Few poets have devoted so large a portion of their oeuvre to subjects which, were it a question of painting, we should classify as still-life. No one has given to the words for common objects so rich a poetical vibra-tion – fenêtre, vitre, console, verrerie, pierrerie lampe, plafond… (Mal-larmé 1936: 305)

Mallarmé and Fry's considerations on him were certainly not lost to Woolf, and her words should be taken at face value when she claims ve-hemently, in a letter to her friend:

You have I think kept me on the right path, so far as writing goes, more than anyone […] I venerate and admire you to the point of worship: Lord! You don't know what a lot I owe you… (Nicolson 1977: 385, 562)

Let us again listen to Virginia when she declares herself unable to discover "whats [sic] the essential difference between prose and poetry. It cracks my poor brain to consider" (Beer 1992: xvii). Let us also bear in mind, with her biographer, that, while writing *The Waves*, Woolf had been thinking continually about the relationship between poetry and fiction (Lee 1996: 609), and let us finally remember that the book was to be "abstract poetic"(Lee 1996: 640): it is easy then to appreciate how much Woolf is involved in the Mauron/Mallarmé/Fry constellation; how much at home she feels when invited by Mauron to "descend to technical considerations and linger in the laboratory", as well as to relegate "sternly to the background the documentary value of a work of literature"; how sympathetic she may be to the idea that "Half-lights are a part of art [that] have their place in the twilight world of poetry where information is at most accessory" (Mauron 1927:14, 70). Woolf is likely to have identified a twin soul in the Mallarmé Fry and Mauron were introducing her to. Like hers, his sentences are full of subtle convolutions (Mallarmé 1936: 29); like her, he bends and dislocates the syntactical order. There is kinship again between the poet and the novelist in their being both fond of "acrobatic feats, rocket-like sallies, glidings upon air"(Mauron 1936: 40); thanks to the rhythmical repetition of single words, the language of *The Waves* "pirouettes" acrobatically, as Jenny does (Woolf 1992b: 79, 132), but is also ready to 'glide' towards an entirely different register, of decomposition and loss. The flow, the ripple and the flutter of life repeatedly give way to the image of a crumbling London (Woolf 1992b: 23, 48, 154), and the whole world appears to be threatened by general extinction: "Our separate drops are dissolved: we are extinct, lost in the abyss of time, in the darkness"(Woolf 1992b: 188).

Following its principle of composition, 'to a rhythm', the novel orchestrates *Leitmotive* as so many threads which eventually combine into a textile metaphor recurring throughout the novel to interrogate its very making: " I have heard threads broken and knots tied and the quiet stitching of white cambric going on and on on the knees of a woman (Woolf 1992: 164). As to the elusive storyline, it is undoubtedly connected to the intricate process of knotting/unknotting: "Love makes knots, love brutally tears them apart. I have been knotted, I have been torn apart" (Woolf 1992b: 179). Weaving may be unsuccessful – "It breaks [...] the thread I try to spin"(Woolf 1992b: 182) – or, on the contrary, it may yield some sort of fabric: "I am wrapped round with phrases"(Woolf 1992b:

181). Sewing, weaving, breaking, knotting: the semantic field unravels its enigmas to the end of the novel, where its aesthetic relevance is, once again, stressed:

> She [Susan] was born to be adored of poets since poets require safety; someone who sits sewing, who is neither comfortable nor prosperous, but has some quality in accordance with the high but un-emphatic beauty of pure style which those who create poetry so particularly admire. (Woolf 1992: 207)

Thus unfolds the poem/novel, sustained by recurring images and sounds which however do not lend themselves to being deciphered into meanings and remain, rather, irremediably obscure. Similarly, the uncertainty never stops as to ways of telling the story; will it be a traditional one, will the writer pretend "that we can make out a plain and logical story, so that when one matter is despatched [...] we go on, in an orderly manner, to the next" (Woolf 1992b: 210); or are we, rather, to believe the narrator who declares himself "tired of stories, [...] of phrases that come down beautifully with all their feet on the ground!" and more attracted by such a language "as lovers use, broken words, inarticulate words, like the shuffling of feet on the pavement" (Woolf 1992b: 199)? Or perhaps it is Bernard's conception of writing that should prevail: "It is the speed, the hot, molten effect, the laval flow of sentence into sentence that I need" (Woolf 1992b: 63). Unceasingly questioning the modes of writing, experimenting fearlessly with the instruments of music and painting, exploring the recesses and crannies of language and its non-intelligibility, *The Waves*, as poetry indeed does, offer a "complex of word-images and their associations tend to set up vibrations which continue in the mind" (Mallarmé 1936: 294). These are some of Fry's many words for Mallarmé, words that, once more, seal the affinity between Mallarmé *l'obscur*[3] and Woolf the obscure, as well as their common attraction for "the mysterious *miroitement en dessous*, imprisoned in the poet's most cryptic verses..."(Woolf 2003: 39)[4]

---

[3] This is the title of the volume Charles Mauron published in 1941.

[4] For a more technical account of the poetic and linguistic intersections Woolf/Mallarmé, see my essay 'Cratylian Turbulence in the Air', in Daniela Carpi ed., 2005.

Miroitement en dessous: Fry's expression resonates with Mauron's invitation to descend under the surface of texts, with the restlessness of language and its mysteries; it is also an indirect reminder that Provence, or the specific triangle delimitated by Les Baux, Arles and Aix, is cut across by various cultural threads and traditions which intersect and contribute to the linguistic and creative turmoil:

> Provence is, perhaps, the most intelligible land that there is [...] who has, inevitably, by one means or another, such continuous references to the Greek spirit. For Provence is even more Greek than Roman, and the bent must have been given to its civilization before Rome came along. Otherwise, how explain the fact that Roman architecture and Roman sculpture are so much finer and more sensitive than in Rome itself? (Fry 1926: 173)

While Greek and Roman, the local culture is also tinged and touched by Oriental influences, as the eagle in the Romanesque doorway of St. Gilles church, "comparable to the very best of Chinese sculpture and to the most beautiful Persian drawings of the twelfth century"(Sutton 1972: 556). And indeed Oriental are the Gypsies who have mysteriously travelled from India and gather annually at the fair of Beaucaire or the Saintes Maries festival. The Gypsies inhabit the country but above all they people the imagination of Provence, from the poems of Frédéric Mistral to the pictures of Vincent Van Gogh, soon to invade British culture, in the paintings of Augustus John and in his philological excursions into the Gypsy language; and, crucially, of course, in Virginia Woolf's writing. Gypsies accompany Orlando in Turkey and turn up unexpectedly in many novels: "I like best the stare of shepherds met in the road; the stare of Gipsy women beside a cart in a ditch suckling their children as I shall suckle my children" murmurs Susan in *The Waves* (Woolf 1992b:79). As it multiplies the dimensions of space, Provence similarly enhances the multiplicity of time, thanks to the palimpsest of presences and traces it is custodian to; St.Rémy, yet again, offers a stage to such unexpected stratification of prehistoric, Greek and Roman relics (Pl. 3). The excavation of the ancient town of Glanum, an archaeological site lying precisely in front of Fry's and the Maurons' *mas*, had started in 1921, in time, therefore, to be contemplated by the Bells and the Woolfs when they visited their friends in 1928 (Fig. 1). It is noteworthy that Van Gogh's picture of the same tract of landscape, 'Olive Trees with the Alpilles in the Background', completed in 1889, should reveal nothing of what was then ly-

Figure 1 – The Ruins of Glanum at St. Rémy, 1999. Photograph.

ing undiscovered under the ground! Whether related to St. Rémy or not, Woolf's archaeologic imagination is as vivid as central to her narrative and linguistic universe, a verbal landscape equally to be excavated and explored in its multiple strata: "Everyday I unbury – I dig up. I find relics of myself in the sand that women made thousands of years ago..." (Woolf 1992b: 104). Sensual and textual, theoretical and anecdotic, austere and enigmatic, this Provence holds a mirror to the anxieties of the English Modernists and offers food for their aesthetic appetites. For once, however, the South was not being consumed, erotically, or culturally; nor was it being 'museified', in the style of Ruskin, or even worse, recreated in the image of British longings and aversions, hopes and fears, as was the habit of Victorians and Edwardians (Pemble 1987: 274). On the contrary, this Provence was being addressed, talked to, interacted with, interrogated; it was made the partner in a dialogue, the pole of a dialectics. In their intimacy with St. Rémy and Mauron, Fry and Woolf had somehow reshuffled the cards of the Mediterranean game.

Bibliography

Banfield, Ann, 2000, *The Phantom Table. Woolf, Fry and the Epistemology of Modernism*, Cambridge University Press, Cambridge.

Beer, Gillian, 1992, 'Introduction' to Woolf Virginia, 1992b, pp. xii-xxxvi.

Bell, Anne Olivier (ed.), [1980] 1982, *The Diary of Virginia Woolf*, Penguin, Harmondsworth.

Carpi, Daniela (ed.), 2005, *Why Plato? Platonism in Twentieth Century English Literature*, Winter, Heidelberg.

Caws, Mary Ann, 1994, *Bloomsbury in Cassis*, Cecil Woolf Publishers, London.

—— *et al.*, 2000, *Bloomsbury and France*, with a Preface by Michael Holroyd, Oxford University Press, Oxford.

Dunn, Jane, [1990] 2000, *Virginia Woolf and Vanessa Bell. A Very Close Conspiracy*, Virago, London.

Fry, Roger, 1926, *Transformations. Critical and Speculative Essays on Art*, Chatto & Windus, London.

——, 1927, *Cézanne. A Study of his Development*, Hogarth Press, London.

Gardes, Joëlle, 2002, *Virginia Woolf à Cassis*, Images en Manoeuvres Editions, Photographies Christian Ramade, Marseille.

Lee, Hermione, 1996, *Virgina Woolf*, Chatto & Windus, London.

Mallarmé, Stéphane, 1936, *Poems*, translated by Roger Fry, with Commentaries by Charles Mauron, Vision Press, London.

Mauron, Charles, 1927, *The Nature of Beauty in Art and Literature*, Translation and Preface by Roger Fry, Hogarth Press, London.

——, 1938, *Esquisse pour le tombeau d'un peintre*, Denoël, Paris.

——, 1941, *Mallarmé l'obscur*, Denoël, Paris.

Nicolson, Nigel, 1977, *A Change of Perspective. The Letters of Virginia Woolf*, vol. 3, 1923-1928, The Hogarth Press, London.

Patey, Caroline, 2005, 'Cratylian Turbulence in the Air', in Daniela Carpi (ed.), 2005, pp. 189-198.

Pemble, John, 1987, *The Mediterranean Passion. Victorians and Edwardians in the South*, Clarendon Press, Oxford.

Reed, Christopher (ed.), 1996, *A Roger Fry Reader*, The University of Chicago Press, Chicago-London,

Silver, Kenneth E., 1988, *Making Paradise. Art, Modernity and the Myth of French Riviera*, MIT Press, Cambridge (Mass).

Sutton, Denys (ed.), 1972, *Letters of Roger Fry*, Chatto & Windus, London, 2 vols.

Vitaglione, Daniel, 2000, *The Literature of Provence. An Introduction*, Mc Farland and Co. Inc. Publishers, Jefferson and London.

Woolf, Leonard (ed.), 1959, *A Writer's Diary. Being Extracts from the Diary of Virginia Woolf*, The Hogarth Press, London.

Woolf, Virginia, [1927] 1992a, *To the Lighthouse*, Oxford University Press, Oxford.

——, [1931] 1992b, *The Waves*, Oxford University Press, Oxford.

——, [1940] 2003, *Roger Fry. A Biography*, Vintage, London.

——, 1942, *The Death of the Moth and Other Essays*, Harcourt Brace and Co., New York.

UNTENDER IS THE NIGHT IN THE GARDEN OF EDEN: FITZGERALD, HEMINGWAY, AND THE MEDITERRANEAN

Mario Maffi

The science of the sea is the study of courses and streams, the chemical analysis of salinity rates, stratigraphic surveys, map of the benthonic and pelagic domain and subdivision into euphotic, oligophotic and aphotic zones, and the measurement of temperatures and winds. But it is also the history of wrecks and the myth of sirens, sunk galleons and primeval Leviathans, humankind's original amnios and cradle of civilization, the Greek form which comes forth from the sea as perfect as Aphrodite, the great trial of the soul which Musil writes of, the encounter with the symbol of the eternal and the persuasion: that is, of life sparkling in its pure, incorruptible present, in its fullness of meaning...

Claudio Magris (1991, 2004)

Foreword

Before analyzing two novels like *Tender Is the Night* by F. Scott Fitzgerald and *The Garden of Eden* by Ernest Hemingway, in the context of this International Conference on "Anglo-American Modernity and the Mediterranean", a few explanations are needed as to the status of the two works. The first was published in 1934 by Scribner's and, with *The Great Gatsby* (1926), was one of Fitzgerald's great successes. *The Garden of Eden*, on the contrary, was published posthumously in 1986 (also by Scribner's), an operation which gave rise to a heated debate. Hemingway had started writing it immediately after the end of World War Two, and when he died in 1961, he left an unfinished manuscript of some 1500 pages: from this the publisher pruned it down to just 250 pages, a questionable decision to say the least. Yet neither was the publishing life of *Tender Is the Night* a straightforward one. After the 1934 publication, Malcolm Cowley put out a new edition in 1951, based upon a drastic revision which had been started but left incomplete by Fitzgerald himself with a view to restructuring the novel along chronological lines (seemingly upon the advice of Hemingway, who thus paid back the debt contracted at the beginnings of his own career, when Fitzgerald had insisted that he cut the opening pages of *The Sun Also Rises*). Several interventions by Cowley himself were also added, up to the point that – as indicated in the 1986 Penguin Edition's final "Note" – "there is no way in which a reader of the revised edition can tell without research whether Fitzgerald actually wrote what he is reading" (Fitzgerald 1986: 339). Over time, and almost unanimously, critics sided against this "revised edition" and the text presently available offers again, with slight variances, the one published in 1934 (see Bruccoli 1963). The fact remains that both novels (although more so Hemingway's than Fitzgerald's) went through continuous variations of status, internal and external interventions, mutations, drifts and shifts. So much so that they contain something fluid and uncertain: this, as we shall see, is hardly serendipitous.[1]

Analogies

The analogies obviously go well beyond the similar status of the two novels. *Tender Is the Night* and *The Garden of Eden* appear to be linked together in a manner which is almost subterranean, subliminal even: perhaps, in the same ambiguous and contradictory (at times even troubled) manner which bound together the two authors as friends and enemies (see Donaldson 2001). Here, I will not delve too deeply into the complex interplay which creates a kind of warp connecting the two novels, made up of cross-references, hints and perhaps even digs ("Once I knew a man who fell in love with his nurse", *TIN*: 170; the continuous references to hair-cutting; several other repeated details…). Certainly, the theme of the fall (of man, of the artist, of "he who creates and acts") is central to both nov-

---

[1] The editions I will use are the 1986 Penguin one for *Tender Is the Night* (hereafter *TIN*) and the 1994 Harper Collins one for *The Garden of Eden* (hereafter *GE*).

els. So is the theme of the difficult relationship with women (seen as a menace, an impending madness, with all the implications which go from homosexuality to androgyny, from fear of castration to insecurity about one's own identity, both cultural and sexual). Another central theme shared by both novels is that of the expatriates in the Twenties, the "colony" of Americans (but not only) on the French Riviera, and the way in which, within such a microcosm (analyzed almost *in vitro*) expatriation, withdrawal within the self, the differences and separation from what lies outside, the non-casual renewal of complicated *ménages à trois*, all work toward creating further tensions, further splits. Money, "to have and have not", consistency with one's own being and doing, the difficulty of redemption after the fall, the threat of failure and the necessity for inner discipline, for rules in creating and behaving, all represent another complex series of analogies. Last but not least, the colours of the coasts, countryside and small villages, the water of the Mediterranean Sea and "what lies beyond" (be it the "offing", the buoy, or the horizon, Africa), the repeated references to the avant-garde painters who were then the interpreters of those landscapes and colours,[2] are another common *Leitmotiv*. In this way, *The Garden of Eden* almost seems Hemingway's answer to *Tender Is the Night*, and *Tender Is the Night* comes across as being in a constant dialectical relationship with Hemingway's work – the work of an artist who, for Fitzgerald, remains a much disputed and contradictory landmark.

The Riviera as garden

In this interplay of cross-references, the Mediterranean Sea (the stretch of the French Riviera situated between Nice and the Camargue) surely plays a central role: with it both novels open, and on it they are struc-

---

[2] A few examples: "Their room was a Mediterranean room, almost ascetic, almost clean, darkened to the glare of the sea. Simplest of pleasures – simplest of places. [...] 'I like this room', she said. [...] '[...] this is a wonderful room, Tommy – like the bare tables in so many Cézannes and Picassos'" (*TIN*: 316, 317). "The room they lived in looked like the painting of Van Gogh's room at Arles except there was a double bed and two big windows and you could look across the water and the marsh and sea meadows to the white town and bright beach of Palavas" (*GE*: 9). These two descriptions of the room in which the characters' most intimate rituals take place belong to the interplay of cross-references between the two novels. Other examples could follow.

tured, with centrifugal chapters, which do however always imply a kind of "return".

Let us read the opening sentences of *Tender Is the Night*:

On the pleasant shore of the French Riviera, about half way between Marseilles and the Italian border, stands a large, proud, rose-colored hotel. Deferential palms cool its flushed façade, and before it stretches a short dazzling beach. Lately it has become a summer resort of notable and fashionable people; a decade ago it was almost deserted after its English clientele went north in April. Now, many bungalows cluster near it, but when this story begins only the cupolas of a dozen old villas rotted like water lilies among the massed pines between Gausse's Hôtel des Étrangers and Cannes, five miles away.
The hotel and its bright tan prayer rug of a beach were one. In the early morning the distant image of Cannes, the pink and cream of old fortifications, the purple Alps that bounded Italy, were cast across the water and lay quavering in the ripples and rings sent up by sea-plants through the clear shallows. Before eight a man came down to the beach in a blue bathrobe and with much preliminary application to his person of the chilly water, and much grunting and loud breathing, floundered a minute in the sea. When he had gone, beach and bay were quiet for an hour. Merchant-men crawled westward on the horizon; bus boys shouted in the hotel court; the dew dried upon the pines. In another hour the horns of motors began to blow down from the winding road along the low range of the Maures, which separates the littoral from true Provençal France. (*TIN*: 11)

And now let us read the opening sentences of *The Garden of Eden*:

They were living at le Grau du Roi then and the hotel was on a canal that ran from the walled city of Aigues Mortes straight down to the sea. They could see the towers of Aigues Mortes across the low plain of the Camargue and they rode there on their bicycles at some time of nearly every day along the white road that bordered the canal. In the evenings and the mornings when there was a rising tide sea bass would come into it and they would see the mullet jumping wildly to escape from the bass and watch the swelling bulge of the water as the bass attacked.
A jetty ran out into the blue and pleasant sea and they fished from the jetty and swam on the beach and each day helped the fishermen haul in the long net that brought the fish up onto the long sloping beach. They drank aperitifs in the café on the corner facing the sea and watched the

sails of the mackerel fishing boats out in the Gulf of Lion. It was late in the spring and the mackerel were running and fishing people of the port were very busy. It was a cheerful and friendly town and the young couple liked the hotel, which had four rooms upstairs and a restaurant and two billiard tables downstairs facing the canal and the lighthouse. The room they lived in looked like the painting of Van Gogh's room at Arles except there was a double bed and two big windows and you could look across the water and the marsh and sea meadows to the white town and bright beach of Palavas. (*GE*: 9)

The two *incipits* contain many of the implications of the two novels. First of all, the sense of "place as a threshold", as a "middle territory", reinforced, in Fitzgerald's text, by the repeated use of such expressions as "half way", "between", "separates", and, in Hemingway's, by the setting of the hotel *between* the "walled city of Aigues Mortes" and the open sea. Then, the repeated situations in which the glance sweeps beyond, towards the horizon, towards faraway landscapes: "the distant image of Cannes, the pink and cream of old fortifications, the purple Alps that bounded Italy" (Fitzgerald), "the sails of the mackerel fishing boats out in the Gulf of Lions [...] the white town and bright beach of Palavas" (Hemingway).

This "threshold", this "middle territory", occupies a fundamental place in the cultural and existential geographies of the American expatriates in Europe in the Twenties. As Malcolm Cowley has convincingly shown in his *Exile's Return* (1934), the flight of this youth – mostly writers, or would-be writers – from the United States towards Europe did not only follow in the footsteps of previous generations, of the Hawthornes and the Melvilles, of the Jameses and the Steins. It also expressed urgencies which were quite different: the "revolt against the village", the "refusal of provincialism and Babbitry", the "reaction against a stifling Puritanism", the "traumatic experience of World War I", the "rifts of modernity", and – perhaps much more trivially, but at the same time more concretely – the "strong dollar" (see Cowley 1994: *passim*). It was a flight from a "stepmother America" and a return to the Old World, a search for identities shaken by so many diverse and barely manageable factors – internal and external, historical, generational, cultural and psychological. All too often, however (as the Parisian works of these authors well illustrate), this attempt to put down roots again in the Old World, to confront European realities, history, and culture, has no happy ending – almost as if these American expatriates had no choice but to carry along, with and within themselves, the very tensions and misunderstandings from which they were running away. The picture Hemingway and Fitzgerald give us of this condition, respectively in *The Sun Also Rises* (1926) and "Babylon Revisited" (1931), is tellingly dramatic: violence, frustration, aggressiveness, disorderly behaviour, drunkenness, fragmentation of the self and of the group, disillusionment and bitterness.

So Southern France and the rivers and mountains of Spain offer two possible alternatives: not so much two escape routes as two middle territories to take off to – America behind (or the American colony transplanted to Europe), the sea ahead (and, beyond it, the African or Spanish "wilderness"), and in the middle the white, almost deserted, beaches, the Grau du Roi, the Corniche, Juan-les-Pins and the Esterel: where one might come to terms with one's self, with one's own identity and with European culture, and thus find one's own voice, mature and modern. *But...*

But this "middle territory" is nothing new to American culture. Leo Marx accurately explored it in his seminal *The Machine in the Garden* (1964), showing how, from the very beginnings of the colonial experience, the New World was conceived and presented (with an aggressive ideological construction enveloping precise material and economical determinations) as a "garden of Eden", where it would be possible to give new birth to human history and culture, with far more complex implications than those of a purely Biblical nature. A "garden", on the other hand, threatened from the very outset: by society, which necessarily encroaches upon it through colonization and civilization, and by a wilderness which is all that lies "beyond the pale", beyond the Frontier (external or internal, real, metaphorical, psychological). What then remains is only that "middle territory", fluid and ephemeral, precariously situated between two extremes, in unstable balance, doomed to be continuously assailed – by machine or by wilderness (see Marx 1999: *passim*).

A great part of the cultural history of the United States between the 19th and the 20th centuries is clearly marked by this precariousness, by this two-sided threat, and by the search for escape routes in order to save the "garden" simply by moving it elsewhere. By the end of the 19th century the Frontier had officially been closed, and American culture thus remained unbalanced, deprived of an element which, from the 17th century on, had functioned as a powerful experiential and ideological safety valve. The flight from the village, the return to the East, the encounter with huge metropolises and triumphant technology, the search for new

routes towards the Pacific Ocean or backwards, across the Atlantic Ocean
to the Old World, are only some of the traces of this unbalance, time and
again deeply anguished. And the generational expatriation of the Twenties also carries with itself some of these connotations.

Once the anguished propulsive push which had led so many youths
to the Eastern metropolises was exhausted, only two roads remained: the
one taken by Nick Carraway in *The Great Gatsby* and that taken by Jake
Barnes in *The Sun Also Rises*, both narrators of generational defeats. The
former, after experiencing New York's turbulent whirlwind, goes back to
the West abandoned years before because it was the "ragged edge of the
universe" (Fitzgerald 1967: 9), thus closing the circle; the latter, with the
complicity of the war, goes East, across the Atlantic Ocean, "over there",
back to old Europe. But neither of the two "types" reaches salvation. The
former will only be able to "beat on, [boat] against the current, borne
back ceaselessly in the past" (1967: 188). As to the latter, Bill Gorton
will tell him: "You're an expatriate. You've lost touch with the soil. You
get precious. Fake European standards have ruined you. You drink yourself to death. You become obsessed by sex. You spend all your time talking, not working. You are an expatriate, see? You hang around cafés"
(Hemingway [1926] n.d.: 120). Uprooted both, and landless.

Once more, here is the double, menacing polarity; and with it, the
need to reinvent a "garden", an (equidistant, neutral) "middle territory",
where it would be possible to find balance and identity again, and perhaps a "separate peace" – far from the uncouth meanness of American
mass society *and* from "fake European standards", those infections cours-
ing through the aging veins of the Old World, which Henry James and
Edith Wharton had often written about. To Fitzgerald and Hemingway,
that "garden", that "middle territory", would be Spain and the French
Riviera, both overlooking a "middle sea", the Mediterranean Sea.

Sea changes

But – as we know – the American Frontier is not a fixed, uninterrupted line. Rather, it is an *oscillating situation*, mobile and uncertain. In
the very same way, the sea appears to offer the necessary element (changeable and undulating) in which to set a painful mutation. We also know
that, again, this kind of mutation is something deeply embedded within

American culture, especially that part of it which has more to do with the primeval essence of the "American Myth": the Western "tall tales" with their "half-horse, half-alligator" characters, James F. Cooper with his partly Indian, partly civilized Leatherstocking, Poe with his "divided selves", Hawthorne and Melville with their recurring images of fearful and uncertain gestures of threshold crossing (not to speak of the modern – and contemporaneous to Fitzgerald and Hemingway – rewritings by Faulkner, or of certain re-readings operated by mass culture, such as the obsessive and significant 1956 movie by Don Siegel, *Invasion of the Body Snatchers*). It is interesting to see how this condition – that of the *homo americanus* caught in the moment of a change of skin – is reintroduced and reinterpreted by our two authors, *in a radically different context*: on a Riviera which, almost as a veritable space-time niche, becomes for a very brief time (as for the Divers) "home".

The sea in which one swims (or on which one travels by steamer) thus really is, metaphorically and symbolically, "humankind's original amnios", with diverse valences and implications: "On the long-roofed steamship piers one is in a country that is no longer here and not yet there. [...] the past, the continent, is behind; the future is the glowing mouth in the side of the ship; the dim, turbulent alley is too confusedly the present" (*TIN*: 224). It is precisely here that the act of mutation can better take place: as occurs in the case of David and Catherine, whose hair and bodies become, on the seashore and in the waves, blonder and blonder, more and more sun-burnt – and ever more similar the one to the other, up to the point of allowing a kind of change of gender, of sexual roles (a dimension already anticipated, in *Tender Is the Night*, by Nicole when – fearing a new crisis – she asks Dick to help her and concludes: "We'll live near a warm beach where we can be brown and young together", *TIN*: 178).

It is precisely here that, in two novels where the theme of change obsessively returns (well expressed by the insistence with which the noun "change" and the verb "to change" resonate), this mutation can take place: "a dark magic" (*TIN*: 181), "the dark magic of change" (*GE*: 28); perhaps, one last chance to reinvent the self, to gain release from the restrictions of the past, so rigid and repressive, so Puritanical and self-referential. At the station in Paris, waiting for the train which will take him to his new "home" on the Riviera, Dick sees (in what is an explicit play of mirrors) a group of Americans, and Fitzgerald gives voice to his thoughts: "Nearby, some

Americans were saying good-by in voices that mimicked the cadence of water running into a large old bathtub. *Standing in the station, with Paris in back of them, it seemed as if they were vicariously leaning a little over the ocean, already undergoing a sea-change, a shifting about of atoms to form the essential molecule of a new people"* (*TIN*: 95, my italics).[3]

An illusion of change and rebirth, then: but one that is destined to last only a very little while.

Other "machines" in the garden

In these two parallel stories of American expatriates on the Mediterranean shore, the early luminosity of sea and shore soon darkens, and the repeated, regenerating and salvific act of diving in the Mediterranean waters and swimming off[4] takes on, in the end, ambiguous implications. It almost becomes an (involuntary) path of soul-searching and unconscious-studying (Dick Diver is a psychoanalyst and David Bourne's writing is quite akin to self-analysis), and as such it brings to light obscure areas – perhaps exactly that "map of the benthonic and pelagic domain and subdivision into euphotic, oligophotic and aphotic zones", which is also (as Magris writes) "the history of wrecks and myth of sirens, sunk galleons and primeval Leviathans", as well as "humankind's original amnios and cradle of civilization" (see Sullivan).

The American expatriates' "last shore" is thus menaced and finally invaded by other "machines in the garden": human ones, this time, and not technological, but no less destructive. What dissolves the utopia of

[3] In "The Sea Change", a short story which Hemingway included in *The First Forty-Nine Stories* (1939), a couple of expatriates splits up in a Paris café: after a summer spent together on the seaside, the girl (*sun-burnt and with very short hair*) leaves the man, having just started a homosexual relationship. At the end, the man will say to the waiter: "I'm a different man, James" (Hemingway 1995: 401). I don't believe that in the complicated Fitzgerald-Hemingway play of mirrors and in the continuous surfacing of such a theme in Hemingway's work, this is casual: it is indeed probable that "The Sea Change" is the initial nucleus of *The Garden of Eden*.

[4] It is important to remember that Fitzgerald's two main characters are called Dick and Nicole *Diver*, better known as "the Divers"; and that David Bourne, Hemingway's main character, loves to dive off the Eden Roc, alternately grazing Catherine and Marita. In the names as well as in the acts, the sexual symbology is quite explicit.

expatriation is something which often surfaces in the work of Fitzgerald and Hemingway, and binds together these two novels more than ever happened elsewhere in the works of their authors: "madness"[5] and women. The entire personal story of Dick and Nicole is ruled by her hovering "madness" and Dick's role (his identity, strong at first and central, then all the more arguable and marginal, finally crumbling to pieces) is that of being also Nicole's doctor. Her "madness" presents itself in a subtle way, in alternate phases, mostly hidden to the eyes of onlookers, but it is a madness which corrodes and empties – and finally explodes when Rosemary Hoyt appears on the scene. For its part, the story of David and Catherine is that of a gradual descent into the abyss of her "madness", which David vainly tries to control: a more turbulent "madness", dotted with sudden crises, acute and explosive, clear and cutting, which hurts others as well as Catherine herself, and also has to do with problems of roles and identity – and finally explodes when Marita appears on the scene. So, while Dick "dives", personally and professionally, into Nicole's subconscious, and emerges defeated from this "voyage to the end of the night", David seems to be rather more engaged in "being borne" and in "bearing" (Bourne = born and borne), through his writing as self-analysis and through the dialectics of arduous interpersonal relationships.

Both kinds of "madness" are then linked together by certain rituals, with hair cutting – and all the symbolic implications it contains[6] – surely the principal one. If in Fitzgerald's novel it is simply alluded to (Baby Warren: "before I knew, almost in front of my eyes, she had her hair cut off, in Zurich, because of a picture in Vanity Fair". Dick Diver: "That's all right. She's a schizoid – a permanent eccentric. You can't change that", *TIN*: 167-168), in Hemingway's it becomes obsessive, the real pivot around which

---

[5] It is quite obvious that, in both novels, the point of view from which Nicole's and Catherine's behaviours are experienced and narrated is a male one, but a discussion of their meaning and character would lead us too far away. I thus put "madness" between brackets, to indicate the ambiguity of the issue (see Sullivan). Besides, in the two authors, the theme of "madness" contains a whole series of well-known biographical and autobiographical implications, which I also chose to leave aside.

[6] It must be remembered that Fitzgerald has his place in the history of American culture and costume as the singer of the "flappers" (so called from their peculiar haircut): a disparaging definition, certainly, but not entirely inappropriate. And that the act of hair cutting is often repeated in Hemingway's writing (Maria in *For Whom the Bell Tolls?*). The feminine model of these two authors is undoubtedly androgynous.

David and Catherine's story revolves, a symbol and an instrument of their crisis through a gradual reversal of roles, and above all of sex and gender: "She said, 'Now you can't tell who is who can you?'. 'No'. 'You are changing,' she said. 'Oh you are. You are. Yes you are and you're my girl Catherine. Will you change and be my girl and let me take you?'. 'You're Catherine.' 'No. I'm Peter. You're my wonderful Catherine. You're my beautiful lovely Catherine. You were so good to change. Oh thank you, Catherine, so much. Please understand. Please know and understand. I'm going to make love to you forever'" (*GE*: 25). It is exactly this "dark magic of change" (*GE*: 28) that will lead to the "disregard of the established rules" (*GE*: 184): perhaps an attack on the Anglo-American Puritanical background, as well as on "fake European standards"?

The second element which links together the two kinds of "madness" (and the two novels) is the role of money and, in a wider sense, of fame and success – a theme often dealt with by the two authors, and by Hemingway in particular ("The Snows of Kilimanjaro"). Money, serving as a springboard to success for the two male protagonists, is firmly in the grip of women: it is *they*, Nicole and Catherine, who "finance" Dick's and David's professional and creative activities, and this – against the backdrop of Puritan ethics and of a burgeoning psycho-analytic interest – cannot but have an influence on the two males, on their self-consciousness, self-esteem and role- and gender-identity.[7] Money is sex and sex is money: but the equation is "dominated" by the female figure, and to evade it one either has to give up money (as Dick Diver seems to do, in a growing self-annihilation, to the extent of disappearing from the scene, buried again in small-town American – perhaps a new and more mature Nick Carraway) or give up sexual identity (as David Bourne is tempted to do, by accepting the "dark magic of change" and penetrating into unknown and disquieting territories, as had already happened to Jake Barnes in *The Sun Also Rises* as a result of a war wound). In both cases, it implies a renunciation, an abandoning of fixed and accepted roles, in favour of a fluidity and uncertainty, of a precariousness which recurs in the two authors' work – a metaphor of modernity which is celebrated (and to which sacrifices must be offered) on the shores of the Mediterranean and in its eternally

---

[7] This does not concerns the two female protagonists only, but seems to be a recurring feature of female characters. Says Mrs. Seer (*nomen omen!*) to Rosemary, at the very beginnings: "economically you're a boy, not a girl!" (*TIN*: 50).

changing waters, which do indeed hide "sirens, sunk galleons and primeval Leviathans".

If therefore the Mediterranean is yet another "middle territory", a "garden" menaced and finally invaded by women as carriers of money and "madness" (i.e., by civilisation as symbolised and embodied by women: from which all the Rips and Hucks of American culture try to run away), the alternative once more remains the "Territory", the "wilderness", what lies "beyond the pale" – in a lesser measure for Dick in *Tender Is the Night*, who experiences it in terms of intoxication, brawls, abandon (especially in nocturnal Rome, where Fitzgerald's Irish Catholic heritage freely and contradictorily mixes with his well-known hostility towards Italians)[8]; in a greater measure for David in *The Garden of Eden*, who finds it, always *à la* Hemingway, in fishing (the long opening scene of the struggle with the fish: as in *The Old Man and the Sea*, in *The Sun Also Rises*, in "Big Two-Hearted River"), and above all in African big-game hunting, lived, imagined, or told: it is hardly a coincidence that David's "African story" (with which the young writer totally identifies, in a *"transfert"* – again the "transference"? – which also involves the structure of the novel itself) ultimately leads to Catherine's climax of "madness", and in the end becomes her "intended victim".

Once again: behind, civilization; in the middle, the garden; ahead, the wilderness. But, at the same time, by now, not even in the Old World, not even on the Mediterranean shore, are these "places" safe and sound. As Dick says: "The pastoral quality down on the summer Riviera is all changing anyhow" (*TIN*: 197). Only desperation remains, and more and more it grips the two authors and their characters: untender is the night in the Garden of Eden. Or, as Leo Marx remarked: "American writers seldom, if ever, have designed satisfactory resolutions for their pastoral fables" (Marx 1964: 364).

---

[8] Elsewhere, Dick will think of the Mediterranean as a "world of pleasure – the incorruptible Mediterranean with sweet old dirt caked in the olive trees, the peasant girl near Savona with a face as green and rose as the color of an illuminated missal" (*TIN*: 214). *À propos* of that initial "incorruptible", one ought to go back to the opening quotation from Claudio Magris.

Mixed techniques

But the Mediterranean contains other implications as well, for the two authors and for the two novels. It is a well-known fact that, to Fitzgerald and Hemingway, the French period spent between Paris and the Riviera in the early 1920s meant an opening towards the avant-garde, towards Post-Impressionism. This opening – for which a key *liaison* was offered by Gertrude Stein, the Paris *salons*, the almost daily encounter with painters and sculptors, the city atmospheres and the colours of the South and of the sea – manifests itself differently for the two writers. In the case of Fitzgerald, what is quite apparent from *The Great Gatsby* onwards (it was finished in the summer of 1924, at Valescure, some two miles from St. Raphaël, on the Riviera), is the influx of modern art techniques, and particularly those of the Cubists. The modes of character construction (and of Jay Gatsby *in primis*) obey the principles of de-composition and re-composition of reality typical of Braque or Picasso: the main character is never "described", but "delineates itself", "materializes" through different points of view (voices, opinions, glances, perspectives) of other characters – in a word, through the aggregation of different and even contrasting features, and thus through the revealing "epiphanies" which emanate from their complex dynamics. Besides, "certain specific features and objects (for instance, 'the eyes of dr. T.J. Eckleburg', the huge, abandoned advertising signpost which dominates the landscape of the 'valley of ashes') openly recall the common features and objects of early Cubist *collages*, especially by Braque" (Maffi 2001: 209), with deeply symbolic implications and valences (almost in an anticipation of Pop Art), which have at their centre the act of seeing, the "vision". In *Tender Is the Night*, it is Dick Diver ("the man in the jockey cap", *TIN*: 19 – quite a Picassoesque title) who is inundated by this Cubist light: he is revealed little by little, through gestures and postures, voices and words, points of view and perspective planes, his own and others'. For instance, in Rosemary's eyes, Dick moves and behaves as if in a *photo montage*, or in a *collage*, or in a portrait by Picasso, or perhaps even better – and not so strangely – in a figuration by Isadora Duncan: "The enthusiasm, the selflessness behind the whole performance ravished her, the technique of moving many varied types, each as immobile, as dependent on supplies of attention as an infantry battalion is dependent on rations, appeared so effortless that he still had pieces of his own most personal self for everyone" (*TIN*: 89) – where the key words are, obviously, *performance*, *technique*, *moving*, *varied types*, *immobile*, *pieces*, and *self*: a veritable fusion of painting, dance, photography, and psychoanalysis.

This is no surprise. The character of Dick Diver was moulded upon the figure of Gerald Murphy (and Nicole's upon his wife, Sarah), the host *par excellence* in the expatriate colony on the French Riviera in the summer of 1924, able to aggregate around himself the most disparate types and personalities, to be the "magnet that attracts" and the socio-cultural fulcrum, and to "be liked" in the widest sense of the expression (something to which Fitzgerald always aspired, throughout his life, in an almost morbid and painful manner; see Bruccoli 1991: 239). Dick thus becomes the point of convergence of all the perspective and narrative lines, and at the same time resolves himself in as many facets as the characters with whom he deals, then recomposing everything through his own specific angle – yet another version of the "eyes of dr. T. J. Eckleburg", which see the world flowing below, but always maintain their own, imperturbable fixity. What is active in these relationships (both the real ones, Fitzgerald-Murphy, and the fictional ones, Dick-others) is a kind of strange element of quasi-psychoanalytic *transfert*. It must not be forgotten that, on the one hand, Dick *is* a psychoanalyst and, on the other hand, those *were* the years in which psychoanalysis began to deeply influence the artistic world (Joyce to Surrealism). The expression "a transference", which echoes throughout the novel (*TIN*: 134, 155, 324), refers both to this background and to that element of fluidity and mutability which is at the core of Fitzgerald's narration: exactly in the same way as the expression "associational fragments in the subconscious" applied to the "refracting objects" in an "inhabited room" (*TIN*: 122) again refers both to psychoanalysis and to the artistic modernity of those Parisian (and more generally French and European) years.

But the Murphys did more than simply offer a *model* for the Dick-Nicole couple:

Both were seriously interested in the arts. They studied painting with Natalie Goncharova and were active supporters of the Russian and Swedish ballets in Paris. Their close friends included Pablo Picasso, Philip Barry, Cole Porter, John Dos Passos, Archibald MacLeish, and Fernand Léger. Between 1922 and 1930 Murphy completed ten paintings that combined minute detail with abstract techniques. His first work was a large-scale arrangement of a safety razor, a fountain pen, and a matchbox; and art historians have credited him with anticipating the pop art school. (Bruccoli 1991: 238)

The Murphys thus gave Fitzgerald a further chance to establish contact with the avant-garde world, thanks to Gerald's charismatic personality and to the indubitable *transfert* Fitzgerald felt for him (not by chance, "In *Tender Is the Night* [Fitzgerald] *transferred* Murphy's 'power of amusing a fascinated and uncritical love' to Dick Diver"; Bruccoli 1991: 238-239, my italics). They were thus the means by which the "transference" took place, from contemporary pictorial techniques to Fitzgerald's writing.

Even more explicit is the link between Hemingway and Post-Impressionism, both in this and in other works of his (see Maffi 2001: *passim*). In reply to George Plimpton's question: "Who would you say are your literary forebears – those you have learned the most from?", in the famous interview given to the *Paris Review*, Hemingway says:

Mark Twain, Flaubert, Stendhal, Bach, Turgenev, Tolstoi, Dostoevsky, Chekhov, Andrew Marvel, John Donne, Maupassant, the good Kipling, Thoreau, Captain Marryat, Shakespeare, Mozart, Quevedo, Dante, Vergil, Tintoretto, Hieronymus Bosch, Breughel, Patinier, Goya, Giotto, Cézanne, Van Gogh, Gauguin, San Juan de la Cruz, Góngora – it would take a day to remember everyone. Then it would sound as though I was claiming an erudition I did not possess instead of trying to remember all the people who have been an influence on my life and work. This isn't an old dull question. It is a very good but a solemn question and requires an examination of conscience. I put in painters, or started to, because I learn as much from painters about how to write as from writers. You ask how this is done? It would take another day of explaining. I should think what one learns from composers and from the study of harmony and counterpoint would be obvious. (Plimpton 1974: 29)

Some of these names reappear in *The Garden of Eden*, where the references to the artistic universe of modernity are direct: the room which looks like the painting by Van Gogh, the painters whom Catherine wants to contact for the illustrations in the book (Marie Laurencin, Pascin, Derain, Dufy, Picasso), the trips by car through "Cézanne country"... But the references are indirect as well, and perhaps they are the most subtly meaningful, because they have to do with the compositional method, the structure and architecture of writing.

In a passage struck out from "Big Two-Hearted River", Hemingway had written:

> He wanted to write like Cézanne painted.
> Cézanne started with all the tricks. Then he broke the whole thing down and built the real thing. It was hell to do. He was the greatest. The greatest for always. It wasn't a cult. He, Nick, wanted to write about country so it would be there like Cézanne had done it in painting. You had to do it from inside yourself. There wasn't any trick. Nobody had ever written about country like that. He felt almost holy about it. It was deadly serious. You could do it if you would fight it out. If you'd lived right with your eyes. [...] He could see the Cézannes. The portrait at Gertrude Stein's. She'd know it if he ever got things right. The two good ones, the ones he'd seen every day at the loan exhibit at Bernheim's. The soldiers undressing to swim, the house through the trees, one of the trees with a house beyond, not the lake one, the other lake one. The portrait of the boy. Cézanne could do people, too. But that was easier, he used what he got from the country to do people with. Nick could do that, too. People were easy. [...] He knew just how Cézanne would paint this stretch of river. God, if he were only here to do it. [...] Nick, seeing how Cézanne would do the stretch of river and the swamp, stood up and stepped down into the stream. The water was cold and actual. He waded across the stream, moving in the picture. (Hemingway 1972: 239-240)

This act of "moving in the picture" is clearly recognisable in *The Garden of Eden*. It takes place every time David turns to his "damned story", to the "African story" which so irritates Catherine that she burns it, thus marking the end of their relationship: the long excerpts in which one passes (another "transference"?) from the French Riviera, across the Mediterranean, to the depths of Africa, while the moment lived today becomes the past remembered, and writing and memory of what happens/happened "beyond the pale" are woven together.

I believe that the point of convergence between Cézanne's painting and Hemingway's writing is precisely here, in this "movement which, within perfectly organized spaces [...], develops in a dialectic manner through three different phases, guiding the observer's eye from bottom to top – a kind of dynamic 'tripartition', which leads from a complexity of details in the foreground (thesis) through a momentary suspension of that complexity (antithesis), in direction of a synthesis which offers the possibility to reach a serenity not disjointed by some disquietude: as can be seen in *Le Golfe de Marseille, vu de l'Estaque* [...] or in the many *Sainte-Victoire* [...], especially of the last period, which Hemingway could have seen in Paris, even if there is no trace of them in his writings" (Maffi 2001: 205).

Certainly, on the Riviera, in "Cézanne country", the traces were many. And the act of seeing – so present and so important in Hemingway, so decisive in this novel, which is structured, even more than *Tender Is the Night*, upon a play of mirrors – fuses together direct experience and artistic references, thus obtaining that effect of flat profundity, which, according to some critics (Hedeen 1985: *passim*; Hermann, in Rosen 1994: *passim*), is common to both artists.

For instance:

> They had three rooms at the end of the long low rose-colored Provençal house where they had stayed before. It was in the pines on the Esterel side of la Napoule. Out of the windows there was the sea and from the garden in front of the long house where they ate under the trees they could see the empty beaches, the high papyrus grass at the delta of the small river and across the bay was the white curve of Cannes with the hills and the far mountains behind". (*GE*: 85)

The cultural tripartition so central to American culture (civilization, the "middle territory", wilderness) is posed again, by Hemingway, even at the formal level, and under direct suggestion from Cézanne: in an exquisite dialectic of thesis-antithesis-synthesis founded upon movement and change. In all this, however, there is no pacification: the movement has nothing teleological in itself, and quietness isn't necessarily a point of arrival, as we know from "The Snows of Kilimanjaro", "Big Two-Hearted River", *The Sun Also Rises*... It is, if anything, "a momentary stay against confusion", as Robert Frost would say. In the anguished vision of American modernism, none of the three stages of such a search (existential, cultural, formal, etc.) is fulfilling or reassuring, stable or sound. Instead, fluidity and precariousness rule, and even upon a momentary radiance Fitzgerald's "touch of disaster" looms. Because, as Hemingway put it, writing to a friend in June 1948, the subject matter of the novel he was writing is "the happiness of the Garden that a man must lose" (quoted in Sullivan).

Epilogues

Exactly that final "must" helps us to understand the diversities (so strictly intertwined with the analogies) between the two authors' points of view. For Fitzgerald, the story set on the French Riviera is yet another chapter in the saga of a generational (and national) defeat, which the author will never be able to overcome or leave aside (and thus his Dick Diver, after occupying the scene and being the point of reference for everybody, will disappear like so many characters before him, perhaps destined to become one of the many "grotesques" peopling America "in one town or another"; *TIN*: 338). For Hemingway, on the other hand, the French Riviera is yet another ordeal to face, in suffering and loss, in the belief that – as he wrote somewhere – "il faut durer". And thus his David Bourne, after going to pieces in the torment of continuous change, will grasp and hold that very "must", an inevitability which is not fatalism but, rather, perception of the ruling laws: in this way he will be able to recover the only certainty which remains and literally sustains him – his *métier*. In the end, we will see him intent on writing: because "there was no sign that any of it would ever cease returning to him intact" (*GE*: 267).

Together, in an unceasing dialogue and dialectic, the two authors, friends and enemies, who throughout the 1920s and 1930s never ceased to cross and question each other, offer us two diverse but converging states of mind of American modernism. And they do so from the shores of the Mediterranean.

Bibliography

Bruccoli, Matthew J., 1963, *The Composition of* Tender Is the Night, University of Pittsburgh Press, Pittsburgh.
——, [1981] 1991, *Some Sort of Epic Grandeur. The Life of F. Scott Fitzgerald*, Carroll & Graf Publishers, Inc., New York.
Cowley, Malcolm [1934] 1994, *Exile's Return. A Literary Odyssey of the 1920s*, Penguin, Harmondsworth.
Donaldson, Scott, 2001, *Hemingway vs. Fitzgerald*, The Overlook Press, Woodstock, New York.
Fitzgerald, F. Scott, [1926] 1967, *The Great Gatsby*, Penguin, Harmondsworth
——, [1931] 1965, "Babylon Revisited", in F. Scott Fitzgerald, *The Crack-Up with Other Pieces and Stories*, Penguin, Harmondsworth.

——, [1934] 1986, *Tender Is the Night*, Penguin, Harmondsworth.

Hedeen, Paul M., 1985, "Moving in the Picture. The Landscape Stylistics of *In Our Time*", *Language and Style. An International Journal*, 18, 4.

Hemingway, Ernest, [1925], "Big Two-Hearted River", in Ernest Hemingway 1995.

——, 1926, *The Sun Also Rises*, Simon & Schuster, New York.

——, [1939], "The Sea Change", in Ernest Hemingway 1995.

——, 1972, *The Nick Adams' Stories*, Charles Scribner's Sons, New York.

——, 1986, *The Garden of Eden*, Scribner's, New York.

——, 1995, *The Short Stories*, Scribner Paperback Fiction, New York.

Jehlen, Myra, 1989, *American Incarnation. The Individual, the Nation, and the Continent*, Harvard University Press, Cambridge, Massachusetts.

Maffi, Mario, 2001, "Cézanne e il 'segreto' di Hemingway", in Giovanni Cianci - Elio Franzini, Antonello Negri (a cura di), *Il Cézanne degli scrittori, dei poeti e dei filosofi*, Bocca Editori, Milano.

Magris, Claudio, [1991] 2004, "Per una filologia del mare", in Predrag Matvejević, *Breviario mediterraneo*, Garzanti, Milano.

Marx, Leo, 1964, *The Machine in the Garden. Technology and the Pastoral Ideal in America*, Oxford University Press, New York.

Plimpton, George, 1974, "An Interview with Ernest Hemingway", in Linda Welshimer Wagner (ed.), *Ernest Hemingway. Four Decades of Criticism*, Michigan State University Press, n.p.

Rosen, Kenneth (ed.), 1994, *Hemingway Repossessed*, Greenwood Press, Westport, Conn.

Sullivan, Robert, "The Marriage of Heaven and Hell: Sexual/Textual Politics in *The Garden of Eden*", Lecture given for the Modernist Journals Project, Brown University (http://www.modjourn.brown.edu/Sullivan/hemlect2.htm).

Part II

EXPLORING THE ADRIATIC

JOYCE AS UNIVERSITY TEACHER
AND THE "SOUTH SLAV QUESTION"

Renzo S. Crivelli

When Joyce joined the staff of the Trieste Revoltella Public School of Business Studies in September 1913, the institute seemed to have become well consolidated within the life of the city [Fig. 1]. It was founded in 1876, seven years after his death, by Baron Pasquale Revoltella, a rich business man of Trieste whose many enterprises had included being a partner in financing the Suez Canal. The School came into being under the guidance of Scrinzi de Montecroce, the executor of Revoltella's last will and testament, flanked by a committee of guarantors made up of representatives from the Town Council and the Chamber of Commerce. Lacking direct experience in the field of higher education, the committee turned to the imperial counsellor Carlo Arenz, director of the Commercial Academy of Prague, and entrusted him with the task of drawing up plans for staff and teaching at what was called the "High School" (Sauer 1882: 10-11). And so the School took its first steps, under the direction of Carlo Marquando Sauer and on the model of the seventeenth century Austrian Commercial Academies which had sprung up at the height of mercantile activity by decree of Maria Theresa of Austria (Vinci 1997: 111).

There were 94 students enrolled in the Revoltella (divided between the 1st and 2nd year course) when Joyce began teaching in the 1913-14 academic year, but this number declined to 74 students by the end of the year. Although it was considered a university-level institute, the School's curriculum meant that it could not confer the degree of Doctor of Business Studies and Economics (but only a diploma), and this undoubtedly

Figure 1 – The 19th century building, with elegant neo-classical pediment, which housed the Revoltella business school in 1913 (Courtesy Civici Musei di Storia e Arte, Trieste). Photograph.

limited its goals. However, it could rely on its local prestige and the reputation it had acquired in the business community in Trieste and neighbouring areas and, as the Archives of the University of Trieste make clear, many firms wrote to the School asking for the names of students they could employ.[1] Beginning in 1908, after a series of requests to the Ministry and within the framework of the reform of Austrian education, an exception was made and Revoltella graduates were permitted to take the exam of the Faculty of Education which qualified them for teaching business studies in the secondary schools. The graduates had to take a supplementary course in practical philosophy, pedagogy and mathematics that was an equivalent of university courses. At all events, in 1913 the Revoltella drew its students from a wide area which included, in addition to Trieste and its province, the Istrian coast, Dalmatia (with students from Zadar and Split), as well as Galicia, Bucovina and Romania. This explains the presence of numerous Slovenian students (from Istria) and Dalmatians of Croatian origin in the year Joyce began teaching at the School, who were all quite sensitive to the "South Slav question" being advocated by "the liberal bourgeoisie and the educated classes of the Slav peoples" (Vinci 1997: 120, 122). As we shall see, these views would come into conflict with the Irredentist aspirations of many of the pro-Italian students.

This was the situation when Joyce assumed his teaching position at the School. He was officially welcomed during the brief "Preliminary Staff Meeting" (composed of Professors Morpurgo, Menestrina, Spadon, Gentile, Mussafia, Grignaschi, Stenta, Wendlenner, Schreiber and Du Ban) and presided over by the Director, Prof. Francesco Savorgnan, which took

---

[1] For example, the *Austrian-American Company* of Via Molino Piccolo 2, Trieste, specialised, according to their letterhead, "in the transport of goods, dispatches and passengers from Trieste to North and South America, Mexico and the West Indies" wrote the School on 20 July, 1914: "Since there will shortly be a vacancy in our firm for a clerk, we would like to know if your School could kindly inform us if any of the students graduating this year would be interested in filling the post. The candidate should have a knowledge of Italian and English, and should be able to do double-entry bookkeeping" (Archivio della Scuola Superiore di Commercio "Revoltella", Università di Trieste 1913-14, File 1/A 19). However, in that same year, the School sent a letter of protest to the Ministry of Commerce, complaining about the disparity in the treatment of Revoltella students and those of the Academy of Exportation of Vienna by the I. R. General Warehouses of Trieste, who gave preference to the latter when hiring (File 1/A 8).

place on 26 September, 1913 and lasted from 6pm to 7.15pm. At the start of the meeting, the Director greeted "the new lecturer of English, James Joyce, in the name of all the teaching staff", and announced a rapid reform of the scholastic programme, for which he invited all of the teachers "to contribute their views regarding their own subject" —something which, as we shall see, Joyce was to do regularly (Archivio Revoltella, University of Trieste, 1913-14: File 1/A 19).

In the autumn of 1913, Savorgnan was the School's undisputed director, a position he had assumed six years before and which he was to leave at the end of the 1913-14 academic year, in order "to remain in Italy". He was at the height of his career as a sociologist and expert in demographic statistics, and was also a town counsellor and member of the Diet of Trieste. Among Joyce's colleagues were two other noteworthy figures: Giulio Morpurgo, who had taken his degree in Graz, was the former director of the University Pharmacies, the director of the prestigious *Giornale di farmacia, chimica e scienze affini* and the architect and guiding spirit behind the School's Faculty of Commodities Studies; and Francesco Menestrina, a jurist from Trento and passionate Irredentist whose 1904 speech in favour of an Italian university in Austria had resulted in the famous Innsbruck student riots, an event which continued to guarantee his prestige in Trieste nearly 10 years later.

Joyce's teaching proceeded smoothly until July, 1914 (the written and oral exams for the diploma were generally held at the end of May) even though he would be involved, involuntarily, in a number of events that were crucial for the School and for the city of Trieste. His attendance at the Academic Council meetings, stipulated in his contract, was not entirely regular: on 5 December, 1913, for instance, he was present and signed his name; on 16 December, however, he was present (his name is given as Yoyce!) but did not sign his name (perhaps he left early?). During this meeting, a case of *harassment* by a certain Professor Grignaschi, who taught Commercial and Financial Calculations, against the "fulltime" student of the first course, Giuseppe Cristoforo Sangulin, was discussed. The episode throws light on the authoritarian nature of schools of the period and, curiously enough, recalls the injustice suffered by young Stephen (and by Joyce himself) in *A Portrait of the Artist as a Young Man* when he accidentally broke his spectacles. Based on the account given in the minutes, Sangulin had spoken to a classmate during the lesson and was ordered by the professor to come up to the blackboard with the words

"Step forward you, the one with the ideas". The student excused himself saying that he "had a headache". The incident ended there, but it was treated as a serious case of insubordination and Sangulin was summoned before the disciplinary committee.

In March, however, Joyce was witness to serious incidents of violence between the Italian and Slav students, fomented by their mutual intolerance. Since 1912, the "conflict between nationalities had been coming to a head in Trieste: everything else was overshadowed while the political factions positioned themselves, without any hope of mediation, along the dividing line of Italian-Slav hatred". While tensions had been building for some time concerning the issue of an Italian university in Trieste, in the period 1912-14 the true nature of this movement had become apparent, manifesting itself as cultural opposition to the Slovenes, who also wanted a university of their own (as well as a new "Yugoslav nation"). The political positions assumed by the Triestine students were all characterised by intolerance – culminating in the incidents of March 1914, whose epicentre was the Revoltella School itself. The political situation had become increasingly confused, given that "among the Italians the scholastic issue had once again reached a fever pitch, and the conflict was carried forward by means of a perfectly orchestrated system of provocations and counter-provocations, accompanied and sustained by a Press campaign of unprecedented harshness" (Vinci 1997: 88). Everything came to a head with the tragic events of 13 March at the Revoltella, which had been designated on various occasions as the most likely nucleus for a new Italian as well as a new Slovene university.

In the very year in which Joyce began teaching, 1913-14, there had been a notable increase in the number of Slovene and Serb-Croat students, leading the Director, Savorgnan, to voice his suspicion in a very concerned letter to the Ministry of Education and Culture of Vienna that this was part of a precise political plan.[2] Suspicions aside, what is certain is that

[2] That this was part of a political manoeuvre has never been confirmed. Given that the School attracted students from as far away as Dalmatia, we can assume that enrolments were not based on political considerations (Vinci 1997: 92). Based on the School's internal inquiry, the initial provocation was to be attributed to the Slavs, while according to the police it was the Italians who created an atmosphere of intimidation, something which was also confirmed by the Triestine newspaper *Il Piccolo*. The director Savorgnan criticised the newspaper for its stance in a note to the editorial office on 25 April 1914 (Archivio Revoltella, University of Trieste 1913-14, File 1/A 19).

on 3 March, during the 11 o'clock interval in the first-year classroom, a number of Croatian students had the *audacity* to speak aloud in their own language, something which the Italian students took as a provocation. A scuffle broke out and the School was closed as a preventive measure for four days. On 7 March, the meeting of the Academic Council resolved unanimously (and thus also with Joyce's approval) to exhort the two factions to show mutual respect, while also acknowledging the "Italian" character of the School in accordance with the intention of its founder, Baron Pasquale Revoltella. But this appeal for moderation did nothing to prevent the even more serious incidents which occurred over the next few days, and this despite the presence of police officers at the School: on 12 March a Slav student was attacked; on 13 March Slav students entered the building after laying down the sticks and cudgels with which they were armed, only to find the Italian students still armed with sticks; a brawl ensued with fists flying and chairs being hurled in the staff room.

However, things did not end there: according to the reconstruction of events by the School, that same day the Croatian student Stefano Sisgoreo fired three pistol shots (some witnesses said they heard four or five shots, but these were probably from other weapons that were then thrown out of the window). One of the bullets hit Giuseppe Cristofaro Sangulin in the left arm (the same student who had been "punished" by the School). The Police intervened and arrested Sisgoreo – who still had the pistol in his hand – and five Italians. They were questioned and immediately indicted on various charges.

Meanwhile, the situation remained confused. According to the minutes of the meeting of the Academic Council held on 28 March, the School had resolved to expel Sisgoreo while awaiting his trial, but could not decide how to proceed with the others. Prof. Grignaschi repeated his reservations about the absence of his colleagues Joyce (who had given no justification for his absence) and Menestrina (who was away from Trieste) and in any case "he did not think it right, in the absence of the two lecturers, to take any decisions regarding the other students who were more or less involved in the inquiry". It was thus resolved to postpone any decision until the next meeting, and to request the presence of Joyce and Menestrina "so that they can participate in the discussion and cast their vote". It was further resolved to suspend the first-year course, while the second-year course would be resumed on 20 April. Further, the School entrance would be strictly controlled to make sure that only second-year students were allowed to enter. At the next meeting on 4 April, in which Joyce was present and expressed his approval verbally, it was resolved to confirm the expulsion of Sisgoreo (who had not even bothered to appear to defend himself) and to wait for the results of the trial before taking any further action. It was also decided to limit the suspension of the first-year course to the second term. Everything was thus deferred pending the outcome of the trial which, as it turned out, never took place due to the start of the first World War.

As for Joyce, even though he was clearly aware of the turbulent situation around him, he seems to have been reluctant to become involved, judging from his "strategic" absences from the School. During this period, the tormented dream of finally seeing *Dubliners* in print was, after a series of frenetic last-minute negotiations with the publisher, about to become a reality (it was published by Grant Richards on 15 June, 1914, in an edition of 1250 copies) while Joyce the writer was completely absorbed in the composition of the second act of the play *Exiles*. But meanwhile the clouds of war were gathering over Europe. On 28 June 1914 the Archduke Franz Ferdinand was assassinated at Sarajevo by Serbian terrorists and, on 28 July, Austria declared war on Serbia in an attempt to curtail the nationalistic aspirations of the Slav peoples. These events filled Joyce with fear and foreboding. Immediately after the declaration of war, he went to the British Consulate of Trieste with his son Giorgio to obtain information on his status as a British subject. He then went to the home of his pupil Boris Furlan, who lived next to the Italian Consulate, in search of additional information. While talking with Furlan, who expressed his pessimism about the future of Europe, a crowd gathered in front of the Consulate and tried to tear down the Italian flag. The result, as Furlan recalled, was disastrous: Joyce took Giorgio by the hand and fled in a state of shock (Ellmann 1983: 380).

In the following weeks, the crisis deepened with Russia, France, Germany and Austria preparing for war, and on 8 August, following Germany's invasion of Belgium, Great Britain also declared war against the Austro-Hungarian Empire. This event was of particular significance for Joyce, who suddenly found himself a British citizen (Ireland was still under British sovereignty) resident in an enemy country.

Despite the Salandra government officially declaring Italy's neutrality on 3 August 1914, interventionist sentiment ran high and it was impossible to know what Italy's position would ultimately be. As a result, numerous pro-Italian families began leaving Trieste and returning to Italy in order to keep their sons from being called up by the Austrians and eventually having to fight against Italian soldiers. Naturally, the School was deeply effected by these events and when it re-opened, a month late, on 1 November 1914, enrolment had fallen off drastically. There were only 30 students in the first year and 10 in the second, with the majority of the School's former or potential students having been conscripted or moved to the Kingdom of Italy, often with the excuse of requiring "urgent medical treatment in specialised clinics" (Vinci 1997: 148, 150). Even the School's Director, Francesco Savorgnan, had left Trieste for Padua in September under the pretext of "temporary sick leave" and was only substituted a year later by Giulio Morpurgo, who had moved to Vienna. However it was Savorgnan, who thought quite highly of Joyce, who had taken the necessary steps to renew his contract as early as the previous July.

Joyce's contract could therefore be renewed, at least officially, even though the problem of his "nationality" still remained. His appointment was confirmed and on 10 December 1914, he sent a letter to the I. R. Lieutenancy requesting the return of his documents, which included his "degree from the National University of Ireland (the original on parchment)", two "newspaper articles on his literary works (copies)" and "two references from the Università Popolare of Trieste".[3] This lengthy process, which was complicated by the war and by Joyce's British nationality, ended on 22 February 1915 when the I. R. Lieutenancy, in response to the Ministry letter dated the 17th of the same month which affirmed that "there is no reason to fear that Joyce will have a negative influence on the students who attend his lessons", informed the Revoltella School that an exception (*deroga*) had been made in Joyce's case and his appointment had been confirmed, but that this appointment was limited to the current academic year and would not be extended in the future.

As occurred so often with his financial loans, this *deroga*, which had no apparent justification, was the result of Joyce's precious contacts with the Trieste establishment, in this case his indirect connection with the I. R. Lieutenant Hohenlohe, who was the highest Austrian official in Trieste. Joyce had given private lessons to his wife, Princess Francesca Ho-

---

[3] The signed document is preserved in the Archivio di Stato di Trieste, I. R. Luogotenenza del Litorale, Atti Generali 1906-1918, Folder 2543.

henlohe-Schillingfürst, and her children and was therefore considered "a serious young man whose only concern is earning his living". It was this judgement, together with the support of the Aulic Counsellor Gelcich, provincial educational inspector and ministerial delegate, which had *made all the difference*.[4] Gelcich, in his role as consultant to the School, played a significant part in Joyce's being confirmed in his post (indeed, all of the above-mentioned documents bear his signature as having been read by him) and most likely he was acting in the School's *real* interest, convinced as he was that Joyce was both an effective teacher and much appreciated by his students.

This appreciation was to be confirmed a month later when Joyce, who had not been paid since November and whose economic situation was becoming critical, received unexpected support from his students, who were becoming increasingly concerned at the delay in starting his course in "English language and correspondence". As regards the economic situation, we know that as early as January, Joyce had been obliged to take out a loan of 600 crowns from the *Consorzio Industriale Mutui e Prestiti* [Industrial Consortium of Loans and Mortgages], and that thanks once again to Italo Svevo he had begun working as a clerk for the Veneziani firm for 100 crowns a month, a sum which enabled him at least to meet his family's basic needs during this difficult period. With respect to the delay in starting his course, on 4 March the students organised a formal protest against the School on his behalf, which resulted in Joyce being officially reconfirmed by Morpurgo (who had become the School's acting head after Savorgnan's departure) on 13 March 1915. In the letter of confirma-

---

[4] As an example of the atmosphere of suspicion surrounding British subjects in Trieste after Great Britain entered the war, we can cite Stelio Crise's reconstruction of the so-called "Berlitz detective mystery", based on documents from the Central Headquarters of the Vienna Police. According to Crise, from October 1914 investigations were carried out into "English enemy agents" in the Berlitz. Indeed an independent inquiry was carried out on the Joyce brothers (referred to as Janez and Stanislaw), who were thought to be the directors of the Berlitz School, which was described as a "hotbed of British subversives". But when it was ascertained that the acting head of the school was, in reality, a certain Charles Joyce (domiciled in Via Belvedere), the latter was interrogated on 9 October, 1914 and the investigation was dropped. The police concluded that Joyce, James, had nothing to do with the matter and noted that he lived in Via Bramante (Stelio Crise, "Sottosopra la polizia austriaca da ipotetici 007 di nome Joyce", *Il Piccolo*, 30 May 1971).

tion, Joyce was told to present himself the following Tuesday at 10.00am in order to resume teaching.[5]

However, due to the critical political situation, the return of his salary was not sufficient to allay all of Joyce's concerns. On 20 May the School decided to bring forward the final exams for the 2nd year course, originally scheduled for the middle of July, because "many students have been called up for military service" and also urgently requested the I. R. Lieutenancy to appoint a new president for the board of examiners to replace Counsellor Gelcich, who had left Trieste.[6] On 24 March, Italy entered the War and the situation became even more complicated. Trieste was a city increasingly in the grip of tensions, departures and public disorders.

The War Ministry made urgent appeals to the young men of Trieste who had not yet been called up for military service – including the pupils of the School – exhorting them to join the *Jungschützkorp* "in order to serve the Fatherland and the city in which they lived". The teaching staff faced the same problems and numerous vacancies opened up at the Revoltella: Francesco Savorgnan, as noted, was in Italy, the acting head of the School Giulio Morpurgo ended up in Vienna, Giovanni Spadon enrolled in the Territorial Militia, Francesco Menestrina was sent to the front as the commander of a *so-called* "Russian battalion" (though he, too, fled to Italy a year later)[7] and Attilio Gentile was called up. As a result, the academic year 1914-15 was a disaster: the teaching staff was decimated and enrolments plummeted. Furthermore, given that Savorgnan was accused of "collaborating with the enemy" (a police report described him as being of "italienisch-radikaler Gesinnung")[8], the Authorities were reluctant to grant the normal funding for the School.

---

[5] Cornell Joyce Collection, Scholes 1293 (See also McCourt 2000: 243).

[6] Archivio di Stato di Trieste, I. R. Luogotenenza del Litorale, Atti Generali 1906-1918, Folder 2569.

[7] In a letter of 19th June to Morpurgo, Menestrina says, with obvious disgust, that he "has to see to the hygiene, de-bugging of the bed linen and emptying of the latrines" for a battalion of 2470 men. Menestrina's departure seems to have been a piece of bad luck, seeing that some months earlier, on 29th January 1915, the School wrote to the I. R. Lieutenancy, pointing out that the lecturer "was exempt from the obligation of being called up since at the end of that year he would have passed the age of 42". Evidently he received his draft papers some months before the expiry of those conditions (Archivio Revoltella, Università di Trieste 1913-14, File 1/A 19).

[8] See Archivio di Stato di Trieste, Tribunali Militari, Folder 20, "Tagebüch über die Strafsache gegen Savorgnan".

There was now no alternative to closing the School, a step taken by the remaining teaching staff on 16 June 1915. That this closure caused considerable distress is evident in the correspondence of some the teachers (and of the caretaker Giovanni Debenedetti, who was left to look after the empty rooms), which contains references to the difficulties of trying to get by without a fixed salary (in a letter to Morpurgo of 28 August 1915, Spadon remarks that he has "lost his only source of income"). The caretaker Debenedetti, who was a veritable institution in the School (and who had a higher salary than Joyce!), writes that "Trieste has become a desert". Among his duties was that of assisting Professor Grignaschi – the only staff member left – in issuing school reports for those students who asked for them. Debenedetti gives an indication of how hard times were in one of his letters to Morpurgo in Vienna:

> Nothing new here; things are going from bad to worse. It's very hard to find foodstuffs even paying sky high prices. Now that I mention it, I should tell you that I took your advice and took advantage of some samples of oil [...] The Commodities Room still has some whale, seal, shark, cod and walnut oils, which might even be good and a great boon for me personally, but they have a very strong fishy smell and so I don't trust them. (Archivio Revoltella, University of Trieste 1913-14: Atti, File 1/A 21)

In another ungrammatical letter in the middle of June we learn that Joyce had turned up at the School asking for a certificate of service and for information about his salary. The School had been closed for a month and there were absolutely no funds available, but given his dire economic straits (which were certainly not helped by the "sky high prices" mentioned by the school caretaker) Joyce still tried, unsuccessfully, to obtain his salary. The debt contracted six months earlier with the *Consorzio Industriale Mutui e Prestiti* for a modest sum not exceeding 600 crowns, fell due on 30 April, but payment had been deferred to 30 July. It is therefore not surprising that Joyce should ask the School about his July salary...

With Italy's entry into the war, life in Trieste had become even more difficult. Anti-Italian demonstrations took place throughout the city. The cafés frequented by the Irredentists (such as the San Marco, the Milano and the Stella Polare in Via Dante where Joyce went) were ransacked and devastated. The statue of Verdi, a symbol of Italian culture and identity, was pulled from its pedestal in Piazza San Giovanni, and the offices of the newspapers *Il Piccolo* and *Il Piccolo della Sera* in Piazza Goldoni were burnt down. The city seethed with hatred and revenge.

In this situation, Joyce had no option but to seek refuge (and a livelihood) elsewhere and, once again borrowing from various friends and pupils – Baron Ralli, Count Sordina, and the Veneziani family –, the Irish writer and his family left for Zurich on 27 June 1915. And the School? The Revoltella weathered the War and reopened, completely renovated, just in time for Joyce – who returned to Trieste in mid-October 1919 – to take up his position once again as professor of English for the 1919-1920 academic year.

But by then Trieste had become Italian and the "South Slav question" had passed under the control of the Italian Government and seemed less urgent. It was to become pressing again during the fascist regime's occupation of Istria and Dalmatia, after Joyce's final departure for Paris in July 1920.

Bibliography

Ellmann, Richard, 1983, *James Joyce*, Oxford University Press, Oxford.
Sauer, Carlo Marquando, 1882, *La pubblica Scuola Superiore di Commercio di Fondazione Revoltella in Trieste* [The Revoltella Public School of Business Studies in Trieste], Tipografia del Lloyd Austriaco, Trieste.
McCourt, John, 2000, *The Years of Bloom*, The Lilliput Press, Dublin.
Vinci, Anna Maria, 1997, *Storia dell'Università di Trieste: Mito, progetti, realtà*, Quaderni del Dipartimento di Storia [Notebooks of the History Department], Università di Trieste, Lint, Trieste.

FROM *EXILES* TO *ULYSSES*: THE INFLUENCE OF THREE ITALIAN AUTHORS ON JOYCE – GIACOSA, PRAGA, ORIANI

Giuseppina Restivo

The hesitant re-evaluation of James Joyce's *Exiles* – subsequent to Harold Pinter's successful staging of the play in 1970 – and its persistent classification as an Ibsenian experiment (which it is not), have long discouraged a thorough analysis of the text. By re-reading it today, we may better establish its role in both the development of its author and the history of modern English theatre, especially considering its impact on – among others – Samuel Beckett, Harold Pinter and T.S. Eliot, all exposed to Joyce's influence.[1]

The 1913-1915 period that gave birth to the play was one of the densest in the history of European culture, and the artistic exchanges which promoted it as the fruit, not of a beginner but of a mature artist composing *Ulysses*, were rapidly transforming Western literary tradition. Within the recently redrawn frame of Joyce's stay in Italy, the analysis of a specific line of these 'exchanges' linking *Exiles* with three Italian writers will be proposed. This will help not only to shed new light on the process and transmission of innovation in a crucial author at a crucial moment, but also to detect what appears as a decisive though unacknowledged influence on the narrative core of *Ulysses*.

The three Italian sources – Marco Praga's *La crisi* (*The Crisis*),

---

[1] Critical controversies have always accompanied Joyce's work, from the international reception of *Ulysses* to the Italian assessment of the author (see, as revealing on the subject, Cianci 1974). In the case of *Exiles* a proper re-evaluation, long overdue, could finally meet Joyce's own self-conscious appreciation of the play.

Giuseppe Giacosa's *Tristi amori* (*Sad Loves*) and a passage from Alfredo Oriani's *Rivolta ideale* (*Ideal Revolt*)[2] – are mentioned by Joyce himself in his *Notes* for the play and have so far been considered in one short article by Dominic Manganiello, "The Italian Sources for *Exiles*: Giacosa, Praga, Oriani and Joyce" (Manganiello 1977: 227-237), at a time when interest in the play had only just been aroused, but the influence of the Italian years and of Trieste on Joyce was yet to be fully appreciated. Interesting hints appeared in the same year in an essay by Michael Mason in "Why is Leopold Bloom a Cuckold?" (Mason 1977), while in barely more than a couple of pages Praga and Giacosa were dismissed as "unquestionably dramatists of second rank" in the later *James Joyce's Italian Connection, the Poetics of the Word*, an otherwise fundamental book by Corinna del Greco Lobner (1989: 10-12). The 1979 *Textual Companion* to *Exiles* by John MacNicholas, on the other hand, offers brief notes on Praga and Giacosa, and mentions Oriani as the author of the 1896 novel *La disfatta* (*The Defeat*), but not of the essay *Rivolta ideale*, which Joyce drew from in his *Notes*.

Joyce lived in Italy, as is well known, for about eleven years between 1905 and 1919 (he spent the 1915/'18 war period in Zurich), staying for most of the time in Trieste, except for a brief initial period in Pola, a few months in Rome in 1907 and three trips to Dublin in 1909 and 1912. His mention of the three Italian authors reflects his absorption of Italian culture, which included not only the great authors of the past (like Dante, Bruno, Vico), but of the present, from D'Annunzio to the 'antithetical' Giacosa, from the Futurist Marinetti to the "Scapigliato" Marco Praga. As was usual for him, in his work literary experience merged with biographical experience, which in *Exiles* referred particularly to his jealousy of his unmarried partner Nora Barnacle.

The two protagonists of the play, Richard Rowan, a writer like Joyce, and his partner Bertha, stage the author's Triestine tensions with Nora and his obsession with betrayal, transposing them to Dublin. The scene opens in 1912, on a couple just back from nine years of voluntary exile in Italy. Richard's old friend Robert Hand, who courts Bertha trying to seduce her, is defined in the *Notes* to the play as "precious, Prezioso", which confirms his link with Roberto Prezioso, a journalist and later director of

---

[2] Translations are mine throughout, unless otherwise indicated.

Il Piccolo della Sera, the most important Trieste daily newspaper. Prezioso took English lessons from Joyce and helped him socially and economically, by commissioning articles for *Il Piccolo* from him. He attended Joyce's house as a friend, to the point of arousing his jealousy, the more so since a previous episode in 1909 had already perturbed Joyce. During a visit to Dublin, his former friend Vincent Cosgrave had insinuated that Nora had frequented him in the same period Joyce was courting her, before they eloped to the continent. This had upset Joyce deeply, until another common friend, John Francis Byrne, convinced him that Cosgrave had lied out of resentment.

Exiles records both episodes, turning them into "portals of discovery": the forging of new ethics for the couple. This implies first of all the right to a free union (as in Joyce's own case) rather than a marriage contract, so as to maintain reciprocal constant individual rights to choose one's partner according to actual feelings, rather than to social and religious constriction, in defiance of both Irish Catholicism and common bourgeois rules. On this basis, the problem of fidelity was no longer a problem of honour, which could disgrace a husband, but of emotions. Since Flaubert's *Madame Bovary,* Joyce argues, the new broad middle class literary public, no longer bound by the old aristocratic concepts of 'honour' and of the husband's 'possession' of his wife, has broken the long literary tradition of contempt for the cuckold. Hence the necessity and possibility of the new outlook on the couple the play engages with, and hence again the author's interest in similar previous plights: "Praga in *La crisi* and Giacosa in *Tristi amori* have understood and profited by this change – Joyce points out – but have not used it, as is done here, as a technical shield for the protection of a delicate, strange and highly sensitive conscience" (Joyce 1992: 345). What does this really mean?

An intellectual and a writer of genius, the male protagonist fearing betrayal in *Exiles* enjoys a position of undisputed superiority, while his journalist rival defines himself as "a common man". Bertha keeps Richard informed of Robert's advances, expecting him to take a stance. But Richard refuses to make claims on her: rather he exhorts her to explore her would-be lover's intentions freely, to choose the man who can better provide her with happiness.

Bertha is worried about this generosity and doubts its causes, the complexities of which are admitted during a dialogue between the two men. Richard suffers owing to his acute jealousy, but at the same time admits his desire to ease his sense of guilt for grossly betraying Bertha in the past, by breaking her ethical superiority in always keeping faithful to him. At the same time, if she betrayed him, he might feel freer for a possible alternative relationship with another woman (such as Beatrice Justice, Robert's cousin and former fiancée), able to understand his literary work as Bertha cannot, because of her lack of education and modest social origin. In this contradictory situation Richard feels prey to a sado-masochist drive: although wounded by jealousy he "wants his dishonour", while his defiance of common ethics and all ties, domestic or of friendship, increases his previous remorse for his past dealings with his dying mother, whose Catholic ethics he had denied. Moreover, Richard is aware of a homosexual drive between Robert and himself, combining with their heterosexual attraction to the same woman. Bertha has become the object of the two men's reciprocal unconfessed desires in what appears to be a form of "triolism" (McCourt 2000: 195).[3] Thus, admittedly, both noble and ignoble drives are entwined in Richard's ambiguous desire to let Bertha choose her own good, to which he is ready to sacrifice his egoistic right to possess her. The doubts and suffering of all involved are openly discussed: in the end Richard proves stronger, the Richard/Bertha couple is confirmed and Robert leaves. But the two men manage to preserve their friendship, in spite of Robert's initial lies and of Richard's final "wounding doubt" as to whether there was a night of love between Bertha and Robert, after he left them alone to let Bertha make her free choice.

The end of the play, while preserving the couple in spite of the daring analysis of human behaviour, is problematic. Lacking action throughout, replaced by a continual play of self-contradictions, *Exiles* closes on an endless suspension of doubt and impossible catharsis, beyond Ibsen's protest (implicit in it), and on the verge of Beckett's future 'absurd'. It develops indeed on "double binds" (of the type "stay with me/ don't stay with me") between Richard and Bertha as between Richard and Robert or Richard and Beatrice, a characteristic which, radicalised, will become pivotal in Beckett's theatre, leading to or verging on "experimental catatonia" (Segre 1974: 274; Restivo 1991: 181, 233). But what does this in-

---

[3] McCourt appropriates Brenda Maddox's reference (Maddox 1988: 156) to the psychological condition called triolism, "in which a homosexual desire for someone is expressed in sharing, or dreaming of sharing, a partner".

tricate situation owe to the Italian sources mentioned in Joyce's *Notes?*

The first of the two plays evoked, Marco Praga's *La crisi* (1907), is set in Milan in 1904. It portrays a new type of emancipated woman, Nicoletta, who has received an unusually free education, considered both of "English" and even of "American" type, and scandalous. She is married to a well-to-do entrepreneur, Piero Donati, so deeply in love with his beautiful and fascinating wife as to be unable to give her up although he suspects she betrays him, which indeed she does, as his brother Raimondo soon finds out. A former colonel with a rigid sense of honour, Raimondo uses a pretext to avenge Piero by challenging Nicoletta's lover, Ugo Pucci, to a duel. But Raimondo realizes that Nicoletta, disappointed with her lover (a mediocre young lawyer who occasionally assists her husband), has decided to leave him in favour of her own husband. Anxious to avoid wounding Piero, who is unaware of anything, Raimondo then withdraws, ready to keep silence on what has passed. To his surprise, though, Nicoletta rejects his connivance and insists on her right to both choose her man and to be sincere: she informs her husband of her decisions as the only way to stay with him. Against moral conformism, she self-consciously points out to Raimondo: "in fondo, sono buona…sono anche onesta…non sorridere…ti dico la verità: sono forte e buona" ("after all, I am good…I am also honest…don't smile…I'm telling you the truth: I am strong and good").

Her polemical stand is the core of the play: not physical fidelity or honour based on duels, but moral integrity as correspondence between social marital status and feelings constitutes real honesty. Hence the new values based on individual freedom and renewed ethics for the couple: a clear antecedent of Joyce's intents in *Exiles,* where the final choice is also left to the woman.

There is therefore more in common between the two plays than Richard's and Piero's choice to live with a woman "not approved of by the society in which they live", or their being "acquiescent cuckolds", as later Bloom in *Ulysses,* or even the presence of words like "wound" and "doubt", characterizing the two male protagonists (Manganiello 1977: 230-231). Nor can Nicoletta's frankness be juxtaposed to the amoral protagonist of another of Praga's plays, Giulia Campriani in *La moglie ideale* (*The Ideal Wife*), who ironically manages to keep both her lover and her unaware husband equally happy: Nicoletta is the opposite of Giulia, and *La moglie ideale* remains rightly unmentioned in Joyce, even though present in Joyce's library. Praga considered this play his best one, but Giorgio Pullini, the author of the most important study on Praga, found more innovation in *La crisi*:[4]

> *La crisi* è forse l'unica commedia novecentesca di Praga, quella che si apre ad una dialettica nuova del personaggio e a cui si riallaccia molta produzione italiana degli anni successivi, Pirandello compreso. (Pullini 1960: 122)

> *La crisi* is perhaps Praga's only twentieth century play, opening out to a new dialectics of character, to which much of later Italian theatre, Pirandello included, is linked.

Pirandello's 'technique of doubt' is indeed implicit in Nicoletta's sentimental experiment and temporary double link.

As Manganiello admits (1977: 229), "Joyce thought so highly of the Milanese dramatist Marco Praga that he expressed the wish that the latter would sign the *Ulysses* protest".[5] Joyce was not overestimating an obscure author: in quoting Praga's moral innovations he was, as usual, lucidly and coldly aware of choices and implications he could build upon, ready to push farther on the same way, being far more complex psychologically in his analysis of what he self-consciously calls "a delicate, strange and highly sensitive conscience" (Richard's, or his own).

In spite of the conclusion left to the woman, *Exiles* shifts moral attention from the woman to the man: her freedom becomes a central subject of *his* claims and not of hers, a fundamental aspect of *his* "battle of the souls" with Robert, which is the only duel to be fought on real moral grounds. Moreover, the friendly relationship of the two males of the 'triangle' comes to the foreground: if in *Exiles* the aspiring lover is as defi-

---

[4] On the same line as Pullini was M. Apollonio's appraisal (*Storia del teatro italiano*:1950).

[5] As is made clear by his letter to his brother Stanislaus of January 8, 1927, Joyce was gathering support for his protest against Roth, a publisher who had reprinted *Ulysses* with modifications and without paying author's rights. Praga did not sign Joyce's protest: as either director or member of the Società Italiana Autori ed Editori (Italian Society of Authors and Publishers) he had become a leader for the defense of authorial rights, fighting difficult battles for the cause (see Pullini 1960, 9-23 and Praga's letters on the subject in Lopez 1990) and probably did not want to meddle with delicate legal problems provoking publishers more than necessary.

nitely inferior (intellectually, not socially) to his rival as in Praga (where the difference is more social than intellectual), the deeper motivations of their common attraction to the same woman are delved into only by Joyce. But Praga's play had been a stimulus.

The point was not that "The Italian theatre appealed to Joyce's penchant for raising relatively unknown names from obscurity" (Manganiello 1977: 227), which sounds like a peevish allusion to Joyce's later launching of Italo Svevo in Paris. At the time of the *Notes* Joyce and his writings were totally 'obscure' and could not benefit anybody, while Praga was well-known in Italy, and Giacosa, the other Italian dramatist alluded to, was not only universally considered the leading playwright of his time in Italy, but an international success, as his four-act *Come le foglie* (1900) had triumphed not only in Turin, Milan, Venice, but in Vienna, Budapest and Prague (the capitals of the Austro-Hungarian world which Trieste belonged to) and then at the Odeon theatre in Paris in 1909, with one hundred repeat performances (Nardi 1949: 818).[6] This was not, at the time, an "undercurrent of Italian drama" (Manganiello 1977: 227) and it would have been difficult to ignore it, especially living in Italy. But Joyce was certainly sharp in distinguishing Giacosa and Praga – useful though second rate authors, whose fame would in time decline – from first rank writers, worthy of more lasting success and memory, as in the case of Svevo's *La coscienza di Zeno*. Quick to grasp useful innovations and combine them into powerful views, Joyce did not enjoy quoting 'unknown names' but rather finding nourishment for his mind wherever good 'food' was available. As Giorgio Melchiori pointed out (1984: 39), "much of Joyce's writing may be seen as a complex game with memory, and personal memory at that; the creative moment for him has always an epiphanic quality: it is a sudden revelation suggested by any banal occurrence or by something that he has just been reading".

Farther reaching, though so far unrecognised, was the influence of Giacosa's theatre. In *Tristi amori* (1887), the play mentioned in Joyce's *Notes*,[7]

---

[6] Comments in the *Figaro* were outstanding: according to Gaston Deschamps "Nul auteur n'a été plus applaudi que M. Giacosa" and *Come le foglie* was "le plus grand succès contemporain de la scène italienne" (Nardi 1949: 818-9).

[7] This three-act play had achieved success with Eleonora Duse in the protagonist's role and in 1892 it was also performed, in the presence of the author, at the Burgtheater in Vienna (Nardi 1949: 590, 751).

Emma is intensely in love with Fabrizio Arcieri, an esteemed lawyer and worthy young man working with her husband in his legal practice. Guido Scarli likes Fabrizio for his many good qualities, ranging from his competence to his patient generosity with his father, whom he maintains by his work and by thrifty administration of his resources. But Fabrizio's father, Count Ettore, a nobleman whose penury is matched only by his willingness to run up debts, is irresponsible to the point of falsifying the signature of a bill of exchange. He then tries to blackmail his son into an economically convenient marriage, which would settle the bill scandal, threatening to denounce Fabrizio's and Emma's love. A victim of a bad father and sincere in his love for Emma, Fabrizio is ready to pay off his father's debts with his own meagre means, but events expose his relationship with Emma, and he decides to emigrate to the States with her. At the last moment, however, as Guido Scarli quietly leaves his wife free to make her choice, she decides to stay, so as not to leave behind her little daughter Gemma. Husband and wife remain together, then, but with a sad pact, shared only for their child's sake.

The influence of this play on *Ulysses* is detected by Manganiello (1977: 229), who points out that:

> Molly states at one point that "I won't forget the wife of Scarli in a hurry supposed to be a fast play about adultery" and remembers that someone in the gallery hissed Emma during the performance. *The Wife of Scarli* (1897) has been identified as an English version by G. A. Greene of Giacosa's play. That Joyce used Emma as a prototype for Molly is borne out by Molly's claim that "its all his own fault if I am an adulteress as the thing in the gallery said". This is a reference to Giulio's indirect admission of his own guilt in his final speech, where he proposes to give his daughter a rich dowry so that she may be free to marry a man of leisure and not one like himself who, in being so busy at earning a living, neglects his wife. Moreover, Molly echoes the reconciliation scene of *Tristi amori* when she says "Why can't we remain friends over it instead of quarrelling?"

But this recognition need not by-pass *Exiles*. "If the fact that Joyce used *Tristi amori* as a model for the 'triangle' theme in *Exiles* is not evident except through the preliminary notes, this does not hold true for *Ulysses*" is an unnecessary comment (Manganiello 1977: 228) if the meaning of *Exiles* is fully considered.

The important point common to Joyce's and Giacosa'a plays is not so much Guido's lack of violence or respect for his wife's rights, since he leaves her alone with Fabrizio to make her choice. This is only a detail in Giacosa, though fully in line with Joyce's final scene. What is central is rather the quality of the 'triangle', based on the relationship between the two rivals sharing the same woman, which in Praga well prepares Richard's intense relationship with Robert. Guido Scarli is so friendly and close to Fabrizio as to secure him personal clients and, though not rich, he generously offers him economic help in his moment of need. It is Fabrizio's apparently unreasonable refusal to play up to this brotherly solution that reveals his secret. The relationship between the two lawyers oscillates between reciprocal attraction or identification and a father/son adoption, in which Guido, older than Fabrizio, is a substitute for his friend's real father, negative or absent. In his letters to Fogazzaro, one of his closest literary friends, Giacosa clearly explains that, in the *Tristi amori* 'triangle', he had meant to draw three equally appreciable characters, each right in his good reasons, each with evident good moral qualities, yet fatally drawn to some compromise and to wounding each other (Nardi 1949: 570). This is indeed the situation linking Joyce's Richard, Robert and Bertha, while the 'adoption theme' is by Joyce deferred to *Ulysses*.

Yet, as pointed out by Joyce himself, the psychological study of the deeper motivations are not 'exploited' in Giacosa as they are in his own tormented approach. Guido and Fabrizio mirror each other and love the same woman out of imitation: they become rivals because they appreciate each other too much. Guido's attachment to his wife, pathetically evident in his sacrifices to obtain the money to buy her a most elegant velvet for a new dress, is a model for Fabrizio's own attraction to the same woman. The two men are similarly hardworking, thrifty and loving, but Giacosa's sentimental range is to a degree closer to that of De Amicis, a good friend, than to Joyce's perturbing implacability. In other words, in exploring the mind's and the heart's contradictions Giacosa starts a process subsequently taken farther by Joyce, but contradictions have already become central in his plays and have begun to change their dramatic structure.

Hence the difficulty to end the play and to produce the traditional catharsis, common to Giacosa's play and to *Exiles*: in *Tristi amori* the couple finds no proper moral solution to personal relationships, but defers the question to the next generation and better economic conditions. Later, considering his *Come le foglie,* Giacosa laments his inability to 'end' his plays properly, since they are 'situation dramas' with no real events or little dramatic action. Lack of action, psychic meanderings with no definite conclusion and catharsis are indeed 'on the way' in Giacosa and typical of *Exiles,* ready to become more radical in Beckett's theatre. What unsympathetic critics saw as poor realistic theatre in *Exiles* was anticipating the early features of the future theatre of the absurd.

Besides the quality of the 'triangle' and the dramatic structure itself, Giacosa's play offers one more important clue to Joyce's play. In a dialogue with Fabrizio, Emma points out a difference between "voler bene", "to wish somebody well", and "amare", love as desire. This difference is resumed by Joyce early in the second act, in Richard's and Robert's dialogue opposing Robert's sensual or 'natural' drive for what is beautiful or for sexual satisfaction to what Richard calls "the spirit's luminous certitude" to pursue the partner's well-being. The same difference is also implied in the "immolation of the pleasure of possession on the altar of love" discussed in the *Notes.*

Joyce's mentions of Giacosa range from 1901 to 1939, from his university days in Dublin to his late Paris days. It is one year after the unprecedented success of *Come le foglie* that Giacosa's name appears in *The Day of the Rabblement,* one of Joyce's earliest critical writings, mirroring his reflections on the writer's and the critic's intellectual role. The text is a protest against the involution of the cultural choices promoted by the Irish Literary Theatre after its promising beginning in 1899, with Yeats' *Countess Cathleen* and a programme to produce European masterpieces, proclaiming "war against commercialism and vulgarity" (Joyce 2000: 50). Disappointingly, the institution had failed to maintain its programme, renouncing all innovations and surrendering to the "rabble" it was supposed to educate by producing plays from the most worthwhile authors. As Joyce points out (*Ibid.*: 50) "the theatre must now be considered the property of the rabblement of the most belated race in Europe".

The list of eleven European authors mentioned in the essay as examples of a necessary renovation includes Yeats, Ibsen, Tolstoy, Hauptmann, Strindberg, Flaubert, Jakobsen, D'Annunzio, and a particular group of three, defined "earnest dramatists of the second rank", including Sudermann, Björnson and Giacosa, who "can write very much better plays than the Irish Literary Theatre has staged"(*Ibid.*: 51): but, Joyce ironizes, "the directors would not like to present such improper writers to the uncultivated, much less to the cultivated, rabblement".

Giacosa's inclusion in this restricted European context is meaningful. It is a fair reflection of Joyce's precise evaluation of the author, well aware that Giacosa is not first rank, but belongs to a 'privileged' group of the second rank, qualifying as a good intermediate 'cure' for the rabble, certainly useful to change the belated or "uncreated conscience" which Joyce was working for.

As late as April 4, 1939, writing from Paris to Livia Veneziani Svevo, Joyce mentions *Come le foglie*; again on May 1, alluding to the 'falling leaves' of a volume by Svevo which needed some glue, he is ready to joke: "come non pensare alla commedia giacosiana?" "How could I not be reminded of Giacosa's play?" Considering Joyce's intermediate mention of Giacosa's *Tristi amori* in the *Notes* to *Exiles,* evidently the memory of Giacosa stayed with Joyce throughout his life.

Come le foglie could but impress Joyce for a number of reasons. He was notoriously eager to identify with other writers and find parallels with his personal experiences, and Giacosa's play could offer him a range of circumstances matching his own life story, starting from the economic ruin which in the play brings a well-to-do family down the social ladder: undoubtedly reminiscent for Joyce of what had happened to his own family in Dublin. The scene opens on an experience frequently repeated in Joyce's life, a family's eviction to satisfy creditors: as Giovanni Rosani faces bankruptcy, his wife Giulia, his son Tommy and his daughter Nennele are forced to exchange their beautiful house in Milan for a modest cottage near Chamonix. But Joyce's English lessons too, economically necessary but fundamentally boring, especially if a systematic teaching of grammar was required, find correspondence in the play, in Nennele's difficulties in teaching English to gain a living. Moreover, Giacosa's involvement with opera – he produced the libretti of Puccini's *Bohème, Tosca, Madame Butterfly* – must have contributed to Joyce's memory of the Italian writer.[8] Joyce's passion for opera and singing certainly added to the range of causes for self-mirroring.

Most interestingly for the Joyce of the period 1913/1915 (busy with the notes and text of *Exiles* as with the first episodes of *Ulysses*) *Come le foglie* once again staged the figure of a tolerant husband. Giovanni Rosani,

---

[8] On 16 October 1909, back in Trieste from Dublin and the 'Cosgrave crisis', Joyce took Nora to see Puccini's *Madame Butterfly*.

engaged in his strenuous battle for economic survival, patiently ignores his wife's flirtation with a Norwegian painter, Helmer Strile, who gives her painting lessons: a situation strikingly similar to that of Molly, the singer who in *Ulysses* betrays her husband Leopold Bloom with her the-atre manager Boylan. But what is even more striking is that the 'betrayal context' occasioned by the wife's artistic education is linked in *Come le foglie* with a father-son psychic adoption, as it was later in *Ulysses*.

In Giacosa's play Giovanni has a poor relationship with his own son Tommy, who, far from following the example of his honest hard work, is incapable of earning a living. He prefers "to take a plunge into millions" ("fare un tuffo nei milioni"), marrying a rich woman with an equivocal reputation, who will allow him to live elegantly as a *rentier*. But, left alone by his natural son, Giovanni finds valid help in his young nephew Mas-simo Rosani, brought up as an orphan and ready to adopt his uncle as a father, on the basis of a common ethics of honesty, self-sacrifice and reci-procal understanding. Theirs is an evident mutual father-son adoption, anticipating the psychic relationship between Stephen Dedalus and Bloom in *Ulysses*. If *Come le foglie* offers the very model of the double anthropo-logical situation on which Joyce's novel is based, the parallel, so far un-heeded by criticism, appears hardly coincidental.

Of course, since the 'betrayal context' had a biographical counterpart for Joyce, so, too, did the father/son relationship: Joyce's friendship with Italo Svevo (or Ettore Schmitz), twenty years his elder, and the image "of the devoted father and husband" (Davison 1996: 164). This relationship has been well pointed out by criticism. Suffice it to recall that "in Zurich it was Schmitz's portrait, and not that of his own father, that hung above Joyce's desk" (Davison 1996: 164). Joyce's need to create paternal role-models was "pivotal to his own masculine self-image", as it embodied his need to "psychologically replace a father whose attention and guidance were erratic" (Davison 1996: 165). This father/son relationship merged in *Ulysses* with other important factors, such as Bloom's Jewishness, in which, along with Svevo's Jewish origin, Joyce's appreciation of another Italian source, Guglielmo Ferrero's work, played a particular role (Nadel 1996).[9]

Giacosa's influence, extending from *Exiles* to *Ulysses,* confirms the cre-

---

[9] During his stay in Rome, Joyce read in particular Guglielmo Ferrero's *L'Europa giovane*, which was certainly influential in many respects.

ative links between Joyce's two works. In particular, the adoptive father-son relationship, already emphasized in *Tristi amori*, could hardly escape a sharp, attentive reader like Joyce and could well appear particularly associated with Giacosa, as in his work it is insistently recurrent. Even Giacosa's facile medieval legend in verse *Una partita a scacchi, A Chess Game* (1873) can be read as a father/son 'adoption story' between Renato and Fernando, rather than as a first-sight love story between Fernando and Iolanda, as is usually the case.

An orphan, Fernando is proud of his personal achievements as a daring young man-at-arms, but he openly regrets his lack of a father, whose influence on his education he has missed. His boasting reflects his unrestrained youth and search for self-assurance. Yet, strong, daring and handsome, he has the basic qualities which Renato, with a glorious past and in possession of many castles, would appreciate in the son he does not have, or in the son-in-law he would like as a husband for his daughter Iolanda. As soon as Fernando, page to the Count of Fombrone, meets Renato, the exchange between them turns into a most intense adoption adventure, though at the risk of Fernando's life. Before he can fall in love with Iolanda, Renato challenges him with an unexpected chance, ambiguously a punishment for his boasting and a prize for his manly qualities: he will become Renato's heir, marrying his only daughter, if he is daring enough to stake his life on a chess game with Iolanda, an undefeated champion. Well aware he cannot win, but dangerously seduced by Renato's offer (which betrays his desire to 'acquire him'), Fernando is unable to refuse and spoil his image in Renato's eyes (it is a real 'love test'). The resulting stalemate is solved by Iolanda, who suggests the chess moves to Fernando, seduced by her chance to marry the man chosen by her father: just before Fernando's arrival, she had invited her worried, aging father to choose a husband for her, thus setting the scene for the father/son or son-in-law 'adoption'. Certainly not a masterpiece, but popular entertainment, *Una partita a scacchi* was so widely known that it could hardly escape notice, confirming Giacosa's intriguing theme.

If Giacosa's contribution to *Exiles* appears no greater than Praga's, his influence on the anthropological core of *Ulysses* makes him outstanding among the Italian authors considered here, in spite of Benedetto Croce's peculiar disparagement of Giacosa: perhaps, indeed, it was Croce who was responsible for the 'cover-up' which has so far misled English criticism. In Italy, Piero Nardi's 1949 monumental and fundamental study of Giacosa redressed Croce's bias, but did not influence opinion abroad, where Croce's authoritative name as a philosopher and critic, also translated into English, prevailed. Yet Nardi's evaluation of Giacosa is undoubtedly much closer to Joyce's, while Croce would not have consented to the circumscribed homage of Joyce's lines in *The Day of the Rabblement* or in the *Notes* to *Exiles*. Joyce was here a better critic, while Croce could surprisingly dismiss *Come le foglie* as a second facile love story of the 'Fernando/Iolanda type' (and a debatable one at that). In fact, as Nardi rightly points out, the only 'lovers' present on stage in Giacosa's most successful play, Giulia and Strile, are aged, hesitant and adulterous, while mutual love and marriage between Nennele and Massimo is an uncertain possibility, taking shape only in the end. Prominent throughout remains the socio-anthropological crisis of a dissolving family: an atmosphere in some respects not dissimilar to Chekov's *The Cherry Orchard*.

If placed in an imaginary gallery of the betrayed husbands considered by Joyce in his *Notes* to *Exiles*, Richard Rowan would appear to enjoy the most undisputed superiority over his rival: Praga's Piero Donati is too oversensitive and emotionally weak; Giacosa's Guido Scarli is a solid lawyer, but his rival Fabrizio is a worthy match, and in the end neither man can win emotionally; Giovanni Rosani, equalled in qualities by Massimo, avoids all confrontation with both wife and rival. Even Shakespeare's Othello, the most important literary study of jealousy mentioned in the *Notes*, has no practical advantage over his real counterpart Iago, and is never a foil to his imaginary rival Cassio.

A clear superiority of the betrayed husband over his rival, even greater than Richard's, may be found in the novel *Gelosia* (1894), by the third Italian author mentioned by Joyce in his *Notes*: Alfredo Oriani. Joyce's quotation evoking Oriani does not actually come from this novel, but from *Rivolta ideale*, a subsequent lengthy socio-moral essay. But the provincial setting of the more or less successful lawyers of *Gelosia* (*Jealousy*) is the same as in *Tristi amori* and might have triggered off a comparison in Joyce's mind. We know from Ellmann that Joyce possessed the book.

In the novel the husband, Filippo Buonconti, rises through a well-deserved career: intellectually strong and morally straightforward, he is admired by all who surround him, and after dealing with great ability with the socio-political problems of his small town, he is elected to Parliament in Rome. His much younger wife Annetta, a plump attractive woman of

22, betrays him with a young, handsome and narcissistic man, but most discreetly, and never injures his self-assurance. Her lover, Mario Zanetti, works as a lawyer in Buonconti's study, but, both intellectually and socially inferior, he is finally destroyed by his affair. He is used by Annetta to conceive a child she could not have with her 47-year-old husband, who is made happier by the birth of a beautiful baby girl.

The jealousy of the title is not the husband's, but the lover's: Mario cannot really possess Annetta or resist his own social decline, while his rival continues to rise, and he is finally left alone when Annetta leaves for Rome with her husband. What may be of interest for *Exiles* is that the betrayed husband is remote from the tradition of the disparaged cuckold. He is not a 'defeated man', but a model of competence and success, which his rival fails to imitate, missing his chance for professional advancement.

Yet it is from the 1906 essay *Rivolta ideale* that Joyce draws for a passage in the second of his extant *Notes*, referring to Bertha: "Her age is the completion of a lunar rhythm. Cf. Oriani on menstrual flow: 'la malattia sacra che in un ritmo lunare prepara la donna per il sacrificio'" ('a sacred malady seems within a lunar rhythm to prepare her for the sacrifice').[10] The ideological meaning implied in this allusion to maternity as 'a sacred illness' – in Oriani's eyes justifying the woman's exclusion from social careers and relegating her to intellectual inferiority and social dependence – is neither referred to nor discussed in Joyce's play. Deriving from the section "Femminismo" (Oriani 1933: 302), it merely contributes a recurrent symbolic link of Bertha to the moon, although in one of the *Notes* Joyce states he does not want to be a feminist, as he is aware he could appear so in the play. Was this, after all obvious lunar reference for the woman the only reason justifying Joyce's mention of Oriani? Or did Joyce take something else from him – and, if so, what? Not the conservative thinking (Oriani's *opera omnia* was later promoted by Benito Mussolini), as Joyce rather sympathized with Ferrero's essays, the socialism and cosmopolitan experience of which were far more influential on him.[11] It was another, longer

---

[10] Manganiello's translation.

[11] Manganiello enlarges on Oriani's anti-feminism as such, implicitly suggesting Joyce's approval, as he reads Joyce's ideological attitude in his debatable essay "Anarch, Heresiarch, Egoarch" along the line its title suggests (Manganiello 1984: 98-115, in Melchiori ed. 1984). Joyce was certainly not ready to exchange the Irish Catholic church for a party church or an ideological church, but in his letters to Stanislaus he insisted

passage in Oriani's book from the section entitled "Delle bassure dell'amore moderno" ("On the baseness of love today"), centred on the intriguing figure of Mary Magdalen, which, as yet undetected and in spite of its rhetorical discouraging quality, contributed to Joyce's reflections.

To Oriani, Magdalen is an important symbol to be opposed to contemporary passionless forms of love. Irritated by current social practices, whereby marriage only meant the matching of "a profession and a dowry" or the union of two patrimonies (Oriani 1933: 305),[12] he evokes this symbol as comprising the two opposing forms of feminine sexuality, the prostitute, ready to accept any man, and Christ's desexualized devotee, or the whore and the nun. A symbol of passion in both cases, Magdalen is repeatedly the object of exhalted rhetorical excess:

> Povera, grande Maddalena di Gesù! La tua anima era rimasta pura, mentre il tuo corpo rimaneva bello come un diamante immerso in un pantano...Tu eri l'ideale di un tempo messianico, nel quale il peccato aveva impeti profetici come la virtù... nessuno amò come te; ecco perché Gesù ti elesse nell'amore umano e ti rivelò, come un segreto ancora ignoto a tutti, la propria resurrezione... tu amasti Gesù da lungi, senza pretendere che fosse tuo, felice nel dolore della tua passione, inconsolabile nello spasimo della sua. Tu eri la donna, tutta la donna, peccatrice e redenta, doppiamente pura nel cuore, che il peccato non aveva potuto corrompere, nel pensiero, sognante l'amore di un Dio. (*Ibid.*: 306-308)

> Poor, great Jesus-possessed Magdalen! Your soul had remained pure, while your body remained as beautiful as a diamond plunged into mire...You were the ideal in a Messianic time, when sin had prophetic urges like virtue... no one loved like you; that is why Jesus selected you in human love and revealed to you a secret still undisclosed to anyone, his own resurrection... you loved Jesus from afar, without pretending he should be yours, happy in the sorrow of your passion, inconsolable in the spasms of his. You were the woman, all of her, both sinner and redeemed, doubly pure in your heart; sin could not corrupt you in your mind, dreaming of a God's love.

in defending his 'socialism' against his brother's hostility, and he was well aware that his ego did not coincide with "the conscience of my race". As a writer his model was Shakespeare's universality.

[12] This was an appealing protest for Joyce, who felt socially disapproved of for having chosen a woman below his social standing, a hotel maid with no education.

Oriani's erotic-mystical thrust could in particular have appealed to Joyce at a time when he had begun his *Ulysses* as a *summa anthropologica,* or a counterpoint to the *Summa theologica* by Saint Thomas. The anthropology of Magdalen's two aspects implies a *coincidentia oppositorum,* the turning of fidelity into betrayal (and viceversa), which lies at the basis of jealousy, the subject of *Exiles.* It is this anthropological intuition, communicated in a striking Christian symbol, that Joyce finds insisted on in Oriani's essay: the 'Magdalen complex' is indeed enlarged on in two of his notes.

Both passages fuse Bertha/Nora Barnacle and Magdalen, while at the same time evoking the author's identification with Shelley, whose grave Joyce had visited in Rome. Shelley's victory over death as a poet and Christ's resurrection here melt into one and the same image:

> She is the earth, dark, formless, mother, made beautiful by the moon-lit night, darkly conscious of her instincts. Shelley, whom she has held in her womb or grave, rises: the part of Richard which neither love nor life can do away with; the part for which she loves him: the part she must try to kill, never be able to kill and rejoice at her impotence. Her tears are of worship, Magdalen seeing the arisen Lord in the garden where he had been laid in the tomb.[...] She is Magdalen who weeps remembering the loves she could not return. (Joyce 2000: 346-7)

> [...] it is a fact that for nearly two thousand years the women of Christendom have prayed to and kissed the naked image of one who had neither wife nor mistress nor sister and would scarcely have been associated with his mother had it not been that the Italian church discovered, with its infallible practical instinct, the rich possibilities of the figure of the Madonna. (Joyce 2000: 349)

Joyce does not here quote directly from Oriani or his subject matter, but he does resume Oriani's Magdalen image, and does so within a vaster system: in the battle of the sexes played out in *Exiles* the mirror-image accusations of Bertha as a "man-killer" and of Richard as a "woman-killer" combine. Richard assumes Christological nuances, as occasionally does Bertha, though she is also Richard's Magdalen. All of which intertwines with the sado-masochist anthropological outlook, implicit in all the relationships in *Exiles* as in its *Notes,* ready to resurface in the encyclopaedic ocean of *Ulysses.*

Bibliography

Apollonio, Mario, 1950, *Storia del teatro italiano*, Vol. IV, Sansoni, Firenze.

Cianci, Giovanni, 1974, *La fortuna di Joyce in Italia*, Adriatica, Bari.

Costello, Peter, 1992, *James Joyce, the Years of Growth 1882-1915*, Pantheon, New York.

Croce, Benedetto, [1908] 1940, "Giuseppe Giacosa" in *La letteratura della nuova Italia*, Vol. II, Laterza, Bari (already published in *La Critica*, VI, 1908).

Davison, Neil R., 1996, *James Joyce, Ulysses and the Construction of Jewish Identity*, Cambridge University Press, Cambridge.

Del Greco Lobner, Corinna, 1989, *James Joyce's Italian Connection*, University of Iowa Press, Iowa City.

Ellmann, Richard, 1982, *James Joyce. New and Revised Edition*, Oxford University Press, New York.

Ferrero, Guglielmo, 1898, *L'Europa giovane. Studi e viaggi nell'Europa del Nord*, Fratelli Treves, Milano.

Giacosa, Giuseppe, 1987, *Teatro*, Mursia, Milano.

Joyce, James, [1901] 2000, "The Day of the Rabblement", in *Occasional, Critical and Political Writings*, ed. by Kevin Barry, Oxford University Press, Oxford and New York, pp. 50-52.

——, [1918] 1992, *Exiles*, in *James Joyce, Poems and Exiles*, ed. by J.C.C. Mays, Penguin, London and New York.

——, [1922] 1960, *Ulysses*, The Bodley Head, London.

——, 1966, *Letters of James Joyce*, ed. by Richard Ellmann, Vols II-III, Faber & Faber, London.

Lopez, Guido, 1990, *Marco Praga e Silvio D'Amico, Lettere e documenti 1919-1929*, Bulzoni, Roma.

MacNicholas, John, 1979, *James Joyce's Exiles, A Textual Companion*, Garland, New York and London.

McCourt, John, 2000, *The Years of Bloom, James Joyce in Trieste 1904-1920*, The Lilliput Press, Dublin.

Maddox, Brenda, 1988, *Nora. A Biography of Nora Joyce*, Routledge and Kegan Paul, London.

Manganiello, Dominic, 1977, "The Italian Sources for *Exiles*: Giacosa, Praga, Oriani and Joyce", in Joseph Ronsley (ed.), *Myth and Reality in Irish Literature*, Wilfrid Laurier University Press, Waterloo, Ontario, Canada.

——, 1984, "Anarch, Heresiarch, Egoarch", in Giorgio Melchiori (ed.) 1984, pp. 98-115.

Mason, Michael, 1977, "Why is Leopold Bloom a Cuckold?", *ELH*, 44, pp. 171-188.

Melchiori, Giorgio (ed.), 1984, *Joyce in Rome, the Genesis of Ulysses*, Bulzoni, Roma.

Nadel, Ira B., 1989, *Joyce and the Jews*, University Press of Florida, Gainesville.

Nardi, Piero, 1949, *Vita e tempo di Giuseppe Giacosa*, Arnoldo Mondadori, Milano.

Oriani, Alfredo, [1906] 1933, *Rivolta ideale*, Licinio Cappelli, Roma.

——, 1993, *Gelosia*, Mursia, Milano.

Pullini, Giorgio, 1960, *Marco Praga*, Cappelli, Bologna.

Praga, Marco, 1907, *La crisi*, Fratelli Treves, Milano.

Rabaté, Jean-Michel, 1991, *Joyce Upon the Void*, Macmillan, London and New York.

Restivo, Giuseppina, 1991, *Le soglie del postmoderno: 'Finale di partita'*, il Mulino, Bologna.

Segre, Cesare, 1974, "La funzione del linguaggio nell''Acte sans paroles' di Samuel Beckett", in *Le strutture e il tempo*, Einaudi, Torino, pp. 253-274.

'A TALE OF TWO SHORES': THE ADRIATIC AND ITS MODERNISMS
On Ivan Meštrović, Adrian Stokes and the Sculptural Imagination[*]

Francesca Cuojati

The basin of the Northern Adriatic has a coastline of stark contrasts: sandy/stony; flat/mountainous; smooth/rough; auroral/crepuscular; 'monolithic'/fragmented in archipelago. The sea which modelled the basin has also carried the rich trade of idioms, ideas and goods, that has endowed the area with a composite and shifting character in history and culture. It has been the home of both Western and Eastern peoples, the theatre of cohabitations, exchanges and diasporas. A province of the Roman Empire, it subsequently became the site of the vicissitudes of the Venetian Republic, the Ottoman, Habsburg, Napoleonic Empires and South Slav states, as well as an important 'actor' (and 'victim') in all the wartime and ideological tragedies, large and small, early and late, of the twentieth century. Now, again, the Adriatic sea and its coasts are witnessing the surge of new, dramatic migrations in the third millennium, while the post-Yugoslav Balkan ethnic and religious communities are precariously engaged in the effort of peaceful mutual recognition and respect. What follows is an attempt to illustrate how and why this self-same Adriatic area – in particular the northern Adriatic region on the fringes of the Gulf of Venice and beyond, including the territory inland – was able to play a positive

* Thanks to Ann Stokes, Philip Stokes, David Carrier and Richard Read for patiently enduring my queries about Adrian Stokes' 1932 travel to Dalmatia. Above all, I am very grateful to Caroline Patey for generously acquainting me with Adrian Stokes and his work. Still more for her 'fearsome' encouragement.

role in contributing Mediterranean languages, images and forms to the international process of shaping the aesthetic features of modernism. A brief history of British Adriatic acquaintanceships, real as well as vicarious or imaginary, will help reveal at which point and through which processes, practices and stages a deep change and a radical renewal were to make their appearance in the aesthetic dealings between Mediterranean localities and traditions and the Northern, international cultural and artistic scene of modernity. In this context, the illustration and investigation of two specific cases of 'lesser', peripheral modernism born out of Anglo-Adriatic exchanges (Ivan Meštrović and Adrian Stokes) bear witness to the possibly still greater relevance of the role of Mediterranean memory and heritage, as re-read and aesthetically re-assessed by early twentieth-century sensibility.

I

1. *Faraway, so close!*

Venice apart, of course, the appearance of the North-Adriatic, and, in general, of both shores of the Adriatic sea on the map of the Anglo-Saxon and Anglo-American traveller is only recent history. The first transoceanic travellers came only with late Victorian tourism, the rush towards the sun of the Nordic émigrés (see Fussell 1980, Triani 1988). The incipient interest in Balkan curiosities was also a factor. Above all, however, the impulse given by rail and steam notoriously fostered tourism, while the increase in Anglo-American globe-trotters was also favoured by the vast commercial travelling enterprise initiated by Thomas Cook (see Pemble 1987, Corbin 1988). Notwithstanding the freshness of the destination and the new facilities in the field of travelling, a substantial continuity may be noticed in the modalities of reception of the Mediterranean landscape involved in such encounters with the Adriatic, with 'ruins' and exoticism as invariable ingredients. In this sense, nineteenth-century and early, pre-war twentieth-century tourism had roots deep in the golden age of the Grand Tour, as if the Adriatic coasts and inlands were merely the last destination added to a travel itinerary to Italy that already encompassed the Tyrrhenian, Florence, and a Venice so far secluded and de-contextualized from its environs, both natural and cultural. It is thus to the eighteenth century of antiquarianism and early archaeology, as also to that of Eastern travelling, as we shall see, that one must turn to come upon those European pioneers who first kindled an interest in the encounter with and exploration of unfamiliar Dalmatia – "very obscure countries" in Edward Gibbon's remark! (2005: 21 n. 86) – and whose experiences set the model and rituals for many an Adriatic journey to come.

2. On the Thames...

In keeping with the English thirst for digging antiquities in the South during the eighteenth century, the first relevant British expedition to Dalmatia had Spalato, or Split, as its destination, and archaeology as its main drive, namely, the surveying and graphic reproduction of the original plan of the palace built for the late Roman emperor Diocletian, after he officially resigned the empire in the year 305 AD. Accordingly, having met Giovanni Battista Piranesi in Rome and learnt the twin practices of excavation and imaginative reinvention of ruins from his work in Campo Marzio, the architect Robert Adam set off to Dalmatia from Edinburgh in 1757. With him were the French artist Charles-Louis Clérisseau, also a Roman connection, and two other draughtsmen (see Pilo 2000: 151-163). Their work, hindered at first, and then hastened by the suspicions of the Venetian governor of Spalatro[1], resulted finally in the publication of *Ruins of the Palace of the Emperor Diocletian at Spalatro in Dalmatia* (1764), a volume "which contains the only full and accurate Designs that have hitherto been published of any private Edifice of the Ancients" (Adam 2001: 30). Given the fact that "scarce any monuments now remain of Grecian and of Roman magnificence but public buildings", Diocletian's palace was chosen because, in Adam's words,

... [nothing] could more sensibly gratify [the architect's] curiosity, or improve his taste, than to have an opportunity of viewing the private edifices of the Ancients, and of collecting, from his own observation,

---

[1] "The Venetian governor at Spalatro, unaccustomed to such visits of curiosity from strangers, began to conceive unfavorable sentiments of my intentions, and to suspect that under pretence of taking views and plans of the Palace, I was really employed in surveying the state of the fortifications." (Adam: 30)

such ideas concerning the disposition, the form, the ornaments, and uses
of the several apartments, as no description can supply. (2001: 26)

Adam's main difficulty on the Adriatic had been that of pursuing his
archaeological endeavour about and around an architectural complex
which, by then, had undergone a drastic metamorphosis. In the seventh
century, while the migration of the Slavs into south-eastern Europe was
changing the ethnographic character of Dalmatia, the nearby centre of
Salona was captured and completely destroyed by the Avars, and its people migrated within Diocletian's building, turning it into the core of the
metropolitan area of today, and partially making use of the palace itself
as a quarry for building stone for their private accommodation, warehouses and shops.[2] By the sixteenth century, Spalato had become an important commercial entrepôt where the Ottoman caravans from Constantinople, entering Dalmatia from Bosnia, met the merchant fleets of
Venice. Thus, "possibly the most serviceable ruin in the world" according to Rose Macaulay (1966: 410), Diocletian's precinct had stopped being a palace and started being a commercial town, and it was actually
from the hybrid language of architecture in the palace-city on the Adriatic harbour that Adam brought the seeds of the renovated, post-Palla-

----

[2] Adam observed: "The inhabitants of Spalatro have destroyed some parts of the
palace, in order to procure materials for building; and to this their town owes its name,
which is evidently a corruption of Palatium. In other places houses are built upon the
old foundations, and modern works are so intermingled with the ancient, as to be
scarcely distinguishable ..." (Adam: 28) In the critical edition of Adam's volume, Marco
Navarra explains that "although it is true that Adam does not expand on how the city
was constructed inside the palace, it is also true that he presents an accurate drawing
of the design of the urban fabric in relation to the ruins and he carefully shows us the
way in which certain buildings, such as the Temple of Jupiter (actually Diocletian's
mausoleum), were transformed for other uses. In his book, Adam offers us a clear image of the transformation of architecture upon itself, suggesting the idea that the city
was born of the palace from whom it takes its name, as he himself explains." (Navarra
2001: 197) According to Aldo Rossi, "l'indagine di Robert Adam su Spalato è uno dei
contributi fondamentali di questo processo dell'architettura (definire un progetto di
architettura e un'ipotesi di progettazione architettonica dove gli elementi sono prefis-
sati, formalmente definiti, là dove il significato dell'architettura è sempre nuovo); ed
uno dei riferimenti più straordinari dove la storia offre, nei suoi eventi, una architet-
tura-città e l'uso alternato di diverse tipologie. Un'idea di architettura riferita ad elementi concreti e non modificabili conforma la città." (Rossi: 373)

dian Neoclassicism he successfully cultivated in Georgian England, selling its 'fruits' at a premium to the wealthy London customers of his studio. Their elegant town houses and the respective interiors profited greatly from Adam's survey of Spalato, and the Adelphi (1758-1762), above all, both a unitary urban 'grand-style' monument and a residential and commercial development built upon a bend of the Thames, seems especially to have derived its modern architectural multi-functionality from the dialectics between palace and city, unity and density that were peculiar to the Adriatic port[3]. Doubtless the style of an ancient Roman imperial palace suited the capital city of a rampant modern empire.

From Spalato there also came part of the oriental side of the eclectic, sometimes whimsical Adam-style: as is well known, Diocletian's splitting up the Roman empire through power-sharing and keeping the East for himself[4] would stand for the dawn of the Byzantine world. Architecturally speaking, some details of the palace, like the so-called "Syriac" arched-trabeation pediment, are said to set the building on the threshold of the Byzantine or Romanesque, and prove the employment of oriental workers in its construction. In Spalato, moreover, according to architectural historians, "Roman architecture, which on the one hand pours into the Byzantine, joins the Arabic world [as] the 'palatia' derived from the 'castrum' and 'castella' (built along the borders in the great examples of Treviri, Salonicco, Antioch and Filippopoli etc.) anticipate the various Islamic castles of the Omeiad princes whose characteristic ruins are spread across the Syrian desert" (Cantone 2001: 229). Besides, in the Spalato volume, Adam lingers with more than one view on the sphinx "which was anciently in the Temple of Jupiter" (Adam 2001: plates I, LV, LVI) and which, sufficiently 'tamed' to suit English interiors, stands out as one of the most recurring elements among the fanciful 'Egyptomaniac' decorations designed by him: a further instance of his preference for the 'hybrid'. The frontispiece of the volume itself "opens on a panorama inhabited by strange figures: a mixture of cultures, ethnological curiosities, architectural fantasies and in-

---

[3] The Adelphi Royal Terrace was demolished in 1936-8 and replaced with the present Deco building. In her 1941 travelogue about and around Yugoslavia, Rebecca West was to write: "…often when we look at a façade in Portman Square or a doorway in Portland Place, we are looking at Roman Dalmatia." (1995: 140)

[4] By promoting Maximian as the Caesar of the Western Empire, establishing subsequently a Tetrarchy.

tellectual adventures" (Navarra 2001: 169): a view tinged with a tranquil exoticism, in which the swirl of the turbans and trousers of the Illyrians in the foreground, the spiral of the broken tortile column and that of the opposite tree behind reverberate agreeably while the draughtsmen's hands and eyes are busy on their drawing papers. The mild oriental side of Adam's Spalato rests serene and free from the radical otherness attached to the identity of the other orients of archaeology. The British architect himself explains that his work was encouraged "by the favourable reception [...] given of late to the Ruins of Palmyra and Balbec" although he was not, like Dawkins, Bouverie and Wood, "obliged to traverse desarts [*sic.*], or to expose [him]self to the insults of barbarians" (Adam 2001: 30).

3. ...as the East

It was only shortly afterwards that the Adriatic turned, or was turned into a modern ethnographic boundary. As a result, Dalmatia 'got' its own 'barbarians' at last. Interesting enough, it was partly because of little known Dalmatia that Europe underwent a reorientation from North to South, to East and West, a remapping which culminated later in the period of the Cold War.[5] Crucial to such rearrangement was the expedition of the Paduan abbot Alberto Fortis, botanist, geologist and anthropologist *ante litteram*, whose 'enlightened' *Viaggio in Dalmatia*, published in Venice in 1774, was soon translated into German (1776), French (1778) and in the same year also into English, under the title *Travels into Dalmatia*.[6-7]

---

[5] "During the course of the Enlightenment, Europeans arrived at a new view of how their continent was structured and divided. Up until the Renaissance it was taken for granted that Europe was divided into North and South, separated by the Alps. This view of the continent dated back to Roman times and was based on the presumption of Mediterranean cultural superiority with respect to northern *barbarism*. Such a perspective was evident in the *Germania* of Tacitus at the end of the first century, and still meaningful at the beginning of the sixteenth century when Pope Julius II rallied Italians against the French with the slogan *Fuori i barbari!*" (Wolff 2005: 1, italics by the author)

[6] *Containing general observations on the natural history of that country and the neighbouring islands; the natural productions, art, manners and customs of the inhabitants: in a series of letters from Abbe Alberto Fortis, to the Earl of Bute, the Bishop of Londonderry, John Strange...observations on the island of Cherso and Ossero. Translated ...with...considerable additions, etc. (Iter Buda Hadrianopolim anno MDLIII exaratum ab A. Verantio), London.*

[7] For a present-day partial travelogue on Fortis' route, see Marzo Magno (2003).

Fortis was sponsored by English patrons from the Royal Academy, which lent his enterprise an international feel and resonance from the start.[8] Still more, the attention paid to Fortis was the consequence of two relevant historical circumstances that from Venice soon echoed all over Europe. 1. Of late, Dalmatia, as a province of the Republic of Venice, had undergone an enlargement of borders at the expense of Ottoman territories[9], hence an inland *nuovo* and a *nuovissimo acquisto* (new and newest acquisitions) had been added to the *vecchio acquisto* (old acquisitions) which consisted of the costal strip of the eastern Adriatic. 2. In the meantime the Mediterranean trade on the Adriatic, already less profitable than it once was, began to suffer the competition of Habsburg Trieste as a free port after 1719. Accordingly, in Larry Wolff's reading of Fortis, investigations into the natural resources and economic potential of Dalmatia were part of the Venetian design to refashion "the rule of centuries across the Adriatic into the form of a modern [Adriatic] empire" (Wolff 2001:3). Considered in retrospect, focusing on the 'obscure' province of Dalmatia was "a final fantasy of imperial resurgence" (Wolff 2001: 5) before the *Serenissima Repubblica* was destroyed forever by Napoleon in 1797. In their turn, Europe and above all Britain took interest in *Viaggio in Dalmazia* and in Dalmatia itself, as the relevance of empire and the issues of imperial rule were topmost on the contemporary international political agenda. In the context of the mixed excitements and anxieties that accompanied the rise of the modern British empire, Edward Gibbon turned to Fortis as well – which won him a reference footnote – when writing about the Roman Province of Dalmatia and declaring his regret for its not having being preserved in its ancient integrity. "The inland parts have assumed the Sclavonian names of Croatia and Bosnia, the former obeys an Austrian governor, the latter a Turkish pasha; but the whole country is still infested by tribes of barbarians." (Gibbon 2005: 21).

---

[8] It was Frederick Augustus Hervey, Bishop of Londonderry, who paid the expenses of his fellow-traveller in Dalmatia Fortis. Hervey was a compulsive eccentric traveller and also a patron of the arts with considerable interest in geology and vulcanology. Fortis' British connections also included John Strange, a Fellow of the Royal Society with archaeological interests living in Venice from 1773 to 1888 as the British Consul in the city, and the third Earl of Bute, John Stuart, who was Prime Minister (1762-63) in Georgian England, a lover of art, architecture and botany above all, and helped to create Kew Gardens.

[9] Peace of Carlowitz (1699), Peace of Passarowitz (1718).

As to the question of who these "barbarians" were, they were identi-fied according to the following taxonomical constellation, not without conspicuous overlapping: Dalmatians, Morlacchi, Illyrians, Albanians, Bosnians, Serbs, Croats, Slavs. Living mostly in the mountains, given to semi-nomadic pastoralism or agriculture, largely Orthodox in religion (though also comprising a Catholic community), their languages were recognized as Slavic. On account either of emigration or of the *nuovo* and *nuovissimo acquisto*, there were also new Dalmatians among them, espe-cially people who lived along the Bosnian border. In Venetian eyes, these people appeared socially 'uncivilized' when compared to the 'civilized' coastal Dalmatians, a contrast which prepared the ground for the future separation of the Italians from the Slavs and the conflicts that were later to follow.[10] Though never in the domain of a literal empire, but in that of a *Serenissima Repubblica* until its end, the relationship between ruling Venice (the *Dominante*) and the dominion beyond the sea (the *Oltremare*) was always basically colonial in nature, and therefore ready to espouse an ideology of empire.

> The perceived asymmetry between the Republic's continental Italian and trans-Adriatic territories was rendered 'legitimate' in the eighteenth century by the ideological articulation of difference in Dalmatia, whether as economic backwardness, anthropological barbarism, or alien nation-ality. (Wolff 2001: 8)

Out of this context and of the pragmatism of the Enlightenment de-rived Fortis' text, which presents a mixture of ideological and scientific issues. While on one hand the administrative question was a matter of discipline (taming and training the Morlacchi so as to secure their loy-alty to the Venetian Republic)[11], anthropologically speaking, it was more a matter of civilization and a civilizing mission, and Fortis at times "was sufficiently anthropological to question whether the Morlacchi really should be civilized" (Wolff 2005: 4). Like Rousseau, he sometimes

---

[10] See Ballinger (2003) for an updated bibliography on such matters and a present-day investigation into the question of memory and exile in the Istria and Quarnaro re-gion.

[11] Wolff reports the phrase *"indisciplinati Morlacchi"* (2005: 4, italics by the author) as the one used by the Venetian Governatore Generale Pietro Michiel in 1765.

seemed to recognize the Morlacchi as *noble savages* [...] and could formulate [...] a critical perspective on the *society that we call civilized*. Whereas Rousseau had speculated philosophically about how a Carib might feel about civilized European life, Fortis actually made empirical observations in his study of the Morlacchi. (2005: 4, italics by the author)

No doubt, the Paduan abbot's 'field-work' was far ahead of its time. Apart from a wide range of contributions on geology (he was the first to explore many caves in the Istrian and Dalmatian Karst – see Shaw 2001), we owe to him precious information about the mores and customs of the peoples of Dalmatia. As for literature and culture, he was a pioneer, for instance, in the field of folklore studies, and in his travelogue he published several Morlachian folk songs as examples of South Slavic oral poetry, including the song *Hasanagitica* that spread all over Europe, finding illustrious mediators like Goethe and Herder, and giving life to the phenomenon of 'Morlacchismo' that was to influence also Madame de Staël in France and Walter Scott in Great Britain. Named elsewhere 'Morlaccomania', the vogue culminated in an article by Giulio (or Julije) Bajamonti from Spalato on the 'Morlacchismo' of Homer (1797) (!), and of course fed the current of Romantic orientalism. Like many experiences and writings of eighteenth century orientalists, explorers and naturalists (Cook included, whose voyages dated 1768-1771, 1772-1775, 1776-1779), Fortis's *Travels* is ambiguous in that it sways between Rousseauist primitivism on the one hand and assimilation *cum* discipline/civilization/subjugation on the other. The inhabitants of the 'obscure', wild, mountainous Dalmatian heartland beyond the Dinaric Alps would continue to fascinate Europe and Britain as utterly 'primitive' in their ruggedness, 'barbarism'[12] and ingenuousness, as much as they would go on inspiring 'civilizing' 'will to empire' missions with the respective practices and rituals of travelling, collecting (data as well as artefacts), classifying, objectifying (conquering). To quote Larry Wolff once more, "[Fortis] was aware that the distance that separated the world of the primitive or *uncorrupted* customs from that of *civilized* society was no wider than the Adriatic Sea." (2005: 4, italics by the author)

---

[12] In 1895, the primitive otherness of the Morlacchi would re-emerge in H.G. Wells' science fiction masterpiece *The Time Machine*: here, subterranean, nocturnal creatures named Morlocks would prey on the Eloi, the graceful, refined inhabitants of the Upper World.

4. *'Delightful Dalmatia'*

> *Viola.* What Country, friends, is this?
> *Captain.* This is Illyria, lady.
> *Viola.* And what should I do in Illyria.
> My brother, he is in Elysium

Notwithstanding Fortis' *invitation au voyage*, Dalmatia continued to be more cherished than travelled to, thus remaining a destination 'off the beaten track' well into the nineteenth and early twentieth centuries. As late as 1906, the Anglo-Indian Lieut-Colonel J.P. Barry would write: "The whole land is quite off the line of the tourist, for I have not seen as yet a Cook's hatband anywhere" (165). And a notice reminding his readers that Ragusa gave Shakespeare's *Merchant of Venice* the "sweet name 'argosies' (*ragusies*) for his flotilla"[13] reveals the substantial immutability in popular representations of Dalmatia dating back to the Illyria of Shakespeare's *Twelfth Night* (note its consonant/assonant vicinity to the word 'Elysium'). Titles like *Dream Cities* (Goldring 1913) and *Delightful Dalmatia* (Moqué 1914) confirm the opinion of Oscar Wilde, who succinctly summed up the popular identification of Illyria with mere fiction when, in a review of an amateur production of Shakespeare's play, he wrote "When there is no illusion there is no Illyria" (XIII: 46). Byron's *Childe Harold,* too, is imbued with Illyrian suggestions in consequence of the poet's travels in the Orient, albeit more ambiguous political undertones have dripped into the text, also on account of Napoleon's re-adoption of the name 'Illyria' for the newly conquered Dalmatian provinces (see Wallace 1998, Wolff 2001: 332ff and Mardešić 1995: 101-105). These, then were the representations of more or less 'dreamful' Romanticism and its exoticism. However, the Victorian period also produced variously mixed replicas of Adam's and Fortis' travel-writing enterprises, verging either more on the architectural/archaeological, like Thomas Graham Jackson's *Dalmatia, the Quarnaro and Istria* (1885) and Mrs Russell Barrington's *Through Greece and Dalmatia. A Diary of Impressions recorded by Pen & Picture* (1912) (not without a touch of Ruskinism on the part of Lord Leighton's friend and biographer), or on the anthropological/ethnological, like John Gardner

---

[13] "dating from the time when the galleys of this stirring little Republic crowded with her commerce all the Eastern Seas" (Barry: 193).

Wilkinson's *Dalmatia and Montenegro* (1848) (see Wolff 2001: 319ff and Mardešić 1995: 206-221; see also Galsworthy 1998 and Allcock *et al.* 2001 as regards women travellers in the Balkans).[14] Though it would be untruthful to speak of British 'looting' of Dalmatia, in the manner of British 'looting' elsewhere in the Mediterranean, starting from the episode of the Elgin Marbles, the politics and poetics of cultural encounter underlying most Victorian and Late-Victorian travelling in and writing about Dalmatia never really went beyond those orientalist practices, in which a whole set of deeper exchanges, aesthetic as well as human, was not contemplated (see Jezernik 2004).

II

1. *Another country*

In the early decades of the twentieth century, accelerations of a geo-political nature turned the Adriatic shores into a labyrinth of borders drafted and redrafted by the downfall of empires and the re-constitution of nations. Such a paroxysm of boundaries was responsible for entangling, even imprisoning, the existence and identity of numerous men, women and children, when not for turning them into the voluntary and involuntary protagonists of migrations and exiles. As Paul Fussell has remarked, it was above all Fiume, once one of the Austro-Hungarian 'pearls of the Adriatic' – Rijeka in Serb-Croat (indeed still Rijeka today, although in Croat only) – which bewildered British citizens travelling after the collapse of the Habsburg Empire. Arriving in the city overlooking the gulf of Quarnaro (Kvarner) in 1924, Osbert Sitwell gets caught in "a thicket of passports and questions" (quoted in Fussell 1980: 35), while Rebecca West, who crosses the border in the company of her husband in 1937, sketches the following description of Fiume as an "impeded" city:

[14] Between 1872 and 1890, also the eclectic traveller in the Orient, translator and explorer Richard Francis Burton contributed essays about the Istrian seaboard and the coast of Dalmatia to the *Journal of the Anthropological Institute of Great Britain and Ireland* and to the *Journal of the Royal Geographical Society of London.* At the time, Burton was British Consul in Trieste.

…a southern port hacked by treaties into a surrealist form. On a round
plan laid out plainly by sensible architects for sensible people, there is
imposed another one, quite imbecile, which drives high walls across
streets and thereby sets contiguous houses half an hour apart by detour
and formality. And at places where no frontiers could possibly be, in the
middle of a square, or on a bridge linking the parts of a quay, men in
uniform step forward and demand passports, minatory as figures projected into sleep by an uneasy conscience.
"This has meant," said my husband as we wandered through the impeded
city, "infinite suffering to a lot of people," and it is true. (1995: 123)

Reporting her husband's words, West denounces the consequences
and moral outrage caused by Fiume's fate onto its own populace, torn to
shreds by partitions, by the imposition of ever changing national identities, and crushed under the perversion and the obliteration of an architectural plan originally born out of humanistic principles (bridges once
built to 'connect', are now 'haunted' by men in uniform). Nonetheless,
aesthetically speaking, the very same awkward scenario is assessed in the
terms and forms of the Avant-gardes ("surrealist form"), and portrayed in
a prose encoded according to the same cipher of ruins which informs the
urban nightmares of most modernism. Highly spatialised, geometrical,
almost abstract in its de-humanized imagery, West's passage about Fiume stands out as a specimen of writing apt to illustrate and subsume
how the dismemberment of reality brought about by the Great War affected the aesthetics of the urban 'modern' in the way we still think about
it now. What is more, however, deliberately late by the time it was written (c. 1938), clichéd, almost parodistic, West's ostentatiously 'graphic'
description is but an episode in *Black Lamb and Grey Falcon. A Journey
through Yugoslavia*, a belated homage to canonized high-modernism, as
contrasted with her fresh rhapsodizing (aesthetically as well as culturally
and politically) over eastern, Byzantine and primitive Old Serbia, i.e. the
Yugoslavian region to which the largest part of her travel narrative and
of her appreciation are committed. If, then, West's encounter with Yugoslavia and the conspicuous travelogue that ensued represent the patent
culmination of a renewed relationship between modern Britain, the eastern shore of the Adriatic and the Balkans behind[15], other practices and

---

[15] For more than one reason, cultural and political as well as aesthetic, West's is a
very controversial book still waiting for a thorough critical evaluation; in the text traces

forms of hybridization, both imaginative and linguistic, are also worth focusing on; those, for example, deriving from minor or forgotten encounters and exchanges between Adriatic expatriates or exiles and their British domiciles and hosts, along with those between Anglo-Saxon travellers of the thirties and the natural and cultural landscape of the Adriatic region. In order to substantiate the idea of the Mediterranean as the engine and pivot of a complex system of diverse modernisms, large and small, sometimes contaminated by the culture and memory of geographically peripheral traditions, two case-studies in particular could be of interest. Although centred around cultural episodes chronologically at variance with each other, and involving intensely different personalities, it is hoped that, in the end, the following attempted recognitions of the aesthetic and intellectual experiences of the sculptor Ivan Meštrović (1883-1962) and the art critic and writer Adrian Stokes (1902-1972) will yield helpful clues for a fuller understanding and appreciation of the multifaceted sculptural imagination that permeated and shaped modernist art and writing, thanks not only to Ezra Pound's undisputed and militant contribution. In order to concentrate specifically on the figures of Meštrović and Stokes, information about the Adriatic legacy pertaining to the so-called middle-European koine, or involved in the creative relationship of James Joyce with Trieste will be purposely left out. As far as those cultural and aesthetic scenarios are concerned, the reader is obviously referred to distinguished critical works respectively by Claudio Magris (1963, 1982, 1986), John McCourt (2000) and Renzo S. Crivelli (1996, 2004), texts that exhaustively explore those subjects, illustrating them in masterly fashion.

Recovering the resonance of Meštrović's experience in London as a Dalmatian expatriate canvassing for a unified, autonomous Southern Slav state, and retracing the impact of his sculpture exhibited at the Victoria

of old primitivist orientalism and of oversimplified manicheism survive and co-exist with novelty and ideological and stylistic daring. As Vesna Goldsworthy points out, however, West's travelogue stands out as a real turning point in travel-writing about the Balkans in that it "reverses the traditional position of British travel writers in the Balkans who usually, before the 1930s, encountered passionate but childish creatures with little understanding of the outside world. In Rebecca West's case, the adventure and the sense of discovery often arose from the encounters with Balkan intellectuals [...]. They turned her journey through Yugoslavia into an inward journey of self-discovery and a spiritual quest" (1998: 180)

& Albert Museum in 1915 will provide a telling instance of the irruption and intricate reception of Balkan politics and Adriatic inheritance in the cultural agenda of Great Britain at the dawn of World War I and beyond. Whereas, following the later, thin thread of Adrian Stokes' scantily documented 1932 visit to the Dalmatian coast, in the art-prose of his *Stones of Rimini* (1934), will help discover the seeds of a different post-war creative awareness: one re-founded on the relationships between artist and corporeality on one hand, artefact and original environmental context on the other. Originally following in the wake of Ezra Pound's *Cantos* and subsequently employing as well as fuelling a Kleinian vocabulary, Stokes' meditation upon the bas-relief sculpture carved by the fifteenth century artist Agostino di Duccio in the Istrian stone of the Tempio Malatestiano in Rimini will be seen achieving a novel sort of Adriatic synthesis. This aesthetic and humanistic fantasy articulates the ideal reintegration of the western Italian and eastern Venetian/Slavonic shores of the Adriatic under the aegis of early-Renaissance artistic practice, and in the light of its deep commitment to the elemental corporeality of the Mediterranean basin, namely limestone and water.

2. Ivan Meštrović: a second coming?

After one look at pictures of the Ivan Meštrović Gallery on the Marjan peninsula in Split, and a glimpse of press reviews about the exhibitions of the artist's works in England, one would be tempted to read the appearance and popularity of the sculptor/architect Ivan Meštrović on the early twentieth century London stage as a sort of second coming. The suggestion is prompted by a pair of intriguing coincidences, and can be easily explained. 1. The Gallery in Spalato was founded on a donation from the sculptor himself, and hosts the largest collection of his works, installed inside the grand neoclassical seafront villa of his own design where he lived with his family.[16] 2. The articles and advertisements about venues

---

[16] The donation was made to Yugoslavia. The opening date was 1952. The building was erected from 1931 to 1939 according to the sculptor's own plans, and was intended as a combination of residential quarters, exhibition halls and ateliers for artists. Meštrović lived there up to the Second World War.
See http://www.mdc.hr/mestrovic/galerija/povijest-en.htm.

for exhibitions of works by Meštrović in Britain include many mentions
of a 1919 event at the Twenty-One Gallery, Durham House-street, Adelphi. Hence, not without the aid of some imaginative enthusiasm comprising also a rather gross rearrangement of dates and characters, another
Diocletian might be spotted, possibly a new designer himself bringing
from Dalmatia fresh creative proposals to be adopted and developed again
on the shore of the Thames, rather as Robert Adam once did with the
Adelphi, only this time right inside it! Unfortunately, however, nothing
pertaining to a similar clear-cut unhistorical conclusion about Meštrović
proves true indeed, starting from the national identity of the artist himself, to finish with the claims of neo-classicism brought forward by his
sumptuous villa in Split.

Ivan Meštrović was born in the Dalmatian highlands (Vrpolje, Slavo-
nia) in 1883, spent his childhood in Otavice, a village of Dalmatinska
Zagora (the Dalmatian inland), spent a year (1900) as an apprentice in the
stonemason Pavle Bilinić's workshop in Split, and then went to study at
the Academy of Fine Arts in Vienna (1901-1906), just as the Secession
movement was reaching its peak. Apart from some years in Zagreb, where
he was appointed rector of the Academy of Art, and in Split (1922-1941),
the rest of his life was spent as an expatriate in Paris (Rodin was among
his connections there), Rome, Switzerland, the South of France and the
United States, where he died in 1962. On account of the last phase and
crisis of the Austro-Hungarian Empire and in consequence of the Balkan
conflicts and the Great War, the question of his nationality is vexed: in
1906, some of his works were on display in the Dalmatian section of the
Austrian Exhibition at Earl's Court, London, whereas, in the 1911 and
1914 World Exposition in Rome and Venice Biennale, he exhibited in the
Serbian Pavilions. Again, in the summer of 1915, he was a Southern-Slav
sculptor at the Victoria & Albert Museum, while, in 1917, he was a Serbo-
Croat artist exhibiting at the Grafton Galleries, to appear, one more time,
at the Twenty-One Gallery as Meštrović, the Serbian sculptor, in 1919.
Not long before the opening of the V&A event, the labyrinthine history
of the national/regional affiliations and labels regarding Meštrović culminated in a rancorous dispute over the title of the show, an episode that involved his promoters and supporters: the denomination 'Southern Slav
Sculptor' on the exhibition poster [Fig. 1] satisfied the anti-Habsburg federal aspirations of the circle of Southern Slav émigrés from Austria-Hungary, whose advisor and spokesman in Britain was the Scottish journalist

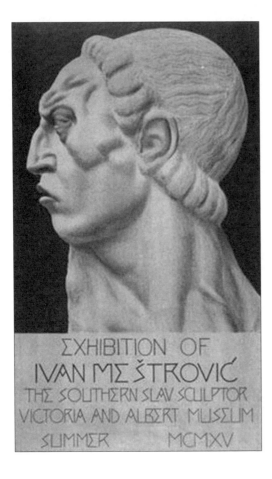

Figure 1 – Poster designed by Ivan Meštrović advertising the 1915 Meštrović exhibition at the Victoria & Albert Museum. Lithograph on paper, 96.5 × 58.4 cm.

and historian R.W. Seton-Watson[17]. Quite the reverse happened with the Serbian Legation and the Serbian Ambassador in London Matija Bošković, who were frustrated in their desire of presenting the sculptor as 'Serbian', in conformance to their identification of Southern Slav unity with the

---

[17] Seton-Watson's travels to and knowledge of the Dalmatian region (see Seton-Watson 1911 and 1915) had begun in the context of the development of modern tourism in "delightful" Dalmatia; Wolff writes that "his challenge to the Habsburgs to attend to their nationalities problems before it became too late made him a hero in Dalmatia when he returned in 1912" (2001: 350).

emergence of a Greater Serbia.[18] After being imprisoned by the Croatian Ustaše (1941), and then becoming a Yugoslavian subject with the constitution of Tito's Socialist Federal Republic (1945), today the *Encyclopædia Britannica* finally describes Meštrović as a "Croatian-born American sculptor known for his boldly cut figurative monuments and reliefs", and devotes only a meagre handful of lines to his career. History undoubtedly pushed Meštrović into the centrifugal machine of crucial European and Balkan conflicts more than once: every time the sculptor emerged from the turmoil of the machine, he was able to recognize and engage in favour of a would-be homeland only 'as displaced', in the shoes of an international émigré, that is from outside ever changing borders and ideologies.

2.1. *The aesthetics of cultural politics*

Linked again to the question of national identity, there was, if possible, even greater cultural and aesthetic confusion over the reception of Meštrović's work following his success on the international scene at the 1915 V&A solo exhibition. If the draughtsman and critic Ernest H.R. Collings opened his 1919 essay on "Meštrović in England" with the following words:

> The Tragedy of war has brought to this country, an island indeed in matters of art, the work of Ivan Meštrović, which has widened our knowledge of contemporary activity and stirred into more vigorous life our understanding of sculpture. (Collings 1919: 48)

in recent times, it has been minutely documented that what Collings had asserted "owed very little to the routine procedures of the pre-War art world and [...] a great deal to the contingencies of wartime cultural-political manoeuvring" (Clegg 2002: 740). As a member of the *Entente*, Britain was in a mood to support the idea and welcome the culture of a unified Southern Slav state, since it would make a most effective barrier against the present and future threat of a Germanic *Drang nach Osten.*

---

[18] Elizabeth Clegg further observes that "Despite [the] apparent victory for Seton-Watson, English linguistic usage was for some time to favour the term proposed by Bošković, be it on account of the familiarity of Serbia as a wartime ally or of continuing confusion as to the identity of the 'Southern Slavs' and, later, of the 'Serbo-Croats'". (2002: 746)

No wonder, then, that at the heart of the V&A show were some figural sculptures and a wooden scale model of the *Cycle of Kosovo*. Not without underlying intentional references to the contemporary unliberated situation of the South Slavs, still under the yoke of Austria-Hungary, the sculptor's "epos in stone" (Collings 1919: 50) was meant to celebrate the "paradoxically inspirational" (Clegg 2002: 740) military defeat of the Serbs by the Turks at Kosovo Polje in 1389, also in the light of the popular cycle of heroic ballads that had gone to form the core of the poetic heritage of Serbia. The architectural-sculptural complex of gigantic proportions was originally conceived by Meštrović when in Paris. The British Authorities saluted it with warmth. As both the artist and his sponsors – though these last diverged in some views – linked the exhibition to the political engagement in favour of the autonomy of the native South-Slav states, his work was promoted accordingly. No matter if "in the last analysis [...] the exhibition at the V&A [sparked] less public and critical interest in Meštrović as a representative of his people and their aspirations than in his achievement as an exceptionally talented individual" (Clegg: 746), it was above all due to a well manoeuvred political press campaign and propaganda that the cultural reaction and the critical assessment of the British readily welcomed or rejected his works in accordance with only limited specific cultural-aesthetic categories. Whereas a mixture of sheer nationalism, barbaric arrogance and racial ostentation stretched and paroxystically contorted Meštrović's desperately proud Serbian stone heroes of the past, other works in a minor key were also on display (*The Annunciation*, 1913; *Shepherd Boy with a Flute*, 1913). They were entrusted with the task of representing the genuine, religious character of the South Slav: an austere rural people of peasants, shepherds and stonecarvers. Therefore, a mythopoeic biographical outline in tune with the sculptor's patriotic mission was what was mainly offered by the titles and content of profiles circulating in the press, one example being a heading in *The Daily Chronicle* that recited: "From Shepherd Boy to Prophet and Leader" (quoted in Clegg 2002: 476). Up to 1919 and beyond, Ivan Meštrović's modest rural origins in the karst were illustrated over and over in England, more often than not in close association with a childhood spent almost in symbiosis with the craggy landscape overhanging the Adriatic. He was "born of peasant parents in Dalmatia, [...], and had [...] little education in the ordinary sense. His genius, strong, pure and direct, [had] been mainly guided by the folklore of the Serbian Croatian peasantry" (*The Globe*, May 13th 1919). Still more, he was trained under "forcible, primitive conventions" (*The Challenge*, May 9th 1919), and it was in the mediums of stone and wood that "as a boy (he) was taught to carve by his father, a Croatian peasant, who had built houses and decorated them with his own carving" (*The Evening News*, July 5th 1919). To the rough scenario of limestone Dalmatia we shall later return. For the moment, suffice it to say that the sculptor's kinship with it suggested also his partaking in the ancient lineage of stonemasons and architects that had once crossed the Adriatic to work in the buildings of Ravenna, Rimini and Renaissance Venice. Above all, Meštrović's Viennese academic training often went unmentioned, perhaps even passed over deliberately by his British biographers, especially at the time of the V&A exhibition.

2.2. A taste for Meštrović (whose taste was Meštrović's?)

The 1915 V&A exhibition ran from 24 June to 25 August [Fig. 2], and had truly resonant echoes. It was a literal sensation[19] which raised an incandescent exchange of views in the letter columns of *The Times*, when a Slade Professor dared to use adjectives such as "wilful, inchoate, amorphous, even monstrous" (quoted in Clegg 2002: 747) in reference to the works by Meštrović on display. In the meantime, pieces by the Adriatic Balkan sculptor were entering English collections, "be it through sales to private individuals, as gifts to friends of the artist, or through acquisition by a public institution" (Clegg 2002: 750). In Italy, the Avant-garde polemicist Giovanni Papini and painter-writer Ardengo Soffici had already expressed their admiration for Meštrović, while Tommaso Marinetti had even gone so far as to write of affinities between Futurism and his works (De Micheli *et al.*: 5) (Marinetti was prone to express all-inclusive enthusiasms, apt to enlist more and more new adherents in his movement). Despite the sculptor's indisputable 'cosmopolitanism', it was rather the Anglo-American Avant-gardes of the period that never really attributed any kind of relevance or even marginal meaningfulness to his works. Neither Pound's Vorticist group, nor the Blooms-

---

[19] The exhibition catalogue "sold so well that numerous reprintings were required, total sales at the Museum being calculated at a remarkable 4,335 copies. [...] In departure from usual V&A practice, the Museum continued to sell Meštrović photographs long after the exhibition had closed ...". (Clegg 2002: 747)

Figure 2 – Exhibition of sculpture by Ivan Meštrović installed in the West Hall,
Victoria & Albert Museum, 1915, London. Photograph.

burians seemed interested in his sculpture, although it is well known
that around the same years both parties were deeply engaged in, and
working hard toward a 'repositioning' of the sculptural imagination in
literature and the arts, experimenting also with a new formal poetics articulated in terms of 'massess' and carved, stylized contours, also on account of their reception of Post-Impressionism, of the newly discovered
African Art, and of the 'Byzantomania' that was in the air (see Bullen
1999, 2003). Parallels matching Meštrović with the Slavic forms and
dynamisms of Diaghilev's coeval *Ballet Russe*, or with the 'archaic'
Mediterranean recovered through the anthropological researches of the
classicist scholar Jane Ellen Harrison never really showed up, although

both trends were quite familiar in Bloomsbury and would have been well worth comparison. On the contrary, when cherishing the aesthetic relevance and beauty of the masterpieces by Henry Gaudier-Brzeska in 1916, Pound compared them favourably with the "glorification of energetic stupidity" of the sculpture by Meštrović (Pound 1916: 151). In a 1921 *Little Review* essay on "Brancusi", the American poet – who, by the way, was the one who could best smell tradition and academicism from afar in spite of any propaganda – again dismissed the South Slav Meštrović in derogatory terms: "no one who undestood Gaudier was fooled by the cheap Viennese Michelangelism of Mestrovic [*sic.*]" (Pound 1980a: 211). As early as 1915, it had been the turn of Wyndham Lewis to 'blast' the artist in his journal of the same name (92). Bloomsbury, meanwhile, remained haughtily silent.

As for the reception of the Avant-garde of Irish origins, though warmer in a sense, in the end they could see nothing new or unconventional in the work of Meštrović. Joyce's strange bent for the sculptor, "whom [he] greatly admired" according to Dario de Tuoni (Crivelli 2004: 118) was of course fuelled by his proximity to the Eastern Adriatic lands, and by his contacts with the South Slavs in Pola and Trieste. As reported by the novelist's pupil in Trieste, the writer owned some framed reproductions of works by Meštrović from the Alinari Institute of Photography in Florence (Crivelli 2004: 118). Paolo Cuzzi and Oscar Schwarz, who visited the novelist in Trieste in his flat on the Via Donato Bramante, also recalled the reproductions. Richard Ellman mentions them in his description of the odd atmosphere that reigned in the room where Joyce gave English lessons. It was a place "perfumed by a burning cone of incense", with a "vellum-covered missal-like volume of *Chamber Music*" "on a reading desk ecclesistical in style" (Ellmann 1983: 381).

To enhance the atmosphere of ritual, three photographs of sculptures by Meštrović took the place of icons. Schwarz recognized them as among the sculptures exhibited at the Expositione Biennale in Venice the previous year. Joyce had cut them from a catalogue, sent them to be framed, and then inscribed his own titles below them. One represented a peasant woman, belly swollen, face contorted with labor pain, her sparse hair half covered by a wretched wig. Joyce's title for this was 'Dura Mater.' The second was a mother and a child, the bony infant hanging from a withered breast, and under this he had written 'Pia Mater.' The third

photograph was of an ugly old woman naked, and under it Joyce had
had engraved [some] lines from Canto V of the *Inferno* [about Helen of
Troy]. (Ellmann 1983: 381)[20]

In Renzo Crivelli's words, this was "a mixture of the sacred and the
profane" (2004: 122), a *mise en scène* in pure Joycean "desecrating, paro-
dic" (124) mode, that, aesthetically speaking, had something in common
with the Irish novelist's fascination for the Catholic and Eastern Greek-
Orthodox liturgical ceremonies then at hand in the churches of Trieste.
If not exactly a full appreciation, Joyce's whim about Meštrović's 'mothers' at least shows the acute awareness of there being a kind of consonance
between the Irish and the South Slav's age-old calvary of mute sacrifice
and submission, a lot that the Irish exile could not abide even in the privacy of his own house, if not by submitting its 'icons' to some form of
subversion. Perhaps, then, it was not fortuitous if, as late as 1927, another Irish writer and intellectual, W.B. Yeats, paid some attention to
Meštrović, this time in reference to institutional matters. As the head of
a Government committee in charge of appointing seven artists who could
participate in a competition meant to select the designs for the new cur-
rency of the Irish Free State, Yeats invited Meštrović (who, in his opinion "expressed in [his] work a violent rhythmical energy unknown to past
ages, and seem[ed] to many the foremost sculptor of the day") to contribute his own design (see Cleeve 1972).

Things went differently in private communications and matters, at
least to judge from the curiosity of attending the V&A exhibition man-
ifested by D.H. Lawrence in his correspondance with Ernest Collings[21].

---

[20] According to what Ivo Vidan wrote in 1974, "It is difficult to identify the three
Meštrović sculptures, but they are definitely not in the catalogue of the 1914 Venice
Biennale, as Schwarz claims. The 'ugly old woman naked' is very likely the famous one
reproduced in the catalogue of the 1911 Rome exhibition, at present in the Ca' Foscari
[*sic*] collection in Venice. The other two are probably also in Italy where the artist had
them stored on the eve of the 1914 war." (1974: 275)

[21] See Lawrence (1981 II: 360 and also II: 157, 570); the misspelling of the sculptor's surname ("Mestrovig") disappears in the ensuing references as if the novelist had
acquired direct knowledge of both work and name. In May 1918, two poems by Lawrence
appeared in the anthology *New Paths*, together with poems by Richard Aldington, F.S.
Flint, Walter de la Mare, Aldous Huxley, Edith, Osbert and Sacheverell Sitwell and
others. Also reproduced in the volume were pictures by Augustus John, Walter Sick-

It should not go unnoticed, moreover, that while in London Meštrović's physiognomy attracted the lens of ultra-modern photographers such as the American Vorticist Alvin Langdon Coburn and the Munich-born Emil Otto Hoppé.

2.3. 'Primitivism reanimated'

What was, then, if any, Meštrović's influence on or consonance with modernism? How to assess his presence in England in the face of such a set of 'vitiate' and contrasting reactions? In the midst of general popular clamour and prevailing silence on the part of the Avant-garde, two observations by *The Burlington Magazine for Connoisseurs* critic Robert Ross and *The Outlook* journalist Charles Marriott seem worth more attention than others. Both expressed ideas contrary to the inclination of focusing on the South Slav primitive exoticism/nationalist rhetoric of the sculptor's works. Ross did not hesitate to stress the cosmopolitanism of Meštrović and to underline its art-historical inevitability; according to him, the sculptures were simply "European", being "no more Serbian than English" (Ross 1915: 206), thanks to the present breaking of cultural isolation among the countries of Europe and beyond. Like other artistic émigrés who were 'carving' the modernist idiom 'out' of tradition, the sculptor was credited by Ross with the aptitude and skill of speaking an international language. Marriott, for his part, insisted on the concept of Meštrović's truth to material; in his opinion his "aim [was] to put the subject before you by convincing your mind through its nice translation into some other material; consequently the treatment of the stone or wood or bronze bec[ame] the most, instead of the last, important part of the business" (TGA 8024). The journalist went on to protest that "Very often a young sculptor [was] blamed for imitating the Easter Islanders when the poor fellow [was] only trying humbly to imitate stone" (TGA 8024). A further observation inhered Meštrović's sculptural-architectural double disposition as especially embodied in caryatids, in which "you ha[d]

---

ert, John and Paul Nash, Mark Gertler and Ivan Meštrović. Besides, in Lawrence's novel *Women in Love* (1921), ch. 30, Meštrović turns up fleetingly as not attuned to the taste of the nihilistic German sculptor Loerke: "He hated Mestrovic [*sic.*], was not satisfied with the Futurists, he liked the West African wooden figures, the Aztec art, Mexican and Central American." See Cianci (1983) for the relationship between Lawrence and Futurism / Vorticism with reference to the novel and the character.

to suggest not only the substance but the structural purpose of the stone, and even when the statue ha[d] nothing to support you ha[d] still to suggest its relation to the building" (TGA 8024). If one adds to this observation the fact that the only two contemporary artists to be mentioned in reference to the Slav sculptor at the time were Jacob Epstein and Eric Gill[22] (Ross 1915: 206), one possible way out of the critical impasse can be attempted by the following considerations. Quite different from one another, the American-born Epstein, the Catholic, late Arts and Crafts and controversial 'guru' Gill[23] and the nomadic Slav Meštrović did share a new sense of 'primitivism'. As far as this supposed new 'primitivism' is concerned, the reader is referred to the coeval switch in ethnographic discourse from what has been termed "an *archaeological* metaphor of the primitive, in favour of a *vitalist* one".

> Unlike the "culture fossils", artefacts or phantasms studied before the closing decade of the nineteenth century, a good deal of what was construed as evidence by ethnographic writers after 1890 was avowed to be *alive* and walking the earth in all the various regions of "savagery", if not at the heart of "civilization". [...] Ethnography began to release the primitive from its bondage to history. (Hoyt 2001: 332, 334).

The exhaustive archival ordering of dead rather than living ethnographic specimens was declining, along with the historical method of evolutionary reconstruction, while more and more voices testified to the "durability, vitality, and potentially subversive agency of the primitive in all its manifestations, whether colonial or metropolitan" (Hoyt 2001: 339). Epstein's, Gill's, Meštrović's close relationship with their materials, their preference for the 'corporeal' as well as their assiduous frequentation of the motives of virility and maternity, though in different degrees, can all be said to strike right at the heart of modern 'reanimated'

---

[22] Not incidentally, John Middleton Murry himself, before championing Epstein as 'the real thing' among contemporary sculptors in 1920, had spoken in favour of Meštrović (see Clegg 2002: 749). In 1925, he would write the introduction of Adrian Stokes' first book *The Thread of Ariadne*.

[23] See MacCarthy for a critical assessment of Gill's life and work as a sculptor and engraver. Almost certainly both Epstein and Gill were influenced by seeing the V&A exhibition or works by Meštrović on other occasions.

primitivism in which artists and writers like Gaudier-Brzeska, Pound, and Lewis also participated.[24] However, for Meštrović in particular, things went differently, as the reception of what was modernist (i.e. vitally primitive) in his work was hindered and complicated by his being a figure of transition in the field of 'ethnographical observation'. If one were to trace him back to one of the eighteenth century declensions of the relationship between the Eastern Adriatic and England, it would undoubtedly be more to Fortis and the underlying imperial ideology than to the neo-classicism of Adam that one should turn. Still encrusted with the relics of Fortis' archaeological palimpsest of classification and data collecting and entangled in the corresponding ideology of deadening orientalist objectification, the work of Meštrović as well as the sculptor Meštrović lent themselves to be generally considered primitive in a by then already 'old-fashioned' or traditional manner. Precisely beause the British authorities' championing of Meštrović in the context of their engagement in favour of the independence of the South Slav states was more a matter of honouring the *Entente* than a true reception of his sculpture as *vital*, the inherent advertising and promotion of South Slav cultural heritage did not go beyond the old proposition of exotic folkloric entertainment. It was because of such geopolitical and cultural circumstances and not because of his youthful academic training in Vienna and Paris that Meštrović was severed from the rank of international modernism and his identity as a sculptor born a 'primitive stonecutter' was misconstrued.

Because it deals with a preference for the primitive and with the concept and practice of 'truth to materials', as it centres around stone carving as well as around the interdependence of sculpture and architecture, and it rests deeply enmeshed in Adriatic corporeality, the work of Adrian Stokes – though later and very different from Meštrović's in terms of style and medium – will perhaps also help to clarify further and shed new light on the South Slav sculptor's position in the context of modernism and its relationship with the Adriatic landscape.

---

[24] Lawrence Rainey, for example, has argued that in Pound's *Malatesta Cantos*, Sigismondo Malatesta is "the historical exemplar of an ahistorical form of life" (1991: 220), no doubt one instance of Hoyt's 'primitive reanimated' as released from the Victorian "bondage of history".

3. *Adrian Stokes'* Stones of Dalmatia[25]

When the London-born 'aesthete'[26] Adrian Durham Stokes left England for a (short?) Dalmatian holiday[27], he was not at all ignorant of things Mediterranean. Nevertheless, because of this selfsame 'trip', it cannot be said that there ever was "nothing particularly unusual about Stokes' geographical and cultural itinerary" (Stonebridge 1998: 109). It is true that, in the Twenties, the young Stokes had got caught in the conjunction between a certain version of modernism and élite tourism, and consequently had become familiar with the gulf of Tigullio, making friends with the Sitwells in Rapallo, and playing tennis with Ezra Pound (1926), whom he also met in Venice (see Patey 2004).[28] And it is also true that Stokes had travelled down the Italian Adriatic coast, from Venice to Rimini via Ravenna, as Pound and Yeats also did.[29] But the choice of Dalmatia as a holiday destination must have sounded simply bizzare to the writer's highbrow friends.[30] Such a peculiar choice, and the scant biographical infor-

---

[25] As regards Adrian Stokes' published works, we have used here the 2002 Penn State University Press edition for *The Quattro Cento* (hereafter *QC*) and *Stones of Rimini* (hereafter *SR*); and the 1978 Thames and Hudson edition for *The Critical Writings of Adrian Stokes* I, II, III (hereafter *CW* I, II, III) other than *QC* and *SR*.

[26] 'Aesthete' according to the definition David Carrier has attached to Stokes, Ruskin and Pater: "In defining an aesthete as someone who gives a central place to the pleasures of visual experience, valued both for its own sake and as a source of self-knowledge, I distinguish aesthetes from art historians and others for whom interpreting and presenting artworks is a profession or a business. Being an aesthete is a way of life." (Carrier 1997: 14)

[27] By then, the Dalmatian inlands and coast were part of the Kingdom of Serbs, Croats and Slovenes, i.e. Yugoslavia.

[28] In the course of his life, Stokes travelled also to Spain and Portugal, the Orient and America, and cruised twice in Greece. Although being fundamentally London-based, he also spent some years in Cornwall and Ascona.

[29] Since the beginning of Grand Tourism back in the eighteenth century, the Italian Adriatic coast had long been 'snubbed' by travellers. The Tyrrhenian coast with Naples was preferred to the opposite unhealthy environment and sandy shores (see Corbin 1988). *Les bords de l'Adriatique* (1878) by Charles Yriarte, and *The Shores of the Adriatic* (1906, 1908) by Frederick Hamilton Jackson sanctioned a change in the travel itineraries of foreign travellers in Italy; both texts promoted and 'advertised' the artistic and naturalistic amenities of the Adriatic shores at a time when also the first establishments of the new 'bathing era' were being developed.

[30] At the beginning of the twentieth century, the Austrian tourist industry trans-

mation about Stokes' holiday itself are all the more reason for 'diving' into his 'unusual' Adriatic and trying to make sense of the aesthete's 'knock-ups' from one shore of the sea to the other.

What is known about Stokes' travel to Dalmatia amounts to its rather uncertain chronology, to the name of the woman that accompanied him on the journey and her identity, to what Stokes was writing and doing at the time, to the behaviour of stevedores in the streets of sunny Yugoslavia (!) (see note 32 below), to what monuments he visited and where, to the 'scientific' motivation behind. Nothing else. No landscape description, no impressions whatever. Although these elements are crucial for an attempt at fathoming the meaning of that journey, they are strikingly few for someone who was the author of two autobiographical narratives imbued with memories of the Mediterranean and a whole body of writings about fifteenth century architecture and sculpture, Adriatic Rimini and Renaissance Venice.[31] What can be done, however, to speculate about Stokes' travel to Dalmatia is to detect and pay attention to indirect clues about the journey in the texts, so as to compare them with factual and biographical data. But first, the essentials:

> I shall be in Venice (on what may subsequently prove to have been a honeymoon) for the first few days of August. Any chance of seeing you? From there I go down the Dalmatian coast, particularly Sebenico. I also want to go to Rimini again. (Quoted in Read 2000: 110)

The abovementioned lines from a letter to Ezra Pound, dated "before August 1932" by Stokes' biographer Richard Read, help establish his departure in August 1932. We also know that it was in the company of the

formed Dalmatia into a curative resort for Viennese vacationers. When the Kingdom of Serbs, Croats and Slovenes was established, Dalmatia was turned into a kind of extraterritorial entity much travelled above all by German tourists who enjoyed swimming and nudist sunbathing despite the strong pro-nazi, anti-Yuogoslav feelings in the air. English tourism in Dalmatia, however, was still only sporadic. Notwithstanding the wealth of Stokes' means, moreover, his travels to Dalmatia cannot match King Edward VIII and Wallis Simpson's 1936 cruise down the Dalmatian coast on the Nahlin, a chartered steam yacht.

[31] *Inside Out* (1947), *Smooth and Rough* (1951), *The Quattro Cento. A Different Conception of the Italian Renaissance* (1832), *Stones of Rimini* (1934), *Venice. An Aspect of Art* (1945).

young sculptress Mollie Higgins (to whom *Stones of Rimini* is dedicated)[32] that Stokes would embark on the journey, starting from Venice and most probably continuing by steamer down the Dalmatian coast and archipelago, as far as the Quarnaro (Kvarner) island of Arbe (Rab), and then to Sebenico (Sibenik), Spalato (Split), the island of Brazza (Brač) to reach finally Ragusa (Dubrovnik) – all this, at least, according to the place names and monuments mentioned in the six pages of annotations entitled "Dalmatian Architecture and Sculpture" in one of his notebooks now in the Tate Gallery Archive.[33] The holiday was probably a short one as Stokes was later to write in *Venice*: "I have visited there only once, and then hurriedly" (*CW* II: 126). At the time, Stokes was undergoing analysis with Melanie Klein[34], his book on *The Quattro Cento* had just been finished and published by Faber, and he was working on a second volume – *Stones of Rimini* (1934)[35] – of what initially was to be a trilogy but never found completion with a third volume.

As a wealthy young gentleman without a profession, Stokes was given

---

[32] Stokes and Higgins were to marry soon, but it actually never happened. Shortly afterwards their return from Dalmatia they split up and she married psychoanalyst Gilbert Debenham. Read's reading of Stokes' break with Higgins is centred around the young writer's bisexuality bearing also a passage from a letter written from Yugoslavia to his old university friend Edward Sackville West as testimony: "I have never seen such extraordinary behaviour in the streets as here. Boys caress themselves and each other. The dirtiest and [foulest?] of stevedores mince along the quay hand in hand. The shoe criterion is non-existent, and the most affectionate dispositions seem to wear layer upon layer of ragged shirt, as too trousers, through which, however flesh shows." (Quoted in Read 1998: 89). Read's *Art and Its Discontents. The Early Life of Adrian Stokes* (2002) gives ample space, documentation and critical relevance to Stokes' juvenile sexual orientation and its import in analysis with Melanie Klein. For an illuminating review of Read's 2002 biography of Stokes' early years, see Carrier (2003) who on his part and advisedly welcomes Read's research and great amount of materials gathered, but maintains at the same time that the visual arguments of Stokes' books deserve more critical attention as much as his extremely inventive art-prose and its genesis call for further consideration.

[33] Apparently, the notes about Dalmatia were more than that, as Stokes was to write in 1945: "The notes I made have been lost" (*Venice*, *CW* II: 126)

[34] Stokes' analysis with Klein had started in 1930 and was to end in 1936.

[35] Through the mediation of Ezra Pound with the editor T.S. Eliot, some instalments from Stokes' work on Rimini and the Malatesta Tempio had been published in *The Criterion*, IX, 34, October 1929 under the title "'The Sculptor Agostino di Duccio', extracted from the third of four essays on the Tempio Malatestiano at Rimini".

to art journalism, but he was much more than that. Positively versed in eclecticism, he was also a 'ballettomane' who reviewed the performances of the *Ballet Russe* and had tried to transfer his own assets to Diaghilev's company in a rush of utter generosity ('fortunately' he had had to see his offer turned down). The core of his field (or fields) of interest, however, were Mediterranean early Renaissance art and landscape, which he perceived as a regenerating aesthetic alternative, a "counter-landscape" (*Inside Out, CW* II: 153) as he called it, to the ugliness of the Edwardian London in which he had spent his childhood days.[36] In the twenties, both in the footsteps of Ruskin and against some of his assumptions, he had been so profoundly affected by his own experience of Venice[37] as to engage in a daring revision of the Italian Renaissance as well as of art in general as centred around Venetian fifteenth century sculpture and architecture rather than on the sixteenth century Florentine masters championed by the school and tradition sanctioned by Vasari[38] and promoted

[36] Stokes was later to employ the following words in describing his first experience of Mediterranean landscape and light in Rapallo: "Existence was enlarged by the miracle of the neat defining light. Here was an open and naked world. I could not then fear for the hidden, for what might be hidden inside me and those I loved. I had, in fact incorporated this objective seeming world and proved myself constructed by the general refulgence. Nothing for the time, lurked, nothing bit, nothing lurched." (*Inside Out, CW* II: 157) See also Kite (2000) about Stokes' aesthetic reaction to the urban landscape of Edwardian London.

[37] Stokes and Ruskin were most discordant in their appraisal of both the Renaissance period and Venice; Stokes perceived Ruskin's championing of Gothic architecture and his dismissal of Renaissance Venice as paradoxical and contrived, as he was later to write in *Venice*: "Not only would he reject the greater part of the stone Venice to which he was most sensible, which inspired him, which surrounded him, but he would strive to prove in terms of this stone that an anterior, for all practical purposes no longer existent Venice, a Venice pre-eminently of wood, brick, stucco, paint, terra-cotta and mud, transcended and shamed the Venice he knew. His own magnificent title is sufficient condemnation of his distorted thesis." (*CW* II: 115). However, as maintained by David Carrier, Stokes' 'Ruskinism' is both evident and undeniable in his participation in a "traditional form of art writing, the art lover's narrative of the grand tour [...] [which] tells of the self-transformation that occurs when the Northern writer goes south" (Introduction to *QC* 2002: 3, 5), and encompasses authors like Hazlitt, Stendhal, Goethe and Ruskin. Paul Tucker, in his turn, underlines their (Stokes' and Ruskin's) "common refusal to isolate aesthetic values [...]. Their concern with visual art [...] as a potent symbol of salvation, a way out of the labyrinths of personal and collective existence. Hence both write of the art of the past with a decided 'eye on the present.'" (2000: 135)

[38] Though proving familiar and well-read in his text, Stokes' opinion of Vasari was

by 'professional' art-historians like Bernard Berenson. In Carrier's words,

> [Stokes] is interested in knowing early Renaissance art through direct looking, mostly without appeal to bookish investigation of its historical context. The oddities of *The Quattro Cento* start with its title, which derives from the Italian *quattrocento*, fifteenth century, written as two words, both capitalized. The Italian identifies a period; Stokes' word picks out the art of highest value regardless of its date. A great deal of *quattrocento* architecture, painting or sculpture is emphatically not Quattro Cento. Conversely, not all Quattro Cento art was made in the *quattrocento*. *The Quattro Cento* focuses on quattrocento sculpture, but later Stokes identifies Cézanne and Henry Moore as Quattro Cento artists. (Introduction to *QC*: 2, italics by the author)

Next, like Pound, Stokes had visited the Tempio Malatestiano in Rimini where of course he had found fertile ground for deep exchanges with his American senior friend and the poetics and politics of the *Cantos*,[39]

---

lapidary: "[The] badly-founded reverence for Florentine achievement is hard to uproot. For it dates back to the excellent writings of Vasari, the most successful booster of all history. How these Renaissance boosters have got away with it!" (*QC*: 128) In the *Vite*, Vasari had not paid any attention to the role and work of both Piero della Francesca and Agostino di Duccio in Alberti's reconstruction of the Tempio Malatestiano at Rimini.

[39] In Caroline Patey's words "nella sua disomogeneità – anima gotica con integrazione albertiana nel Quattrocento – il Tempio si configura presto come un possibile correlativo oggettivo dei *Cantos*. [...] La stratificazione del tempio e le mani diverse succedutesi nella edificazione di un insieme incoerente, nel quale de Pasti costeggia Agostino di Duccio, entrambi sottoposti, dal 1450 in poi, alla 'ricomposizione' di Alberti, fanno dell'edificio un incongruo teatro della memoria che accoglie le ceneri del neoplatonico Gemisto Pletone e le ombre di Sigismondo e Isotta; la strana e ibrida creatura di Rimini si lega così all'interesse ossessivo di Pound per il ruolo della committenza nella costruzione dell'identità dell'artista, e alle numerose contrattazioni, Tiziano, Vittore Carpaccio e altri, registrate nei *Cantos* (XX, XXI, XXIV, XXV), come si presta a offrire al poema lo specchio della propria commissione. [...] Più forse dell'architettura e del disegno complessivo, [a Pound] importa il gesto dello scultore, che, a Rimini, si esplica nei meravigliosi rilievi di Agostino di Duccio [...] Così, per alcuni anni, penna e cesello convivono nei *Cantos*, all'insegna di un'arte alchemica capace di fondere la fluidità dell'acqua e la dura precisione dello scalpello ..." (2004: 77, 78). See Davie (1965), Rainey (1991) and North (1983) for in-depth recognitions of Pound's work with reference to the Tempio in Rimini. See Davie (1965), Carrier (1986), Rabaté (1988) and Read (1998, 2000, 2002), on the relationship between Ezra Pound and Adrian Stokes,

but, even more than that, he had realized the importance of bas-relief carved 'into and out of' limestone slabs. Stokes' 'scientific' bent for Dalmatia, thus, no doubt partook of all that, and especially of the attention Stokes was paying to architecture and sculpture built by the work of architects from Dalmatia like Luciano and Francesco Laurana, Giovanni da Traù and Giorgio da Sebenico[40], or the Florentine stone carver Agostino di Duccio[41]:

> Luciano and Francesco Laurana came from Dalmatia. Their roots were in the local art and in Venetian art. [...] They both came from the neighbourhood of Zara in Dalmatia. [...] The art of Dalmatia belongs to the Venetian subject. [...] Agostino spent a year or two in Venice before he appeared in Rimini". [...] So it will be finally in Dalmatia and the Veneto that we shall seek out the origins of Quattro Cento static sculpture. (*QC*: 196, 151n., 187n., 196)

As Richard Read has kindly suggested me, it might also be possible that Stokes was prompted to research in Dalmatia by someone else, as happened at times, perhaps by Pound himself, or the Sitwells, or even Leigh Ashton, who had played some role in the purchase of the di Duccio *Virgin and Child with five angels* relief for the Victoria & Albert Mu-

---

on mutual influences and on their break in the Thirties. Like Pound's, Stokes' interests in the Tempio, went through a lot of reading and research which included *Rimini: Un Condottiere au XVᵉ siècle* by Charles Yriarte (1882), *La sigla di Sigismondo Pandolfo Malatesta* (1904) by Giovanni Soranzo, *Un tempio d'amore* (1912) by Antonio Beltramelli, *Il Tempio Malatestiano* (1924) by Corrado Ricci, i.e. those texts that since the second half of the nineteenth century had helped bring the Tempio Malatestiano, its history and its artists to the attention of international tourists and intellectuals. See Neri 2004, Introduction: 5-41.

[40] Stokes seems to have anticipated what André Chastel, Aldo Rossi e Federico Zeri would later hope for, that is, further research into the complex relationships between the centres of the Western coast of the Venetian State and the Eastern ones: "Un ulteriore studio di questi rapporti dovrebbe comprendere le città della costa orientale (Zara, Spalato, Trogir, Dubrovnik) e altre città d'oriente; ma in questo caso i parametri di riferimento andrebbero allargati perché altrimenti si rischierebbe di ridurre un complesso campo di analisi a un meccanico rapporto di influenza e di soggezione." (1982: 379). See also Dempsey 1996, Introduction: 7.

[41] "Before coming to Rimini, Agostino probably spent some time in Venice". (*SR*: 99) About Agostino di Duccio see Campigli (1999).

seum (see MacLagan 1926). However, as is made clear and reiterated over and over in Stokes' writings, from Dalmatia came not only the talent and craft of the best Quattro Cento architects and stone carvers he admired so much, but also the material itself in which their creations were carved and built, that is Istrian limestone,[42] which to Stokes and Stokes only was "the greatest instrument of [...] instant revelation" (QC: 15). After being entranced by the Malatesta Tempio, and especially by the bas-reliefs by Agostino di Duccio decorating the chapels inside, it was as if Stokes saw Dalmatia as the quarry to provide the stuff on which his dreams of an early-Renaissance uncorrupted Mediterranean aesthetics were made (which it really was!): "Venice became an offshoot of Istria" (SR: 47); "Aemilia and Romagna are poor in stone" (177) [Fig. 3][43]; "One of the secrets of Venice is the bleaching effect that sun and salt air has on exposed Istrian stone, while on unexposed surfaces lichen spreads (QC: 11); "…the second step of the Procuratie Nuove under whose portico you sit, is made of white Istrian alternating with Verona marble…" (12);[44]

---

[42] Giorgio Vasari wrote thus in the chapter "De le diverse pietre che servono a gli architetti per gli ornamenti e per le statue alla scoltura": "Cavasi ancora in Istria una pietra bianca livida, la quale molto agevolmente si schianta; e di questa sopra di ogni altra si serve non solamente la città di Vinegia, ma tutta la Romagna ancora, facendone tutti i loro lavori, e di quadro e d'intaglio. E con sorte di stromenti e ferri piú lunghi che gli altri la vanno lavorando, e massimamente con certe martelline, e vanno secondo la falda della pietra, per essere ella tanto frangibile. E di questa sorte pietra ne ha messo in opera una gran copia Messer Iacopo Sansovino, il quale ha fatto in Vinegia lo edificio dorico della Panatteria et il toscano alla Zecca in sulla piazza di San Marco. E cosí tutti i lor lavori vanno facendo per quella città, e porte, finestre, cappelle et altri ornamenti che lor vien comodo di fare; nonostante che da Verona per il fiume dello Adige abbino comodità di condurvi i mischi et altra sorte di pietre, delle quali poche cose si veggono, per aver piú in uso questa. Nella quale spesso vi commettono dentro porfidi, serpentini et altre sorti di pietre mischie che fanno, accompagnate con esse, bellissimo ornamento." (1991 I: 28)

[43] In a note Stokes explains that "Some of the blocks [with which the Tempio Malatestiano encasement was built] were lifted from the graveyard. Tomb inscriptions can be deciphered on the façade and flanks. Sigismondo was always hard up and always on the prowl for stone. He lifted some of the port of Rimini itself. His most extensive haul probably was the Istrian stone that had been intended for a new bridge over the Metaurus at Fano." (SR: 177-178n.)

[44] Contrast the liveliness ascribed to Istrian limestone to Stokes' description of the kind of Tuscan stone available to Florentine sculptors: "Pietra morta, 'dead stone', they

Figure 3 – Giovanni di Bartolo Bettini da Fano, the building of the Temple. Miniature from *Hesperis* Libri XIII by Basinio da Parma, Cod. 630, c. 126r, post 1457-ante 1468. Bibliothèque de l'Arsenal, Paris.

Still more, Istrian limestone, its aesthetic potentialities and the culture it literally exuded, had come to mean so much to Stokes that, possibly, he had decided to visit the other shore of the Adriatic because somehow, consciously or unconsciously, he was also looking for more ancient myths about it, for instance for that of overseas primitive stonecutters and lapicides (sometimes called *maestri da muro* or *tagliapiere* working the *piera de mar* – sea-stone), like Marino and Leo who were the rightful an-

call round Pisa way the various livid Florentine sandstones and arenaceous limestones. [...] that is a good title for the whole lot, deficient as they are in grain and fibre, unreflective of light. So I shall use the expression *pietra morta* as a generic term, one that is no more inaccurate, and far more apposite, than the usual expression *pietra serena*. [...] the acute deadness of this material treated for decoration. [...] For this stone cannot be seen as welling up gradually as indicating some core within. [...] How could the Florentine masons love their dead stone?" (*QC*: 71, 72, 74)

cestors of his Quattro Cento stone carvers. Both born of humble families in Lopar, on the island of Arbe (Rab), they were apprenticed to limestone cutting and cut stones in Arbe up to around 257, when they crossed
the sea and went to Rimini to help rebuild the city after it had been destroyed by the fleet of Demosthenes, the king of the Liburnians. Perse-
cuted by the emperor Diocletian, Marino and Leo preached the Gospels
in the Montefeltro region where they spread Christianity, until they
found shelter in the mountains behind the sea and gave themselves totally to hermitage and prayer. To these stonemason saints is owed the
foundation of the villages of San Marino and San Leo, a fact which also
explains the presence of the bell-tower of the Church of Santa Maria Mag-
giore in Arbe in the background of a painting portraying Saint Marino
by the seventeenth century painter Guercino (San Marino, Museo di
Stato). The limestone bell-tower is the most beautiful of four that can
still be seen when approaching the Quarnaro archipelago from the sea,
and confers a special beauty to the island.[45] Stokes' few notes about "Dalmatian architecture and sculpture" include precisely words about Istrian
limestone and Roman, Gothic, Renaissance stone architecture in Rab:
"Famous from Roman times the Lopar stone of Arbe"; "One sees doors
like Venetian Gothic except they have a lintel"; "Have seen in Arbe, a
very plain door; almost Laurana" (TGA 8816.14), and in *Venice*, he would
later write: "Some of the Venetian palace doorways in Dalmatia are very
beautiful, particularly in Rab. The Dalmatian sculptors are notable for
depth of feeling" (*CW* II: 126). So much for the reasons, manifest and
supposed, behind Stokes' Dalmatian expedition.

3.1. *Of stone and water: corporealities*

Stokes's prose in *Stones of Rimini* (written during the time of the journey in Dalmatia) developed a poetics of limestone – always Dalmatian or
Istrian limestone – that was also the 'cornerstone' of a theoretical aesthetics based on the distinction of 'carving' from 'modelling' which championed static, Quattro Cento 'primitive' Venetian sculpture like Agostino's

---

[45] In R. West's 'Euclidean' portrayal, "The city [...] is built of stone that is sometimes silver, sometimes, at high noon and sunset, rose and golden, and in the shadow
sometimes blue and lilac, but is always fixed in restraint by its underlying whiteness.
[...] It achieves expressiveness: a grey horizontal oblong with four smaller vertical ob-
longs rising from it. Euclid never spoke more simply." (West 1995: 129)

as superior to that – dynamic and monumental – of works by the Do-
natellos and Michelangelos in Renaissance Florence. Instead of writing
in terms of bookish historicism or evolutionism in the arts, Stokes based
his *Different Conception of the Italian Renaissance* on an old distinction between artistic techniques revived of late by Eric Gill in his 1917 essay on
Sculpture, an essay from which "it seems that around 1932 [Stokes] profited greatly", an essay moreover, "which had inspired the cult of 'truth to
materials' among Stokes' artist friends of the 1930s" (Read 2002: 217).
According to the principles of 'truth to materials' and to the practice of
'direct carving' (see Zilczer 1981), Stokes maintained that "Carving is an
articulation of something that already exists in the block" (*SR*: 114),
"Quattro Cento effect comes when the stone [...] leads the sculptor" (*QC*:
212), and he let the word reverberate in a hundred lexical variations. What
"is in the block", and "leads the sculptor" is the accumulated force of the
Mediterranean, the ancestral shapes and beings that form the geological
and biological matrix of limestone the carver is able to draw back out of
the stone, realize and, let "stone-blossom", and fix on its surface (a shell,
a fish, a crab [Fig. 4]).

> In the first instance,[however] it was the waters that fed those *organisms*
> whose remains formed the *nucleus* of limestones: it was the waters that
> carried the *calcium carbonate* which cemented those remains into rock.
> (*SR*: 33, italics mine)
>
> Limestone [...] is the link between the *organic* and *inorganic* worlds....
> (*SR*: 40, italics mine)
>
> ... the men who obtained nourishment from this environment soon con-
> ceived those many *aspects of life and death* which, when forming some
> kind of 'objective' whole, we name culture. (*SR*: 33, italics mine)
>
> Marine decoration of every kind is abundant in Quattro Cento art: *dol-
> phins, sea-monsters, as well as the fruits of the earth and the children of men en-
> crust* the stone and *grow* there. (*SR*: 42). The Tempio Malatestiano at Ri-
> mini is an ideal quarry whose *original substances* were renewed by the
> hand of the carver to express the abundant seas collected into solid stone.
> (*SR*: 43, italics mine)[46]

---

[46] In Jean-Michel Rabaté's words "Les pages très poétiques que Stokes consacre à
l'interaction de l'eau et de la pierre sont des morceaux d'anthologie, dans lesquels le
savoir du géologue se combine magistralement à la fantaisie du rêveur." (1988: 26)

Figure 4 – Agostino di Duccio (and workshop?), the crab over Rimini (Sign of Cancer). Bas-relief, Chapel of the Planets, Tempio Malatestiano, Rimini.

Having learned Pound's lessons in 'the poetics of stone carving' very well, in the text, Stokes elaborates upon the permeability of limestone and water, having them oscillate from one to the other and finally reach permutation one into the other.[47] In sculptures like Agostino's bas-reliefs in the temple, moreover, thanks to its carver, limestone is seen betraying its watery origins, thus creating the basis for a daring *vitalist* metamorphic poetics ("shifting and indeterminate" in Stonebridge's words – 1998: 111) which figuratively dissolves the difference between the organic and

[47] As Pound had reached in the *Cantos*. See Patey (2004: 78-79), Rabaté (1988: 25-29).

the inorganic, the natural and the cultural and which culminates in Stokes' meta-narrative thinking and questioning about his own inventive prose narrative as a further instance of carving the Mediterranean out of words and into *ekphrasis*[48]:

> I *write of stone*. I write of Italy where stone is habitual. Every Venetian generation *handles the Istrian stone* of which Venice is made. (*SR*: 15)

> The interaction of limestone and water is always *poetic*, always appealing to the imagination. [...] The *story* of limestone and water has many further *chapters* that are palpable to the senses, many *variations*. (*SR*: 33, italics mine)

> Each *cadence* and each *break* will be a transposition into *rhythm* of what these stones mean, each fact and each idea will correspond to a carved emblem; and the *literary form*, with its *self-conscious come-and-go*, with *protagonists* and *drama*, must emulate the white certainty of Alberti's encasement. (*SR*: 171, italics mine)

> ...it seemed unusual that stone memorials by themselves should feed popular romantic misconception: *in comparison with poetry*, that is; especially post medieval building. Could it be the Tempio itself which was causing the Riminesi to evoke the names of Sigismondo and Isotta in an art encounter with foreign bathers? And I would have known – so I thought – the Sigismondo and Isotta story *if it had been extensively 'written up'*. Surely there was no gesticulation, no eloquent statuary. *And how did a more compact architecture suggest a story?* (*SR*: 77, italics mine)

Finally, Stokes celebrates the "transfusion" of water into stone and of stone into water in the fantasy about the influxion of the moon relief in the Chapel of the Planets. Here, Stokes has Cesare Clementini's chronicle of the tremendous August 1442 storm in Rimini[49] merge with as-

---

[48] According to Stanko Kokole, Agostino di Duccio's figurative carvings in the Chapel of the Planets at Rimini can in their turn be read in an ekphrastic key (for example, by retracing the influence of the *ekphrasis* of the Palace of Syphax in the *Africa* by Petrarch on the sculptured triumphal arch) so as "to shed some light on the function of *formae* in relation to the presumed 'philosophical' program of the whole ensemble" (1996: 204). Since their original conception, Agostino's bas-reliefs seem bound to articulate the permutation of the fluidity of words into the static condition of stone and *vice versa*.

[49] *Raccolto istorico della fondazione di Rimino e dell'origine e vite de' Malatesti*, 2 vols, Rimini, 1617 and 1627.

trology, iconology and Isotta's legendary attracting influence on Sigismondo to create a unique Mediterranean landscape and its inhabitants:

> This court of dolphins, maidens and infants, unsuppressed by centuries
> of watery weight, have broken through the ocean, have dried themselves
> by penetration of the marble, and now fill hoarse unaccustomed lungs
> with breeze. […] On an island of jungle land surrounded by the dol-
> phin-populated flux, the elephant and the lion have walked down to
> join the inspired flood and to vouchsafe the secret places of the wood.
> Mountains jump up in mid-ocean, scattering the sea. (*SR*: 253, 254)
> [Fig. 5]

Stokes' experiments in 'prose-carving' reflect what would later become the "total absence of drawing or dependence on drawing" Richard
Wollheim pointed out in his painting[50]. Elusive, elliptic and lacking in
general design, his books of the Thirties privilege short sentences, expressions and words put side by side rather like objective masses or *tesserae*
in a mosaic. They tell 'stories' that aspire to be directly visual, to be ap-
prehended instantaneously, that is displaying time "laid out as ever-pre-
sent space" (*Inside Out*, *CW* II: 156).[51] As much as Quattro Cento art
brings forth and objectifies fantasies associated with its medium (Istrian
limestone), Stoke's prose strives to do the same with language, putting
single words in relief and carving out unique, fresh expressions like the
characteristic "stone-blossom". The following example, therefore is emblematic of Stokes' prizing and aspiring visually as well as linguistically
after concreteness and materiality: turning once more to the power of the
Southern landscape to make things immediately present to the senses, in
1947, he remembered seeing "purple earth, terraces of vine and olive,
bright rectangular houses free of atmosphere, of the passage of time, of
impediment, of all the qualities which steep and massive roofs connote

---

[50] Adrian Stokes took up painting in 1936, while staying in a hotel in the seaside
town of St Ives, Cornwall. He painted until the very last days of his life in 1972.

[51] "*The Quattro Cento* and *Stones of Rimini* are, almost, travel books, and Stokes' ideal
of presentness is implicit in their very structure. […] Stokes replaces the historical narrative with an account which can enable us to see the artwork as it is, a text meant to
be read while looking at the works it describes. The carved work as Stokes describes it
is visually self sufficient, meaningful without reference to art's history or some theory
of art." (Carrier 1986: 757)

Figure 5 – Agostino di Duccio (and workshop?), influxion caused by the moon (the Flood). Bas-relief, Chapel of the Planets, Tempio Malatestiano, Rimini.

in the north" (*Inside Out*, *CW* II: 156) on his first arrival in Italy, but, most interestingly, he also remembered realizing the same thing happening in language with the Latin word "*mensa*" he had once learned to decline as a child, a word which, all of a sudden, he saw materialized everywhere in infinite Mediterranean declensions:

> At the age of seven […] I was fascinated immediately by Latin. I knew one word the first day – *mensa*, a table – and how to decline it. […] Not that I have any gift for languages; yet I possess the image of this declension of the word '*mensa*' on the first day of Latin, […] Of the table, for the table, by the table, each expressed by one simple word. The genitive case was the possessiveness of a simple love. […] The *mensa* table […] was a revolution in my life, an image the 'feel' of which corresponds with an adult image of a simple table prepared for an al fresco meal, the family mid-day meal under a fig tree, with a fiasco of wine on the table,

olives, a cheese and bread. […] As the train came out of the Mont Ce-
nis tunnel, the sun shone, the sky was a deep, deep, bold blue. I had
half-forgotten about my table for more than ten years. At once I saw it
everywhere… (*Inside Out* , *CW* II: 153-154, 156, italics mine)

Tellingly enough, an episode almost identical involving the Latinate
word "*Zisterne*" together with the Eastern coast of the Adriatic is reported
in "Dalmatinischer Aufenthalt" by the German writer Ernst Jünger who,
like Stokes, was a lover of the Mediterranean, and was also travelling in
Dalmatia with his brother in 1932. In addition, Jünger didn't omit to
invoke a precedent in a passage in De Quincey's *Confessions of an English
Opium-Eater*:

> Es gibt Lagen, in denen uns der tiefere Sinn von langvertrauten Worten
> mit einem Schlage sichtbar wird. De Quincey beschreibt in seinen
> "Bekenntnissen" ein solches Erlebnis in Bezug auf die Worte *Consul Ro-
> manus*; ich hatte ein ähnliches mit dem Wort *Zisterne* bei diesem
> köstlichen Trunk inmitten der ausgeglühten Einsamkeit. (Jünger 1982:
> 32)

Then, mentioning, like Stokes, roofs (flat in the South, steep in the
North), the author goes on pondering the metaphors whereby the differences between his Northern world and the Mediterranean one have been
articulated: "Form und Bewegung, Sonne und Nebel, der Zypresse and
der Eiche, dem flachen und dem spitzen Dache" (32) adding his own con-
traposition between "der Zisterne und der Quelle" (32) and expanding
thereon. In the foreword to his travel diary, Jünger would write that "Dalmatinischer Aufenthalt" was meant as an exercise of the eye, and that language is like a tool, the use of which is learned through exercise in the
description of (and one might add 'inscription in') its object.

3.2. *It's raining stones!*

In Stokes, the carver's realization of low-relief, moreover, is male gendered, and thus acted upon the limestone, onto which a motherly character is bestowed.

> All the fantasies of dynamic emergence, of birth and growth and phys-
> ical grace had been projected within the stone. The stone is carved to
> flower, *to bear infants*, to give the fruit of land and sea. These emerge as
> a revelation or are encrusted there. (*QC*: 15, italics mine)

> But only sculptors with a passion for the material, stone, will keep so close
> to this primary fantasy that on their low relief they create for the stone
> her *children* in the image of male *human infants*. (*QC*: 132, italics mine)

Stokes' philosopher friend Richard Wollheim enjoyed seeing in the
carver a "lover of stone" or even a "stone-struck artist". For Stokes, consequently, "Polishing stone [was] also like slapping the new-born infant
to make it breathe." (*SR*: 112), a statement verging on the ridiculous, if
one has never been to the Tempio and has never watched the *putti* carved
by Agostino playing with water and riding dolphins [Fig. 6]. Similar remarks, and further insistence on the maternal corporeality of Istrian and
Dalmatian limestone (its literal identity of "matrix") remind us that
Stokes, by the time he was in Dalmatia, had already dived deep into Kleinian psychoanalysis, and, on his return from there was not to meet Pound
in Venice as he had wished in the August 1932 letter. In point of fact,
Read observes that "The organic and naturalistic qualities attributed to
the Tempio reliefs in *Stones* are polemically anti-Poundian" (2002: 222),
adding that Stokes' employment of the word "vortex" (*SR*: 157) when illustrating that "Agostino gave back to marble its primeval eddies" (*SR*:
157) "is at the furthest remove from Pound's and Gaudier's definition of

Figure 6 – Agostino di
Duccio (and workshop?),
putto on dolphin over tym-
panum of the door to a
chapel. Bas-relief, Tempi-
oMalatestiano, Rimini.

a masculine in-rush of intellect upon material [...] by contrast, [his] 'vortex' is something organic to the mythological origins of stone 'cut by Agostino to show its original liquidity and condensation in swirling whirlpools" (Read 2002: 223).

If some motifs and vocabulary derived from Kleinian object relations theorizations are interspersed in the text, however, one should also be alert to the fact that Klein herself may have been influenced by the art-critic in her own "account of reparation, in which art is produced from a loving attempt to repair and restore objects damaged through destructive anxiety" (Stonebridge 1998: 16).[52] Therefore Stokes's mapping of the opposition between 'modelling' and 'carving' art onto Klein's paranoid-schizoid and depressive positions, at least in this phase of his writing career, is more the fruit of exchanges and mutual recognition than a mere mechanical reading of the Tempio in psycho-analytical terms. Leaving the question of Kleinian positions aside by referring the reader to Lyndsey Stonebridge's book *The Destructive Element. British Psychoanalysis and Modernism* (1998) and to Read's biography (2002), what is interesting as far as keeping close to the question of Stokes' modernism and Dalmatia is concerned, however, is Stokes' troping the tight relationship between the individual and the Mediterranean (Adriatic) local environment in the same 'reparative' or 'restitutive' terms as those which bind sculpture to its own matrix in bas-relief: "The carved form should never, in any profound imaginative sense, be entirely freed from its matrix. In the case of reliefs, the matrix does actually remain: hence the heightened carving appeal of which this technique is capable." (*SR*: 114). This brings us to understand the Rimini temple, with its Albertian white encasement, its 'stone-blossoming' and encrustations in the terms of a gigantic bas-relief carved out of and into a block of Istrian limestone, out of and into a piece of the craggy Dalmatian coast itself. Such a block might be called a "motherstone" and Stokes bestows upon it the capacity of embracing all things, animate and inanimate, Mediterranean, thus confirming Predrag Matvejević's definition of the Adriatic as a "sea of intimacy" (1999: 14). To Stokes, the Istrian "motherstone", moreover, becomes a paradigm for the understanding and evaluation of sculptural or even pictorial works by his

---

[52] Read suggests that in the course of Stokes' analysis Klein had access to drafts of *Stones of Rimini* (2002: 211-212).

Hampstead, and later Cornwall friends Barbara Hepworth, Ben Nicholson and Henry Moore[53], all of them as close to Melanie Klein as Stokes
himself was; and all of them, together with Stokes and Klein, employing
'carving' techniques on different mediums so as to respond (or 'repair')
aesthetically to the tragic, disrupting events that had marked their youths:

> One way of understanding the "mother-stones" of Hepworth and oth-
> ers of the generation born around 1900 is as quasi-magical objects of-
> fered in generative redress, to repair the violence done by the Great War
> – the event that, as children, formed the background to their lives. Even
> as their progressive subjects turned from infants to fullness and holes,
> they revitalized sculptural tradition while instancing a nature and or-
> ganicism to which it has become harder and harder to lay claim. (Wag-
> ner 2005: 191)[54]

In the review "Miss Hepworth's Carving" (1933), Stokes lingers on
pebbles and their shape before expressing his appreciation of an artwork
by the sculptress entitled *Composition* [Fig. 7][55]:

> What are the essential shapes of stone? Pebbles are such shapes since in
> accordance with their structures they have responded to the carving of
> the elements. They are nearly always beautiful when smooth, when they
> show an equal light, when that light and the texture it illumines con-
> vey the sense of all the vagaries of centuries as one smooth object. [...]
> Nothing, it would appear, should be attempted for a time on the moun-
> tain and mother-and-child themes in view of what Miss Hepworth has
> here accomplished. The stone is beautifully rubbed; it is continuous as
> an enlarging snowball on the run; yet part of the matrix is detached as
> a subtly flattened pebble. This is the child which the mother owns with
> all her weight, a child that is of the block yet separate, beyond her womb

[53] See Adrian Stokes, *Reviews of Modern Art* (*CW* I: 305-316). The reviews were published in *The Spectator* in 1933 and 1937.

[54] Stokes' eldest brother Philip, a geologist, was killed in the First World War. "At
Rugby School which Stokes attended from Michaelmas term 1916, the names of no less
than 686 boys who perished at the front were tolled in chapel on an almost daily basis
during the course of the war." (Read 2002: 8)

[55] Hepworth's Hampstead studio was destroyed by bombings during the war, and
so were several of her sculptures of the Thirties including *Composition*, also called *Figure*
and *Figure (Mother and Child)*. See Wagner (2005: 142).

Figure 7 – Barbara Hepworth, *Composition*, 1933. Grey Cumberland alabaster in two parts, 66 cm. Destroyed.

yet of her being. So poignant are these shapes of stone, that in spite of the degree in which a more representational aim and treatment have been avoided, no one could mistake the underlying subject of the group. In this case at least the abstractions employed enforce a vast certainty. It is not a matter of mother and child group represented in stone. Miss Hepworth's stone *is* a mother, her huge pebble its child. (*CW* I: 309, 310, italics by the author)

Such reparative 'praise of pebbles', however, is only an adjunct to those recurring already in *Stones of Rimini*, where pebbles and their "gradually rounded shapes" (116) "where complete roundness is avoided" (67), or "rounded forms that are yet altogether one with the matrix and with the building" (128) are celebrated as the formal key elements in Agostino's bas-reliefs.[56] Actually, it is through stone 'ovoid[s]', 'flattened sphere[s]',

---

[56] In presenting the book to the public, Stokes had pebble shapes put in the foreground thanks to a dust jacket of Ben Nicholson's design: "... in recent times Stokes had been luring Nicholson away from Continental biomorphic abstraction towards his own preference for organic carving. Stokes played enthusiastic games of tennis with

or 'thinned sphere[s]' that tridimensionality[57], perspective[58] and permutation of the stone into water and back again are realized in the limestone:

> Refracted light through clear water throws marble into waves, tempers it with many dimensional depths. [...] For the Agostino reliefs in the Tempio have the appearance of marble limbs seen in water. From the jointure of so many surfaces as are carved in these reliefs, from the exaggerated perspective by which they are contrived, from the fact that though bas reliefs they suggest form in the round, we are reminded of those strange elongations of roundness, those pregnant mountings up and fallings away of flatness, those transient foreshortenings that we may see in stones sunk in clear waters, [...]; we experience again the potential and actual shapes of the stone in water, changing its form, glimmering like an apparition with each ripple or variation of light. But whereas we pick the stone out of the tide [...] Agostino's forms never cease to be potential as well as actual. (*SR*: 97-98)

It might seem far-fetched, but also as far as pebbles are concerned, turning again to Ernst Jünger and his "Dalmatinischer Aufenthalt" can be of help in establishing a possible further proof of the importance of Stokes' own travel to Dalmatia and of his direct experience of that landscape, stony beaches included. The German author describes the pebbles

---

Nicholson as he had with Pound. Inspired by their shared enthusiasm for 'primitive games' as well as for pebbles, the collaboration on the cover and jacket was a process that eventuated in Nicholson's *October 2 1934* (*white relief – triplets*) that, he admitted, Stokes had 'predicted' and that came to stand as a premonition because it was completed 'four hours before the triplets [to Barbara Hepworth] were born (one of course being expected)'. (Read 2002: 224)

[57] "The reliefs are for the most part low, yet their forms possess many values of sculpture in the round: while the quickened mass of a human shape between wind-strewn films of drapery, the delicious torture of hair and clothing by an unseen, evocative wind upon the outer and intermediary surfaces of a relief, give to its body the effect of vitality, of that stone-blossom we prize so high." (*SR*: 105); "...three dimensional form may become all the more significant from being represented by the compressed shapes of low relief...(*SR*: 116-117)

[58] "The ovoid is the perspective appearance of the sphere. [...] Throughout the work [Agostino] has given us this key. Thus, for instance, when he represented clouds, they are cut to pure ovoidal shapes. Fingers, of course, and fish were dear to him. So too, globular hair-locks and the elongated contour of breast or stomach, buttock or thigh, beneath tight strands of transparent drapery." (*SR*: 150)

rocked by the waves on the Dalmatian beach where he used to go for a swim, like Stokes, lingering precisely on the infinite archive of ovoidal shapes it offers, as well as on the recurring 'avoidance' of prefect round-ness in each and every pebble observed:

> Die hohen Geröllhalden konnte man als das unerschöpfliche Archiv eines Bestrebens betrachten, alle Arten der Rundung darzustellen, die denkbar sind. Sehr selten stießen wir jedoch unter den unzähligen scheibenförmigen, ovalen und zylindrischen Schliffen auf die Kugelform. (*Sämtliche Werke* VI: 21)

In its capacity to integrate and bridge the distance between the Eastern and the Western shores of the Adriatic[59], in its turn, the Tempio is compared by Stokes, this time in a fantastic 'major' key, to yet another stone, a legendary, holy "aerolith" (*SR*: 247), that is to the house of the *Blessed Virgin* herself, which from Palestine was first miraculously transported to Fiume to become the Church of the Madonna del Tersatto, and later on crossed the Adriatic, again 'by air', to land south of Ancona and become the Sanctuary of Loreto (see Scotti 2005: 81-84) (all of them, are it goes without saying, built in the whitest of Istrian limestones):

> Behold an aerolith. Out of the blue, stone has poured in torrents over the blunt Gothic pins and needles, flooding with crystallizing foam the grey Germanic flanks. In 1291 Palestine was the aerodrome and upon a Syrian carpet the House of the Blessed Virgin rose over the Dead Sea, landing at Tersalto [*sic*] near Fiume. In 1295 the Casa Santa took to the air again, this time by night. A landing was made near Recanati in a laurel grove. The miracle was recognized immediately, the laurels cleared, and the shrines of Loreto crept near the wooden, holy house. [...] Whence

---

[59] Fabio Fiori's words of today confirm Stokes' when he states that "In writing now of Venice, I have not in mind Venetian sculpture nor marble palaces reflecting the waters between them. I refer to the less signal yet vast outlay there of the salt-white Istrian stone.." (*SR*: 15): "Banchine portuali, fondamenta e murazzi compongono un'ideale prosecuzione della bianca *orlatura* orientale verso occidente. Un continuum adriatico, anch'esso rappresentativo dello stretto intrecciarsi di elementi naturali e storici. [...] Milioni di metri cubi di calcare, un'isola istriana andata alla deriva, naufragata nella laguna veneta. La bellezza senza tempo delle architetture veneziane *galleggia* sopra una zattera calcarea, un'ulteriore testimonianza dei rapporti tra le due sponde dell'Adriatico." (Fiori 2005: 112, 113)

came this other building, from whence fell this encasement that fits like a strong and knotted gauntlet? (*SR*: 247, 248).[60]

In the end it must be said that probably Dalmatia, or rather Dalmatian Venetian architecture and sculpture, disappointed Stokes; his few 'technical' notes on the cathedrals in its city-ports do not seem to contain any particular enthusiasm (apart from that for palace doorways in Rab), often interspersed as they are with expressions like "disappointing", "exceedingly disappointing", "dull", "little stone-blossom". Something else, however seems to have struck his imagination enormously, and that was the white limestone from which the Dalmatian coast and its islands were themselves carved. 'Weathered ', permeated, and carved by the water of the Adriatic, these craggy places, which were also the quarries from which the stone for Renaissance carving was hewn, were the primitive, original matrix of Mediterranean art and culture, in concreteness. In 1932, however, to affirm the substantial organic and cultural homogeneity of the limestone coast of the Eastern Adriatic shore and the sandy beaches of the Western coast, without including politics, also racial politics, into the calculation, was pure fantasy.[61] Still, it was a completely absorbing fantasy which Stokes formulated "for its own sake", as well as "as a source of self-knowledge" (Carrier 1997: 14);[62] it was a rêverie of reparation and also a way of keeping 'true to material' and safe from the wounds of history. A quotation from Claudio Magris' *Microcosmi* is maybe the best illustration of the relationship Stokes seems to have established with the coastal and insular landscape of stony Dalmatia. It is also the best commentary on its topicality:

---

[60] "Quasi a *concretizzare* il trasferimento dei sacri muri da una parte all'altra dell'Adriatico ci sono i documenti settecenteschi riguardanti le difficoltose operazioni di rifornimento litico dei cantieri lauretani. Per qualche anno infatti, in relazione alle ostilità tra lo Stato Pontificio e la Repubblica di Venezia, i *paroni* istriani dovettero letteralmente contrabbandare la pietra, scaricandola direttamente sul lido di Recanati." (Fiori 2005: 115, italics by the author)

[61] See Richard Read (2002: 163-192) about Stokes' Sephardi Jew origins and for a reading of some of his early works as influenced by the racial theories of his time and connections (Pound above all).

[62] "Auden's poem 'In Praise of Limestone', written in Italy in May 1948, reads as a critical elegy for Pound's and Stokes' pre-war enthusiasm for limestone's poetic and totalizing potential…" (Stonebridge 1998: 120, for a comparative reading of Stokes' and Auden's texts, see 119-129).

Ma ogni volta che si arriva sull'arcipelago – raramente dal mare, in barca, molto più spesso in macchina, prendendo il traghetto a Brestova, sulla costa orientale dell'Istria, e sbarcando a Porozine, sull'isola di Cherso – ogni riferimento a una Storia presente in tante cicatrici ancora fresche si dissolve, svanisce come foschia nei riverberi del sole sul mare e sulle candide rupi ciclopiche ai bordi della strada, paesaggio epico e omerico in cui non c'è posto per la tortuosità della psicologia e dei risentimenti. La Storia viene assorbita, come la pioggia o la grandine nelle fessure delle rocce carsiche, nel tempo più grande e incorruttibile di quella luce estiva e di quelle pietre di un bianco abbagliante; le ferite e le cicatrici ch'essa ha inflitto non vanno in suppurazione, ma si asciugano e rimarginano, come graffiature sulla pianta del piede nudo che si taglia sbarcando sull'isola e posandosi su quei sassi aguzzi. (1997: 152)

3.3. *From palimpsest to bas-relief*

> The temple ⊥⊥⊥ is holy,
> because it is not for sale.
> (*Cantos*, XCVII)

Thanks in part to Stokes' free, un-academic, non-professional attitudes,[63] a new vision of Anglo-Mediterranean relations would finally emerge to be later reflected in certain episodes of post-World War II cultural politics.[64] No longer a palimpsest archaeologically dug and ima-

---

[63] "Stokes' wish to point to the artwork itself, his refusal to treat it as an object that must be placed in a historical context; his frank interest in the role of fantasy in aesthetic contemplation: all place him outside the world of professionalized art history." (Carrier 1986: 765)

[64] See for example Jim Ede's Kettle's Yard in Cambridge. The result of the migrant experiences of the collector and his intellectual friends during the *entre deux guerres* and beyond, Ede's house-museum (established 1957) is an anti-traditionalist, anti-taxonomist proposal of integration and conservation of cultural and life experiences that were issued and fertilized by a plurality of diasporic waves received in London (Gaudier-Brzeska's, Melanie Klein's, Mondrian's, Naum Gabo's sojourns in London, Hampstead in particular) as well as experienced at home (Wales, Cornwall) and in the Mediterranean (France, North Africa). Ede's formal and living exhibition adventure intertwined with Adrian Stokes's friendship and promotion of primitive idioms, be they *Quattro Cento* ones or that of the painting developed by the Cornish mariner Alfred Wallis. Kettle's Yard shares also in Stokes' and the other Hampstead artists' 'pebblemania' and was once ironically defined "the Louvre of the pebble" by concrete poet and gardener artist Ian Hamilton Finlay.

ginatively 'looted', collected and museified according to the long wave of British antiquarianism and to the nineteenth century principles of appropriation and deadening classification[65] (see Jezernik 2004 about the Victorians in the Balkans), neither an heritage made up by attributions, nor one 'filed' in institutions by Berenson, Warburg[66] or Panofsky,[67] the Mediterranean context seems finally to have been entitled to a new, though extremely difficult, exemplary role in inspiring and promoting projects of geographical and artistic integration and reparation based on equal cross-cultural exchanges and on the enhancement of locality as a fundamental feature of the artwork.

> When we visit an ancient stone building of no particular merit, we might yet find it so far as it makes us aware of its *locality*. For the building, if constructed of *local materials*, is an expression of its neighbourhood. We may even see upon its opposite hill a disused quarry from which the stones were taken. *The building is part of the landscape carved by man*: there exists the connection, though the building be an 18th-century palace, with the cave dwelling and the rock tomb. (*SR*: 44, italics mine)

[65] In his 1932 review of *The Quattro Cento*, Ezra Pound wrote that Stokes: "ha[d] for a number of years ransacked Italy not as archaeologist, but as a looker. He ha[d] carried his eyes about and made them work. He ha[d] very definitely scrutinized the shapes" (1980b: 222).

[66] Plate 25 of Warburg's *Mnemosyne*-Atlas (2002) is a *montage* of pictures regarding the Tempio Malatestiano: Leon Battista Alberti's unfinished encasement and bas reliefs by Agostino di Duccio. The Tempio is shown as an example of the *coincidentia oppositorum* and transition between two different modes of apprehending antiquity, the Apollinean and the Dyonisian, and the *montage* foregrounds the superimposition of antique themes, images and forms onto Christian ones. Plate 25 is part of a series of *montages* (plates 20-27) thematically meant to illustrate the tradition of daemonic-astrologic culture in the West.

[67] "Stokes did not, like Bernard Berenson, do attributions; nor did he, like Aby Warburg, study art's social history in the archives; nor, unlike Panofsky, was he an iconographer. Those pursuits find an institutional home in an artworld where the museum depends upon attributions, and the social history of art and its iconography are studied in art history departments. The theory of art's presentness is more personal; Stokes' analysis actively resists institutionalisation. [...] Stokes' relation to Pater and Ruskin is more like that of a creative writer to his precursors than of an art historian to his teachers." (Carrier 1986: 765)

The thousand-and-one marble figures, the Hellenistic statuary, [...] were yet a stone display amid limestone temples, amid grove and court-yard and flowering tub, as they stained themselves with pure water above the fountain, or stood on the sky-line striking the blue with posed yet marble arm. No doubt they are almost *meaningless in our northern museums*: not so when they stand throughout the sun, or pale against the moon. (*SR*: 124, italics mine)

Going back to quarried stone in 1951, and quoting from Lewis Mumford's critique of mining as the basis of industrialism, Stokes would reaffirm the maternal relationship of intimacy the stonecarver, or architect can establish with the natural environment:

There is evidence for an early fantasy of burrowing into the mother's body and robbing the wealth she is supposed to contain, particularly her babies. I have glanced at architecture from the side of a reparation for these and other fantasies. Quarried stone, rock carved thoughtfully and indeed the quarry itself, are as love compared with the hatefulness, the wastefulness and robbery that could be attributed to mining; compared with the process, 'Mine: blast: dump: crush, extract: exhaust.' (*Smooth and Rough*, *CW* II: 248)

It is not to be forgotten, moreover, that the carving technique championed by Stokes contrasts "the cutting away of excess material by the sculptor of stone to the additive procedure of the bronzeworker" (Carrier 1986: 754). Parallel and accessorial to that of restitution as articulated by Stokes himself, such a negative function of carving, its "cutting away", somehow points to the bas-relief as a modernist alternative to the palimpsest in the representation of Mediterranean memory. Removal and subtraction take the place of antiquarian accumulation and excess, therefore a 'creative' oblivion, 'conquered' and embodied in limestone, comes to the fore as fundamental in order to better remember, celebrate and interact with the Mediterranean natural and cultural heritage. As in Marc Augé's "métaphore marine", "les souvenirs sont façonnés par l'oubli comme les contours du rivage par la mer":

L'océan, durant des millénaires, a poursuivi, aveuglément, son travail de sape et de remodelage: le résultat (un paysage) doit bien dire quelque chose [...] des résistances et des faiblesses du rivage, de la nature des ses roches et des ses sols, des ses failles et des ses fractures, que sais-je?

...Quelque chose aussi, naturellement, des poussées de l'océan, mais la force et les sens de ces dernières dépendent aussi des formes du relief sous-marin – ce prolongement du paysage terrestre...quelque chose donc, au total, de la complicité entre la terre et la mer, qui ont contribué toutes deux au long travail d'élimination dont le paysage actuel est le résultat. [...] L'oubli, en somme, est la force vive de la mémoire et le souvenir en est le produit. (2001: 29-30)

Conclusion

An expatriate in London from the Adriatic and a British traveller in the Adriatic of some fifteen years afterwards, Meštrović and Stokes brought its shores out of obscurity and Mediterranean marginality and onto the international scene of twentieth century culture and aesthetics. One striving after independence from the crumbling world of empires had to negotiate his own identity as an artist with World War I and the turbulent Balkan politics of his time. The other looked back at fifteenth century Venetian Renaissance and at the Rimini of the Malatesta to find an alternative to the murkiness of Edwardian London and the war, and finally project it in a fantasy of elemental integration bound to affect and fertilise the aesthetics and cultural politics of the Nineteen-thirties and beyond, albeit understatedly. Even if sometimes perceived as merely a-historic and un-academic, or even fantastic, the strength of Stokes' inventiveness strikes as unique in its apprehension of Mediterranean corporeality and integrity between image and art-word, as much as Meštrović's sculpture can yield interesting intimations of modernist 'primitivism' once the complexities of its production and reception have been taken into account and contextualised. Though, in a sense, both the 'children of a lesser modernism' (and wide apart from each other as to their cultural vicissitudes and achievements), tellingly did Meštrović and Stokes share their experience of the Adriatic with those intellectual personalities bound to impress a lasting imprint on the twentieth century and its relationship with Mediterranean culture: James Joyce, Rebecca West, Ernst Jünger, and Fernand Braudel above all. It was indeed thanks to his 1935 Adriatic sojourn and to the researches in the archives of Venice and Dalmatian Dubrovnik (Ragusa) that the great historian was later to articulate the radically new understanding of the Mediterranean he illustrated in his masterwork.

Bibliography

Adam, Robert, [1764] 2001, *Ruins of the Palace of the Emperor Diocletian at Spala-
   tro in Dalmatia*, Marco Navarra (ed.), 2001.
Allcock, John B. - Antonia Young (eds), 2001, *Black Lambs and Grey Falcons.
   Women Travelling in The Balkans*, Berghahn, London and New York.
Augé, Marc, [1998] 2001, *Les formes de l'oubli*, Éditions Payot & Rivages, Paris.
Ballinger, Pamela, 2003, *History in Exile. Memory and Identity at The Borders of
   The Balkans*, Princeton University Press, Princeton.
Barry, Lieut.-Col. J.P., 1906, *At the Gates of the East: A Book of Travel among His-
   toric Wonderlands,* Longmans G., London, New York, Bombay.
Bullen, J.B., 1999, "Byzantinism and modernism 1900-14", *The Burlington Mag-
   azine*, Vol. 141, No. 1160, pp. 665-675.
——, 2003, *Byzantium Rediscovered*, Phaidon, London.
Campigli, Marco, 1999, *Luce e marmo. Agostino di Duccio*, L.S. Olschki, Firenze.
Cantone, Ugo, 2001, "Postscript", in Marco Navarra (ed.), 2001, pp. 209-252.
Carrier, David, 1986, "The Presentness of Painting: Adrian Stokes as Aestheti-
   cian", *Critical Inquiry*, Vol. 12, No. 4, pp. 753-768.
—— (ed.), 1997, *England and Its Aesthetes: Biography and Taste*, G&B Arts International, Amsterdam.
——, 2003, Review of *Art and Its Discontents: The Early Life of Adrian Stokes* by
   Richard Read, *The Burlington Magazine,* Vol. CXLV, No. 1204.
Cianci, Giovanni, 1983, "D.H. Lawrence and Futurism/Vorticism", in *AAA –
   Arbeiten aus Anglistik und Amerikanistik*, Band 8, Heft 1, Gunter Narr Ver-
   lag, Tübingen, pp. 41-53.
Cleeve, Brian (ed.), 1972, *W.B. Yeats And the Designing of Ireland's Coinage.* Texts
   by W. B. Yeats and Others, Dolmen Press, Dublin.
Clegg, Elizabeth, 2002, "Meštrović, England and the Great War", *The Burling-
   ton Magazine*, Vol. CXLIV, No. 1197, pp. 740-751.
Collings, Ernest H.R., 1919, "Meštrović in England", in Milan Ćurčin (ed.),
   1919, pp. 48-54.
Corbin, Alain, 1988, *Territoire du vide: l'Occident et le désir du rivage (1750-1840)*,
   Aubier, Paris.
Crivelli, Renzo S., 1996, *James Joyce: Triestine Itineraries,* Mgs Press, Trieste.
——, 2004, *A Rose for Joyce*, Mgs Press, Trieste.
Ćurčin, Milan (ed.), 1919, *Ivan Meštrović: A Monograph*, Williams and Norgate,
   London.
Davie, Donald, 1965, *Ezra Pound. Poet as Sculptor*, Routledge and Kegan Paul,
   London.
De Micheli, Mario *et al.*, 1987, *Ivan Meštrović*, Vangelista, Milano.
Dempsey, Charles (ed.), 1996, *Introduction to Quattrocento Adriatico: Fifteenth-Cen-
   tury Art of the Adriatic Rim*, Nuova Alfa Editrice, Bologna, pp. 5-12.

Ellmann, Richard, [1959] 1983, *James Joyce*, Oxford University Press, Oxford.

Fiori, Fabio, 2005, *Un mare. Orizzonte adriatico*, Diabasis, Reggio Emilia.

Fortis, Alberto, 1774, *Viaggio in Dalmazia*, presso Alvise Milocco, Venezia, Eng. transl. 1778, *Travels into Dalmatia*, J. Robson, London.

Fussell, Paul, 1980, *Abroad. British Literary Traveling Between the Wars*, Oxford University Press, Oxford.

Gibbon, Edward, [1776-1788] 2005, *The Decline and Fall of the Roman Empire*, ed. by Hans-Friedrich Mueller, Modern Library, New York.

Goldring, Douglas, 1913, *Dream Cities*, T. Fisher Unwin, London.

Goldsworthy, Vesna, 1998, *Inventing Ruritania. The Imperialism of the Imagination*, Yale University Press, New Haven and London.

Hoyt, David L., 2001, "The Reanimation of The Primitive: *Fin-de-Siècle* Ethnographic Discourse in Western Europe", *History of Science,* Vol. 39, Part 3, No. 125, pp. 331-354.

Jackson, Frederick Hamilton, 1906, 1908, *The Shores of the Adriatic: An Architectural and Archaeological Pilgrimage… with plans and illustrations from drawings by the author and from photographs,* 2 vols, John Murray, London.

Jackson, Thomas Graham, 1885, *Dalmatia, the Quarnaro and Istria*, 3 vols, Clarendon Press, Oxford.

Jezernik, Božidar, 2004, "Lovers of Art on the *missions civilisatrices*", *Ethnologia Balkanica*, Vol. 8, pp. 178-193.

Jünger, Ernst, [1934] 1982, "Dalmatinischer Aufenthalt", in *Sämtliche Werke*, 22 Bänden, Klett-Cotta, Stuttgart, Tagebücher VI, Band 6 Reisetagebücher, pp. 11-35. First published in *Blätter und Steine*, Hanseatische Verlags-Anstalt, Hamburg.

Kite, Stephen, 2000, "The Urban Landscape of Hyde Park: Adrian Stokes, Conrad and the *topos* of negation", *Art History*, Vol. 23, No. 2, pp. 205-232.

Kokole, Stanko, 1996, "*Cognitio formarum* and Agostino di Duccio's Reliefs for the Chapel of the Planets in the Tempio Malatestiano", in Charles Dempsey (ed.), 1996, pp. 177-206.

Lawrence, D.H., 1979-2000, *The Letters of D.H. Lawrence*, ed. by James T. Boulton , 8 vols, Cambridge University Press, Cambridge.

Lewis, Wyndham, 1915, "Blasted", *Blast,* II, p. 92.

Macaulay, Rose, [1953] 1966, *Pleasure of Ruins*, Thames and Hudson, London.

MacCarthy, Fiona, 1989, *Eric Gill*, Faber and Faber, London and New York.

MacLagan, Eric, 1926, "A Relief by Agostino di Duccio", *The Burlington Magazine for Connoisseurs*, Vol. 48, No. 277, pp. 164+166-167.

Magris, Claudio, 1963, *Il mito asburgico nella letteratura austriaca moderna*, Einaudi, Torino.

—— - Angelo Ara, 1982, *Trieste: un'identità di frontiera*, Einaudi, Torino.

——, 1986, *Danubio*, Garzanti, Milano.

——, 1997, *Microcosmi*, Garzanti, Milano.

Mardešić, Ivo, 1995, *Croatia/Great Britain. The History of Cultural and Literary Relations*, Most/The Bridge, Zagreb.

Marriott, Charles *et al.*, Tate Gallery Archive, *Twenty-One Gallery London*, Two press cutting albums and loose cuttings about the Twenty-One Gallery, London and its exhibitions 1913-1935, TGA 8024.

Marzo Magno, Alessandro, 2003, *Il leone di Lissa. Viaggio in Dalmazia*, prefazione di Paolo Rumiz, il Saggiatore, Milano.

Matvejević, Predrag, 1987, *Mediteranski brevijar*, Grafički zavod Hrvatske, Zagreb; 1999, *Mediterranean. A Cultural Landscape*, trans. Michael Henry Heim, University of California Press, Berkeley, Los Angeles, London.

McCourt, John, 2000, *The Years of Bloom. James Joyce in Trieste, 1904-1920*, The Lilliput Press, Dublin.

Michaud, Yves (ed.), 1988, *Adrian Stokes. Les Cahiers du Musée National d'Art Moderne*, No. 25, Centre Geoges Pompidou, Paris.

Moqué, Alice Lee, 1914, *Delightful Dalmatia*, Funk and Wagnalls Com, London and New York.

Navarra, Marco (ed.), 2001, *Ruins of the Palace of the Emperor Diocletian at Spalatro in Dalmatia*, Biblioteca del Cenide, Reggio Calabria.

Neri, Moreno (a cura di), 2004, *Visitatori celebri nel Tempio di Rimini*, Raffaelli Editore, Rimini.

North, Michael, 1983, "The Architecture of Memory: Pound and the Tempio Malatestiano", *American Literature*, Vol. 55, No. 3, pp. 367-387.

Patey, Caroline, 2004, "Lungomare, Rapallo: poetiche angloprovenzali e politica culturale", in Edoardo Esposito (a cura di), *Le letterature straniere nell'Italia dell'entre-deux-guerres*, 2 vols, Pensa, Lecce, Vol. I, pp. 65-82.

Pemble, John, 1987, *The Mediterranean Passion: Victorians and Edwardians in the South*, Oxford University Press, Oxford.

Pilo, Giuseppe Maria, 2000, *"Per trecentosettantasette anni". La gloria di Venezia nelle testimonianze artistiche della Dalmazia*, Edizioni della Laguna, Mariano del Friuli (Go).

Pound, Ezra, 1916, *Gaudier-Brzeska: A Memoir*, London and New York.

——, [1921] 1980a, "Brancusi", in Harriet Zinnes (ed.), 1980.

——, [1932] 1980b, Review of *The Quattro Cento* in *Symposium*, 3, reprinted in Harriet Zinnes (ed.), 1980, p. 222.

——, [1934] 1980c, Review of *Stones of Rimini*, *The Criterion*, 13: 52, reprinted in Harriet Zinnes (ed.), 1980, pp. 167-169.

——, 1985, *I Cantos*, a cura di Mary de Rachewiltz, Mondadori, Milano.

Rabaté, Jean-Michel, 1988, "Adrian Stokes et Ezra Pound", in Yves Michaud (ed.), 1988, pp. 25-38.

Rainey, Lawrence S., 1991, *Ezra Pound and the Monument of Culture: Text, History, and the Malatesta Cantos*, Chicago University Press, Chicago.

Read, Richard, 1998, "The Letters of Adrian Stokes and Ezra Pound", *Paideuma*, 27: 2-3, pp. 69-72.

——, 2000, "The unpublished correspondence of Ezra Pound and Adrian Stokes: modernist myth-making in sculpture, literature, aesthetics and psycho-analysis", *Comparative Criticism*, Vol. 21, pp. 79-127.

——, 2002, *Art and its Discontents. The Early Life of Adrian Stokes*, The Penn State University Press, University Park, Pa.

Ross, Robert, 1915, "A Monthly Chronicle: Meštrović", *The Burlington Magazine for Connoisseurs*, Vol. 27, No. 149, pp. 205-207+210-211.

Rossi, Aldo, [1969, 1970] 1982, "L'architettura della ragione come architettura di tendenza"; "I caratteri urbani delle città venete", in *Scritti scelti sull'architettura e la città 1956-1972*, a cura di Rosaldo Bonicalzi, Clup, Milano, pp. 370-433.

Russell Barrington, Mrs, 1912, *Through Greece and Dalmatia. A Diary of Impressions recorded by Pen & Picture*, Adam and Charles Black, London.

Scotti, Giacomo, 2005, *Fiabe e leggende del Mar Adriatico*, Santi Quaranta, Treviso.

Seton-Watson, Robert William, 1911, *The Southern Slav Question and the Habsburg Monarchy*, London.

——, 1915, *The Balkans, Italy and the Adriatic*, Nisbet.

Shaw, Trevor R., 2001, "Bishop Hervey at Trieste and in Slovenia, 1771", *Acta Carsologica*, 30/2, 20, pp. 279-291.

Stokes, Adrian, [undated], Tate Gallery Archive, *Adrian Stokes Papers*, Dalmatian Architecture and Sculpture, TGA 8816.14.

——, [1932, 1934] 2002, *The Quattro Cento* and *Stones of Rimini*, forward by Stephen Bann, ed. with introductions by David Carrier and Stephen Kite, The Penn State University Press, University Park, Pa.

——, 1978, *The Critical Writings of Adrian Stokes*, 3 vols, ed. by Lawrence Gowing, Thames and Hudson, London.

Stonebridge, Lyndsey, 1998, *The Destructive Element. British Psychoanalysis and Modernism,* Routledge, New York.

Triani, Giorgio, 1988, *Pelle di luna. Pelle di sole. Nascita e storia della civiltà balneare 1700-1946*, Marsilio, Venezia.

Tucker, Paul, 2000, "Adrian Stokes and the 'Anti-Ruskin Lesson' of British Formalism", in Toni Cerutti (ed.), *Ruskin and The Twentieth Century: The Modernity of Ruskinism*, Edizioni Mercurio, Vercelli, pp. 129-143.

Vasari, Giorgio, [1550] 1991, *Le Vite dè più eccellenti architetti, pittori, et scultori italiani, da Cimabue insino a tempi nostri*, nell'edizione per tipi di Lorenzo Tormentino, Firenze, 2 voll., a cura di Luciano Bellosi e Aldo Rossi, Einaudi, Torino.

Vidan, Ivo, 1974, "Joyce and the South Slavs", in Niny Rocco Bergera (ed.), *Atti del Third International James Joyce Symposium*, Università degli Studi, Trieste, pp. 166-133.

Wagner, Anne Middleton, 2005, *Mother Stone. The Vitality of Modern British Sculpture*, Yale University Press, New Haven and London.

Wallace, Jennifer, 1998, "A (Hi)story of Illyria", *Greece & Rome*, 2nd Ser., Vol. 45, No. 2, pp. 213-225.

Warburg, Aby, [2000] 2002, *Mnemosyne: l'Atlante delle immagini*, a cura di Martin Warnke, edizione italiana a cura di Maurizio Ghelardi, Nino Aragno editore, Torino.

West, Rebecca, [1941] 1995, *Black Lamb and Grey Falcon. A Journey through Yugoslavia*, Penguin, London.

Wilde, Oscar, [1908] 1969, "*Twelfth Night* at Oxford", *First Collected Edition of the Works of Oscar Wilde 1908-1922*, ed. by Robert Ross, 15 vols, Barnes & Noble, New York, Vol. XIII, p. 46.

Wilkinson, John Gardner, 1848, *Dalmatia and Montenegro: with a journey to Mostar in Herzegovina, and remarks on the Slavonic nations: the history of Dalmatia and Ragusa, the Uscocs, etc.*, 2 vols, London.

Wolff, Larry, 2001, *Venice and the Slavs. The Discovery of Dalmatia in the Age of the Enlightment*, Stanford University Press, Stanford.

——, 2005, "The Adriatic Origins of European Anthropology", *Cromohs*, No. 10, pp. 1-5, http://www.cromohs.unifi.it/10_2005/wolff_adriatic.html.

Wollheim, Richard, 1978, "Adrian Stokes, Critic, Painter, Poet", 4th William Townsend lecture, Slade School of Art, extended version published *Times Literary Supplement*, 17 Feb. 1978, p. 207, reprinted in Stephen Bann (ed.), "Adrian Stokes 1902-72", supplement, *PN Review*, 15, Vol. 7 No.1, 1980, p. 37, also available at Philip Stokes (ed.), http://adrianstokes.com/.

Yriarte, Charles, 1878, *Les bords de l'Adriatique et le Monténégro. Venise, l'Istrie, le Quarnero, la Dalmatie, le Monténégro et la rive italienne*, Hachette, Paris.

Zilczer, Judith, 1981, "The Theory of Direct Carving in Modern Sculpture", *Oxford Art Journal*, Vol. 4, No. 2, pp. 44-49.

Zinnes, Harriet (ed.), 1980, *Ezra Pound and the Visual Arts*, New Directions, New York.

"MAKING GEOGRAPHY A ROMANCE": JOYCE'S MEDITERRANEAN FEMININE

Laura Pelaschiar

Do you consider, by the by, he said, thoughtfully, selecting a faded photo which he laid on the table, that a Spanish type? (U: 533)

Mr Leopold Bloom, in the "Eumeus" chapter of *Ulysses*, produces a photograph of his wife Molly in order to enquire of Stephen whether he thinks she is the Spanish type. Even at 1 or 2 o'clock in the morning, after much travelling, wandering and waking, Bloom once more shows his obsession with his wife's Mediterranean features and southern colours, to which he seems unconsciously to attach a special significance. And rightly so, as we shall see.

This essay will try to analyse the origins and significance of the Mediterranean Feminine in James Joyce's work and to show how the protracted contact with the Southern/Middle-European female identity of Trieste was fundamental for the reformulation of Joyce's idea of womanhood and the feminine in *Ulysses*.

By and large, although in different and somewhat opposite ways, both

Abbreviations

CW   Joyce, James, 1989, *The Critical Writings of James Joyce*, ed. by Ellsworth Mason - Richard Ellman, Cornell University Press, Ithaca, New York.
D   Joyce, James, 1993, *Dubliners*, ed. by John Wyse Jackson - Bernard McGinley, St. Martin's Press, New York.
U   Joyce, James, 1986, *Ulysses*, Vintage Books, New York.
SH   James Joyce, 1977, *Stephen Hero*, A Triad Grafton Book, London.

the pre- and post- Trieste female figures that Joyce creates in his art are at odds with the unflattering ideas and opinions that he professed about women in his life. His attitude to women was perhaps the only element of his Irish cultural heritage which he never felt the need to question or put in doubt. His female protagonists – especially the most powerful of them all, Molly Bloom – challenge and transcend the cultural limitations of the patriarchal female construct which Joyce brought with him when he left Ireland. What he never really rejected as a man, he denounced and deconstructed as an artist. The multiethnic Mediterranean feminine which he encountered in Trieste and which expressed a completely different female identity from the Irish one he had known in Dublin was the basis of this artistic enterprise.

As Sheldon Brivic, one of the most accomplished readers of Joyce's work from a feminist perspective, writes in his *Joyce's Waking Women*, Joyce's cultural relationship with women and with concepts of womanhood is a complex matter:

> As an egotistical man who grew up in Victorian Ireland, Joyce had serious limits in his view of women; but he worked against those limits so remarkably that, like his early master Ibsen, he deserved a place in the history of feminism. Perhaps he should be put in a line of male feminists whose contribution must be weighed against the ambiguity of their positions as men, including several figures I will refer to: Sophocles, John Webster, John Stuart Mill, and Jacques Lacan. (Brivic 1995: 3)

Joyce would probably have been pleased to find himself in the company of figures such as Sophocles, Webster and Stuart Mill, but would have bridled at being labelled a feminist writer. He was indeed an egotistical male who grew up in Victorian – and Catholic, it should be added – Ireland; a male who more than once expressed prejudiced ideas about women, especially in so far as their intellectual capacities were concerned. Joyce's dismissive opinions and deep scepticism on this subject are scattered throughout his letters and in the recollections and memories of his friends and acquaintances, as Richard Brown explains in his ground-breaking study *James Joyce and Sexuality* (Brown 1985: 89-91). What's more, these are opinions which Joyce retained throughout his life and which remained unaffected even by the presence of intellectually active women of letters such as Harriet Shaw Weaver, Sylvia Beach and Adrienne Monnier, with whom he lived in close proximity when he was in Paris and who helped him most generously in his literary career.

Even when he defends Nora from the poor opinions of people like his friend Cosgrave in Dublin or of a colleague of his at the Berlitz School in Trieste, who had expressed his perplexity about the *liaison* not on the basis of any moral reservation but in the belief that Nora was not worthy of him, Joyce cannot help betraying his chauvinistic beliefs. In a letter to Stanislaus of July 12 1904, on the one hand he justifies Nora's depression and state of prostration and accuses her critics of "self-stultification" for forming such a verdict, while on the other he stresses the stupidity of this notion since – he adds with disheartening candour – after all it is only Skeffington and fellows like him who believe that woman can be man's equal (Joyce 1966: 92).

Francis Skeffington was a good friend of Joyce's at University College Dublin. He became a fervent feminist and later married Hannah Sheehy, another friend of the Joyce brothers who belonged to a prominent upper-middle class Dublin family. The Sheehys were active nationalists who held an 'open house' and organised, every second Sunday of the month, intellectual and literary debates which Joyce often took part in. To honour their feminist beliefs, Francis and Hannah chose to combine their two names after they married and called themselves Sheehy-Skeffington, thus rejecting the tradition which forced the female member of a couple to renounce her family name after marriage. In 1901, Joyce and Skeffington published together two polemical essays: Joyce's "The Day of the Rabblement", on the Irish National Theatre, and Skeffington's "A Forgotten Aspect of the University Question", on women's right to be admitted to university. That Joyce thought Skeffington's feminism his most evident trait becomes obvious when one considers that in the surviving fragment of *Stephen Hero* he appears as "the young feminist McCann" (SH: 43). In spite of his friendship with Skeffington (which waned when Joyce left Ireland with Nora), Joyce never showed any interest in or respect for feminism and the feminist cause his one time friend espoused.

The reading of Stanislaus Joyce's unpublished *Book of Days*, as Stanislaus himself called his 1907-1909 Trieste diary, amply confirms Joyce's reputation as that of a man who did not much care for women. First, a few words on the diary itself. Stanislaus Joyce's Trieste diary consists of fifteen 48-page closely written notebooks which Stanislaus, James's younger brother, kept after he came to live in Trieste in 1905 on James's insistent invitation. The diary begins on 1 January 1907 and ends abruptly on 11 February 1909. The original handwritten document is in London, in the possession of the Stanislaus Joyce Estate, and scholars do not have access to it. But a photocopied version of the entire document is part of the Richard Ellmann Archive at the University of Tulsa's McFarlin Library, in Oklahoma. The eminent American critic and biographer was given permission to read it by Stanislaus's wife, Nelly Joyce, when, as a widow, she was trying to sell the documents and papers which had come into her possession after her husband's death in 1954. This precious and rich collection survived thanks to Stanislaus's careful hoarding and now forms the bulk of the famous Cornell University Joyce Collection. The version of the history of these papers – documents, letters, manuscripts – which Brenda Maddox gives in her acclaimed biography of Nora Barnacle (1988: 505-526) is interesting but totally biased, and a good example of how dangerous a biographer's personal aversion for a character or an event may be (especially if the biographer is a famous journalist like Maddox), and how his/her distorted and at times even openly fictional version of events can treacherously become received wisdom. Brenda Maddox – who, with no evidence whatsoever that the reader is made aware of, accuses Stanislaus of having destroyed or censored to his own advantage papers and documents in his possession, and in the best of cases of having kept them out of sheer greed – is the main architect of the anti-Stannie stance which for decades has been fashionable among Joyceans interested in biographical matters. Her description of the Trieste Diary "as covering the turbulent years in Trieste until the brothers were separated by the First World War"(510) is totally wrong, since the Trieste diary ends in 1909, long before the beginning of World War I, and her claim that Stanislaus manipulated the letters or other documents in his possession is an invention.

In spite of Maddox's antipathy for its author, the Trieste diary is an amazing document in which Stannie recreates with the narrative skill of a novelist the daily life of the extended Joyce family during those first difficult years in Trieste. Stanislaus's focus of attention is almost always his brother. A precious compendium of new information, the diary gives detailed descriptions and records of all that took place in Trieste: reports of lessons, narrations of events, descriptions of students, conversations on all sorts of topics, details on the books the two brothers were reading, the progress and evolution of James's writing, comments on Nora and the children, reports on their opinions about events both in Dublin and Trieste.

Among this varied range of subjects, women do appear from time to time, and Stanislaus's reports do not leave much space for a more encouraging version of Joyce's attitudes. In April 1907 Joyce and Stanislaus attended a performance of Ibsen's *A Doll's House*. This event inspired some reflections on women and their nature. It is a well known fact that Joyce, since his early days in University College Dublin, had nurtured (unlike Yeats) a great admiration for the Dano-Norwegian playwright, whose art he once declared was greater than Shakespeare's. His criticism of the Irish National Theatre derived partly from his huge admiration for Ibsen's art, which he had already articulated in an essay, *Ibsen's New Drama*, published in 1900. One of the things Joyce admired most in Ibsen was what he defined as his deep knowledge of women. In the essay he writes:

> Ibsen's knowledge of humanity is nowhere more obvious than in his portrayal of women. He amazes one by his painful introspection; he seems to know them better than they know themselves. Indeed, if one may say so of an eminently virile man, there is a curious admixture of the woman in his nature.[1] His marvellous accuracy, his faint traces of femininity, his delicacy of swift touch, are perhaps attributable to this admixture. But that he knows women is an incontrovertible fact. He appears to have sounded them to almost unfathomable depth. (*CW*: 64)

Joyce would hold with his opinion that Ibsen's art was an unparalleled study of women and of their condition in society throughout his life. Arthur Power's *Conversations with James Joyce*, a memoir in which Power, an intellectual who met and spent a lot of time with Joyce in Paris in the 1930s, provides ample evidence of this, reconstructing Joyce's defence of Ibsen's art. According to Power, Joyce praised the Norwegian dramatist's research into and analysis of "new psychological depths which have influenced a whole generation of writers".

> The purpose of *A Doll's House*, for instance, was the emancipation of women, which caused the greatest revolution in our time in the most important relationship there is – that between men and women; the revolt of women against the idea that they are the mere instruments of men. (Power 1999: 44-45)

---

[1] Leopold Bloom in the "Circe" episode of *Ulysses* becomes "the womanly man".

This is a much-quoted passage by all those critics who want to show that Joyce's supposed dislike for women was not so all-comprehensive as not to leave space for more acceptable and politically correct considerations on the matter. It must be said that, as far as one can gather from the reading of the material now available to critics, this is as good as it gets. It is true, on the other hand, that, in spite of his admiration for Ibsen's "knowledge" and understanding of women's psychology and for his powerful and assertive female characters, Joyce is deeply sceptical about the possibility of these Ibsenian women actually existing in real life. Joyce's unflattering disbelief is recorded by his brother Stanislaus. After the performance of *A Doll's House*, the young expatriate reiterates once more his admiration for Ibsen's treatment of women. Yet he admits that what Ibsen found in them is for him at least "hidden". In other words the strength, courage and independent-mindedness of a woman like Nora Helmer, the heroine of *A Doll's House*, was for Joyce at best a "plausible hypothesis". In reality – he continues, before a rather embarrassed audience of colleagues – for men woman is just an "aperture" and in the end the difference between a spouse and a prostitute can be measured in terms – so to speak – of time, since the prostitutes one has only for a few minutes, while a wife one has for an entire life.

Joyce admired Ibsen's women as literary creations and as artistic symbols rather than as explorations of human beings. Such women do not exist in real life, he bluntly claims, while sipping wine in a bar and contemplating without much enthusiasm a group of particularly silly looking/sounding female customers. He also admits to having an "Irish way" – as he calls it – of looking at women and consequently to being unable to take them seriously in anything they do.

Yet, as early as 1907, Joyce was already becoming aware of the fact that there were indeed other ways of considering womanhood and that the "Italian way"(meaning, it must be remembered, the way women were treated in Trieste) could be considered more satisfactory for both sexes because there was more equality between them. Nevertheless his belief is that the Irish way is "higher". What seems to disturb him is the confusion between the traditional gender roles assigned by society to men and women which this equality necessarily implies or at least encourages. He finds the interest that men "here" (meaning Trieste) take in cooking and eating and in domestic affairs annoying, a dislike which his Aunt Josephine would have shared, if we can judge from what she wrote to Stanislaus in

Trieste *à propos* of Nora's complaints about her life there.

> My poor Jim what has got over him. I am afraid he has got so many re-
> buffs from Fortune lately that it had left him too tasteless for anything,
> but honestly Stannie I can't understand Nora surely it is a monstrous
> thing to expect Jim to cook or mind the Baby when he is doing his ut-
> most to support both of them. (Joyce 1966: 139)

Joyce cannot consider women as intellectual and spiritual equals. In
his opinion things are as straightforward as this: women simply do not
count. According to Stanislaus, perhaps because of the not always easy relationship with Nora and of the difficult conditions of life in Trieste,
James's disparaging attitude towards women becomes more pronounced
as time passes until, on April 21 1908, he writes that his brother's dislike for women has become almost "savage", so much so that this hatred
has come to form a crucial part of his personality.

Even if we concede that Stanislaus might have infused in his brother's
statements an excess of misogyny and chauvinism, little space is left for
doubting that Joyce was not an estimator of women's intellectual potential.

What is interesting to notice is that Joyce – the young artist, the intellectual who fled his own country, rejecting all the oppressive elements
of its backward-looking nationalism, of its ultra-conservative Catholicism and of its repressive social system – did not challenge but rather accepted the male-centred, reactionary, anti-female aspects of that culture.
His attitude is nevertheless ambivalent. More than once he showed an
awareness of the cruelty towards women intrinsic in the Irish social and
cultural system. On August 29 1904 he writes to Nora in an attempt to
explain why he does not believe in (and hence cannot offer) the traditional
way of life:

> My mind rejects the whole present social order and Christianity – home,
> the recognised virtues, classes of life, and religious doctrines. How could
> I like the idea of home? My home was simply a middle-class affair ru-
> ined by spendthrift habits which I have inherited. My mother was slowly
> killed, I think, by my father's ill-treatment, by years of trouble and by
> my cynical frankness of conduct. When I looked on her face as she lay
> in the coffin – a face grey and wasted with cancer – I understood I was
> looking on the face of a victim and I cursed the system which made her
> a victim. (Joyce 1966: 48)

In *The Book of Days* he expresses similar remarks about the subordination of women in his country. On April 17 1905 he admits that in Ireland, far more than in England, the husband is a despot and the wife is never allowed to think for a moment that she has any duties towards herself or any life of her own. Yet this awareness never brought him any closer to even a vague consideration for women's rights. For all of his teachings, Ibsen's celebrated feminist sensitivity (Scott 1984: 48) remained completely lost on him.

The fact is that, as for so many other things, Joyce was a product –even though a very original one – of that very Ireland he left in 1904, of its anti-female intellectual, social, cultural and religious *milieu*. All the female members of his numerous family, his mother and his six sisters, belonged to the category of female subjects which women's studies would call 'silent women', their silence being an inevitable heritage of a social system which taught them to submit to patriarchal authority in all its various embodiments – fathers, husbands and priests – and to dedicate/sacrifice their life to motherhood and housekeeping.

Yet it is also true that if in life he did not express any political interest for women's predicament, in his art Joyce gave voice to an implicit, perhaps even unintended criticism of the limits imposed upon female roles and aspirations in early twentieth century Ireland. Most women in *Dubliners* are in a position of submission and servitude to the male characters they are related to and, when they are not, they are portrayed as rapacious and castrating wives and mothers ("The Boarding House", "A Mother"). One need only mention the opening story – "The Sisters" – to get to the point. The title seems to announce a plot built around two female characters; instead the reader soon finds out that the narration is focussed upon two male figures, the boy who is also the narrator and Father Flynn, the dead priest. This somewhat puzzling title choice is in fact a meaningful detail and an implicit critique of a male social practice: like the lives and identities of Nannie and Eliza, Father Flynn's two sisters which give the story its title, were totally sacrificed for and annulled in the life of their brother, so now "their" story is entirely occupied by his illness and death, while they are consigned to irrelevant or at least secondary roles.

In *Stephen Hero* Joyce gives life to a type of subdued and pre-cooked female identity in the figure of the protagonist's sister Isabel.

> She had acquiesced in the religion of her mother; she had accepted everything that had been proposed to her. If she lived she had exactly the

temper for a Catholic wife of limited intelligence and of pious docility
and if she died she was supposed to have earned for herself a place in the
eternal heaven of Christians. (126)

Isabel is a representative of the type of womanhood which her creator
would have seen in Ireland and in his household. Joyce did not feel much
sympathy for these women because their silent acceptance of their role
made them collusive with male authority which, from a cultural point of
view, had elevated their prone passivity and pale abnegation to the status of indispensable and laudable domestic virtue. As Stanislaus Joyce
writes in his memoir *My Brother's Keeper* the Irishwoman (as represented
by his mother) was for Joyce the accomplice of the Irish Church "the accomplice, that is to say, of a hybrid form of religion produced by the most
unenlightened features of Catholicism under the inevitable influence of
English Puritanism" (Joyce 1958: 238), a deadly cultural cocktail which
precluded any possibility of even a vaguely autonomous female identity.
Joyce's acceptance of the 'Irish way' of looking at women was therefore in
part due to the repressive male dominant culture of his country, but also
caused by the total absence of female resistance to, and even participation
in, that very culture. Even politically committed and university-educated
women such as the Sheehy sisters never really questioned the traditional
roles assigned to them by society.

Yet these beliefs and ideas were to be slowly but deeply put to the test
in Trieste, where a totally different type of femininity existed and had, to
an extent, existed for a long time. The female world which Joyce encountered in the Austro-Hungarian/Mediterranean city port presented a
cultural and religious texture which could never have been found in Joyce's
Dublin. This new universe was the product of the social, political and
economic history of the city and of its more modern origins. As such it
stimulated, if not Joyce's political ideas, certainly his artistic creativity
in the invention of the "womanly man" Leopold Bloom who, like the men
in Trieste whose 'womanly' behaviour Joyce did not approve of, does have
an interest in cooking and eating and domestic affairs, and of the sexu-
ally active, highly seductive, hyper-physical Mediterranean-Irish Molly
Bloom.

Women in Trieste had a long story of public visibility and social as-
sertiveness which nevertheless did not clash with their traditional roles
of wives and mothers but rather seemed to reinforce them. Even in its remote past, Trieste presented an unusual culture of female autonomy. In

the Middle Ages, for example, there was the rare phenomenon of two exclusively female confraternities, the most famous of which was that of
Santa Maria. The members were all women, they belonged to all strata
of society (which meant that there was a consistent net of female solidar-
ity), they had their own confraternity meetings, their own altar in the
Cathedral and if they were married to prominent members of the City
Council they could sit in the seats of the high choir where they were visible, honoured and respected by the community – a privilege which local Bishops sometimes found difficult to accept (Paolin 2003: 5).

Trieste became the multicultural, multiethnic, multireligious, poly-
glot Mediterranean port of the Austro-Hungarian Empire in the second
half of the Eighteenth century when, after the establishment of the Free
Port, waves of immigrants from all Europe but especially from the East
arrived to start new lives there, attracted by the policy of religious tolerance initiated by Joseph II of Austria.

> Many of them were 'luxury immigrants', who were welcomed by the
> authorities and by the religious community to which they belonged.
> Yet there were also poor people used to surviving by their wits. [...]
> During the nineteenth century the city became the Austrian Eldorado,
> the adoptive motherland to Jews, Greeks, Protestants, Serbs, Italians,
> these latter arriving from the peninsula, some attracted by the possi-
> bility of improving their material and financial situation, others flee-
> ing their birth-places because of religious persecutions. This is the pe-
> riod when the famous Triestine bourgeoisie was born, a social class which
> would later be proudly described as "cosmopolitan"(composed as it was
> of various religious groups) and which would also control the economy
> of the city. The male members of this powerful elite were helped in their
> social presence by wives, mothers, daughters and sisters who were all
> educated to this purpose.
> It is in this cultural and religious *koiné*, characterized by a deeply lay
> spirit, which Trieste's distinctive female identity found the space to de-
> velop and take root in society. It is a female identity which owes much
> to its many and various cultures of origin, and here lies the secret of its
> so often praised specificity. These were women who worked in the ed-
> ucational field and pioneered *avant-garde* pedagogical systems such as
> the Froëblian method; women capable of managing the family business
> with expertise when the husbands were away; women who loved the sea
> and the mountains, and who smile in old yellow photographs without
> any shyness, as they pose in front of the camera in their bathing cos-

tumes or in their climbing gear, near their nimble male companions, with whom they share outdoor hobbies and entertainments; even then "physicality was a natural and fundamental component of the relationship between the sexes". (Catalan 2003: 15)

This is the multiethnic society with which Joyce was in daily touch during his Triestine years and this is the strong, self-assured 'un-Irish' feminine which he came to be confronted with: a type of female identity against which he would be forced to pitch his received ideas about womanhood and which – more importantly – would impact enormously on the creation of his female characters.

Once more, the pages of Stanislaus's diary are precious to understand how it was in Trieste that Joyce came across a new physical and cultural version of woman which he further hybridized and poured into his own Spanish-born, half-Jewish Irish Penelope. The *Book of Days* offers some fascinating female portraits which are described with a mixture of attraction, fascination and surprise in their mentality, habits, culture and even physique. The female students at the Berlitz School were the young and well-educated daughters of the rich Triestine merchants and businessmen. They come across as highly sexualised, independent girls who, to Stanislaus, looked foreign and exotic. They evoke and create what he calls an "oriental illusion". Like Stendhal before him, who came to Trieste in 1831 as French consul (Adamo 2004: 66), Stanislaus clearly perceives in Trieste the presence of what Europe then conceived as the Orient, so much so that he nicknames one of the girls in his class the Odalisque. Some habits he finds strange and even "barbaric", like the wearing of long nails; some others he just finds surprising, like the fact that these girls are not disgusted at the idea of menstruation but rather amused when the subject is mentioned in class. He admires their freedom from embarrassment which in his opinion can only be the happy outcome of having been educated in many lands. Their multi-linguistic competence and their talent for foreign languages make him feel limited, their lay spirit strikes him as completely new and a-typical. He discusses religion and religious practices with them and explains how in Dublin people believe it is a mortal sin not to go to Mass on Sunday. One of them – a young Lutheran who is studying to become a school teacher – replies that maybe, if the weather was good, Dubliners would be much better off to go for a nice walk; then she explains that she goes to church only twice a year and that when she is independent she will not go at all because, after much thinking, she has decided that she does not believe in the church. This sounds excitingly new for Stanislaus, who admires these women's autonomous and lay ways of thinking.

Stanislaus's memories are further evidence that the Mediterranean/Triestine type of feminine offered a new vision of women which could not have been found in Dublin. It is also for this reason that Trieste became for Joyce a creative workshop in which to conceive – among many other things – new ideas of womanhood. The depiction of Molly as the female protagonist of *Ulysses* owes much of its power to the Triestine feminine which in fact becomes a foundational part of her deep complexity. Molly is not given much space in the novel – of the eighteen episodes that form *Ulysses*, only one is completely reserved for her, while Stephen and Leopold Bloom comfortably occupy the previous seventeen. Yet her voice is unforgettable and in the space of a chapter she manages to acquire mythic dimensions as consistent as those of her male counterparts.

There are only two moments in Joyce's texts when a female consciousness is allowed to dominate the narration from beginning to end, without the mediation or perception of a male character: these two moments are "Eveline" in *Dubliners* and "Penelope" in *Ulysses*. The gap existing between these two female spaces is indicative of the distance which Joyce covered in his conception of the feminine during his Trieste experience.

Although they do share the common destiny of having lost their mother at a very early age, the differences between Eveline and Molly are many and highly significant. Eveline is a young unmarried woman of nineteen who has been confined to "incestuous entrapment" (Henke 1990: 22) by a promise made to her dying mother and who since then has acted as her surrogate in the house. Eveline is three times trapped, at work, in her father's house and in Dublin: when at the North Wall she decides not to get on board the ship that would take her away from "that life of commonplace sacrifices closing in final madness" (*D*: 31), she fails to escape from her multiple state as a prisoner. Eveline constructs her identity in terms of erasure, abnegation and self-annulment, and this is also why no physical detail is given about her throughout the story. As a physical presence, Eveline remains vague, undefined, impossible to see and describe: we know nothing of her aspect expect the paleness of her cheeks, the cold of her lips which move in a silent fervent prayer, her white face, her expressionless eyes at the end. Her body simply fails to materialize as her life fails to happen.

Exactly the opposite can be said of Molly Bloom, whose Mediterranean features are described and commented upon time and again by various characters, not least by Molly herself in the monologue. Molly is very aware of her generous physicality and especially of her un-Irish looks, which she estimates higher than the Irish ones, since she disparagingly refers to Kathleen Kearney and other more successful singers in Dublin as "Irish homemade beauties"of little or no significance (*U*: 627). Molly's intense physical presence is a symbol of her visibility not only as a female object, which she has often been accused of representing, but also as a female subject empowered by her sexual allure and proud of her power over men.

Molly is a mature matron of thirty-four, clearly in charge both in her job (even though she does depend on a male impresario for her performances) and at number 7 Eccles Street, where she is not at all the woman of the house who serves the patriarch (or at least not on June 16 1904) but rather the nymph who in "Calypso" is served breakfast in bed by her husband (even though the roles, as we are given to understand at the beginning of the monologue, might be inverted the following morning).

Molly is the only married woman and mother in Joyce's fiction with a profession which is pursued not because of pressing monetary needs or because of a violent and aggressive patriarch but because it is her own choice to do so. She is an opera singer and in this she also subversively transforms singing – one of the many female talents traditionally reduced to "parlour accomplishments" (Scott 1984: 57) and valued only in so far as they could be privately enjoyed by the male members of the family – into a public profession. It is also a profession which allows Molly a great deal of independence, as it gives her the possibility to travel on her own and perform in public.

Opera singing was also a type of occupation which would have been much more obvious in Trieste than in Dublin. It should also be remembered that Joyce had a deep passion for Italian opera (he assiduously attended the Opera House – the *Teatro Verdi* – in Trieste) and more importantly that *il bel canto* was the only other artistic profession he tried to pursue – by enrolling at the *Conservatorio Tartini* in Trieste – in the rather unrealistic hope of becoming a professional tenor. For this career he was ready to give up his writing, much to Stanislaus's exasperation. Molly's profession, therefore, has to do with talent (a talent she shares with Joyce) and self-realization and not with financial needs.

Molly is not as literate or sophisticated as the young women of Tri-

este that the Joyces came to know, but she certainly possesses their freedom from embarrassment, which Stanislaus rightly connected to their
multicultural life-experience. Born in Gibraltar of an Irish father and of
a Spanish Jewish mother, she – like her Triestine counterparts – has been
educated in many lands, and she proudly carries the traces of her mixed
blood in her father's Irish accent and her mother's Spanish eyes and figure (U: 627). As for many of Stanislaus's students, religion for Molly is
more a matter of personal choice than a dogmatic imposition: she goes to
confession, but she questions the priest's authority, while she remembers
that old "cantankerous" Mrs Rubio, the woman who looked after her in
Gibraltar, was unhappy with her "because I didnt run into mass often
enough in Santa Maria to please her" (U: 624).

But perhaps the most significant difference between Eveline and Molly,
and one which is even more directly connected with Trieste, is their opposing relationships with and reaction to alien spaces, foreign locations,
faraway lands. Eveline, who presumably has never left Dublin, travels to
foreign places thanks to Frank's exotic and possibly fictional travelogues.
He – like Othello with Desdemona – tells her about his adventurous life
as a sailor and talks about the strange lands he has visited.

> He had tales of distant countries. He had started as a deck boy at a pound
> a month on a ship of the Allan Line going out to Canada. He told her
> the names of the ships he had been on and the names of the different
> services. He had sailed through the Straits of Magellan and he had told
> her stories of the terrible Patagonians. He had fallen on his feet in Buenos
> Ayres, he had said, and had come over to the old country just for a hol-
> iday. Of course her father had found out the affair and had forbidden her
> to have anything to say to him. (D: 31)

Exoticism, connected to the fear of abduction, had been already implicitly evoked in the story in relation to Frank when, before remembering his "tale of distant countries", Eveline had mentioned the time when,
early in their relationship, he had brought her to see *The Bohemian Girl*,
a very popular opera by the Dubliner Michael William Balfe which had
been performed for the first time in London in 1843 and had immediately become wildly popular all over Europe.

The libretto of *The Bohemian Girl* came from the pen of Alfred Bunn,
who derived his idea of it from a ballet called "The Gipsy," written by
Saint Georges in Paris in 1839 (the Italian version of the opera in fact re-

tained this title, *La Zingara*). Saint Georges had taken his story from the Spanish "Novelas Exemplares" of Cervantes. The choice of this opera is relevant in the signifying structure of "Eveline" as *The Bohemian Girl* tells the story of Count Arnheim, the Austrian Governor of Presburg, and of his six-year-old daughter Arline, who is kidnapped during a party by the Gipsy Chief, with the threatening name of Devilshoof. Although *The Bohemian Girl* ends happily, its tale of a little girl who is taken away from her widowed father's house by a man who has travelled in foreign lands enacts the hidden fears which brood in Eveline's heart and which will prevail over her desire to escape, paralysing her at the North Wall and deleting, with Frank himself, any possibility of significant change from her life.

Eveline comes across as one of the most paralysed of all the inhabitants of the Dublin of *Dubliners* (which is very different to the Dublin of *Ulysses*) because she is the one who comes closest to the possibility of escape. It is also useful to remember – as Katherine Mullin has explained in her *James Joyce, Sexuality and Social Purity* – that "Eveline" was published first in *the Irish Homestead*, the journal of the Irish farming movement that was edited by George Russell and characterised by a very strong nationalistic spirit. One of the aims of the *Irish Homestead* was to slow or discourage the unstoppable flow of Irish emigration by publishing fictional stories which portrayed emigration as the wrong choice since it "ruptured the natural bonds of lovers, families, community and nation" (Mullin 2003: 58). Accordingly "Eveline" interprets emigration from the Irish female perspective and sees it as a betrayal of female duties to home, family, father and hence fatherland. That home is a cage and father is a violent, unforgiving, exploitative patriarch is Joyce's way of articulating his subversion of *The Irish Homestead*'s patriotic goal. But for Eveline overseas lands and foreign people are somehow attractive as long as they remain "distant tales": domestic imprisonment is bad, yet not so bad after all.

Molly has a totally different approach to and relationship with geographical alterity. Although on June 16 1904 she does not leave her home, she is the most successful traveller of all the Dubliners we encounter in Joyce's work. Of her, one could easily say what Stanislaus in his diary says about Trieste and the foreign place-names he hears everyday in his adoptive city: that they "make Geography a romance". In her monologue Molly certainly makes geography a romance, both for herself, for the men she flirts with and for the reader. While Eveline's memories and dreams are contained and confined within the narrow space of her street and her house, Molly's location for memories, dreams and desires is her Mediterranean land of origin: Gibraltar, that very obvious Joycean surrogate for Trieste.

Like Eveline's, Molly's first boyfriend is a sailor called Jack Jo Harry Mulvey. Harry is actually highly reminiscent of Frank himself as seen through Eveline's eyes " his peaked cap pushed back on his head" (D: 31). Molly too pictures Mulvey "with his peaked cap on that he always wore crooked as often as I settled it straight H M S Calypso"(U: 626). But Molly, like the girls in Trieste and unlike Eveline, is not scared of geographical romances simply because she has them within herself, and her monologue is made up of Spain, Gibraltar, the Mediterranean, voyages and travels as much as it is made of Ireland.

While in "Eveline" the narrator of exotic stories is – as we have seen – Frank the sailor and the seduced listener is Eveline, in "Penelope" it is Molly, not Mulvey, who becomes the storyteller. She is the exotic native who invents fascinating stories to capture the young man's attention and seduce him ("what did I tell him I was engaged for fun to the son of a Spanish nobleman named don Miguel de la Flora and he believed me" – U: 625) and explains to her Irish sailor, a newcomer to Gibraltar, the mysteries of an unknown land: Spanish money, thunder and lightning, the strange monkeys which – even though they are not swimmers – live in Spain and on the opposite African shore, all the frightful and lovely sights of Gibraltar:

> he couldnt count the pesetas and the perragordas till I taught him [...] up on the tiptop under the rockgun near OHaras I told him the tower was struck by lightening and all about the old Barbary apes they sent to Clapham without a tail [...] the galleries the casemates and those frightful rocks and Saint Michaels cave with the icicles or whatever they call the hanging down and ladders. (U: 625)

She has very obviously adopted a similar 'Othello-like' technique of seduction with Leopold, she has told him the same fascinating stories, and many more we must presume, because when earlier in the day, in the "Nausicaa" episode, Bloom remembers Molly's girlhood in Gibraltar, the very same examples came to his mind.

As he remembers his wife's words about her land of origin, he also recalls the reasons she gave him when he asked her why she had chosen to marry him. These are crucial words for a proper understanding of Molly's nature and her attitude to foreign and geographical otherness.

> Brings back her girlhood. Gibraltar. Looking from Buena Vista. O'Hara's tower. The seabirds screaming. Old Barbary ape that gobbed all his family. Sundown, gunfire for the men across the lines. Looking out over the sea she told me. Evening like this, but clear, no clouds. I always thought I'd marry a lord or a rich gentleman coming with private yacht. *Buenas noches, senorita. El hombre ama la muchacha hermosa.* Why me? Because you were so foreign to the others. (U: 311)

Joyce gives Molly her un-Irish and Mediterranean/Triestine background, childhood and looks because he wanted to rid his heroine of that Irish female heritage of silence which taught women to submit to patriarchal authority. Mediterranean Molly is completely free from those Irish female shackles. And if she married "Poldy" because he was so "foreign to others", or "not Irish enough" (*U*: 616) and if she is in turn seduced by his own fascinating "tales of distant lands", it is because she is much better travelled than any other Dubliner in the novel and she is not scared of distant lands and of the people who inhabit them. This is also why – unlike all the others – she has no need nor desire to escape. Molly has her own foreign stories to tell, her own vastly horizontal geography to put to use in memories and desires, her rich and exotic places to evoke whenever she wants: like her husband and unlike any other character in the novel, she possesses that mental ability to visit alien spaces, to expand, imagine, invent foreign geographies, to make (as Stannie so beautifully put it) geography a romance which Joyce had seen, admired and learnt in cosmopolitan, multicultural, 'oriental' Trieste.

Bibliography

Adamo, Sergia, 2004, *Ritratti di una città, Trieste tra scritti di viaggio e immagini retrospettive*, Istituto Giuliano di Storia, Cultura e Documentazione, Trieste.

Brivic, Sheldon, 1995, *Joyce's Waking Women, An Introduction to* Finnegans Wake, The University of Wisconsin Press, Madison.

Brown, Richard, 1985, *James Joyce and Sexuality*, Cambridge University Press, Cambridge.

Catalan, Tullia, 2003, Introduzioni – "Un lungo cammino: lo sguardo dello storico", in Carla Carloni Mocavero - Marina Moretti - Elvira Prenz - Graziella Semacchi Gliubich - Marina Silvestri - Marina Torossi Tevini, *Trieste, la donna e la poesia del vivere*, Ibiskos Editrice, Empoli (FI).

French, Marilyn, 1993, *The Book as World: James Joyce's* Ulysses, Paragon House, New York.

Henke, Suzette, 1990, *James Joyce and the Politics of Desire*, Routledge, New York and London.

Joyce, James, 1966, *Letters of James Joyce*, ed. by Richard Ellmann, Vol. II, Viking Press, New York.

Joyce, Stanislaus, 1958, *My Brother's Keeper: James Joyce's Early Years*, Viking Press, New York.

Kershner, R. Brandon, 1989, *Joyce, Bakhtin and Popular Literature*, University of North Carolina Press, Chapel Hill and London.

Maddox, Brenda, 1988, *Nora*, Minerva, London.

Mullin, Katherine, 2003, *James Joyce and Social Purity*, Cambrige University Press, Cambridge.

Paolin, Giovanna, 2003, Introduzioni – "Un lungo cammino: lo sguardo dello storico": in Carla Carloni Mocavero - Marina Moretti - Elvira Prenz - Graziella Semacchi Gliubich - Marina Silvestri - Marina Torossi Tevini, *Trieste, la donna e la poesia del vivere*, Ibiskos Editrice, Empoli (FI).

Power, Arthur, 1999, *Conversations with James Joyce*, The Lilliput Press, Dublin.

Scott, Bonnie Kime, 1984, *Joyce and Feminism*, Indiana University Press, Bloomington.

"ALL THE SEAS OF THE WORLD": JOYCEAN THRESHOLDS OF THE UNKNOWN. A READING OF THE MARINE AND WATERY ELEMENT FROM *DUBLINERS* TO *ULYSSES*

Roberta Gefter Wondrich

Of all the great writers in the English language of the modern age, James Joyce could be said to have wonderfully combined the quintessentially Modernist focus on the city with a universal, all-embracing perspective that fully includes other locations and natural elements, among which water and the sea play a major part. As a deflated epic version in prose of the Odyssey, the most important poem about a hero's marine wanderings of Western civilization, *Ulysses* celebrates an urban navigation that is both accurately physical and ceaselessly imaginative. In *Ulysses* (and in *Finnegans Wake* too) there is indeed room for "all the seas of the world", but the semantic import of the sea – and of water imagery – is mainly identifiable in two contrasting ideas of it, as the following analysis will try to illustrate: one "insular", threatening, disabling, associated with Ireland, the other essentially fabulous, mythical, adventurous, associated with the Mediterranean and the dimension of travel.

The sea and its related features – from climate to temperament and civilization – imply the fundamental role of water itself as a primary symbol and motif. Joyce's work literally overflows with water and sea imagery: as early as 1959 W.Y. Tindall pointed out its importance and, among the most recent studies, Katharina Hagena has devoted an entire volume to the subject of the sea in Joyce's masterpiece[1] and Robert Adams Day a perceptive essay on Joyce's "acquacities", in which he reminds the reader that "the sea embraces the text of Ulysses". (Day 1996: 12)

It is not my intention to probe into the biographical sources that may be at the root of Joyce's marine imagery, yet it is worth remembering that his was a "decidedly acqueous existence", as Day puts it (*Ibid.*: 5), since Joyce spent all of his life close to water, either by the sea or in cities crossed by rivers, from Dublin to Pola, to Trieste, Zurich and Paris. The most decidedly – although idiosyncratically – Mediterranean of these was Trieste, then the Adriatic port of the Habsburg Empire, where the Irish writer in his self-imposed exile of about ten years (1904-1915 and 1919-1920) experienced a unique mixture of cultures, ethnicities and languages that was to exert a powerful influence on his work, and most notably on the characters of Bloom and Molly. The Triestine years, the city "as a crossroads of competing cultures" and their great importance have been brilliantly reassessed by John McCourt in *The Years of Bloom*. McCourt demonstrates how Trieste "was his principal source [...] for much of the Oriental, Jewish and Greek elements of Ulysses, for much of the multilingual chaos of 'Circe' and *Finnegans Wake*" (McCourt 2000: 5).

The centrality of the sea in Joyce's life and work suggests that it may be interesting to attempt a comparison between different versions of the idea of a threshold of sea or water, especially in relation to a gap separating land and sea, earth and water as multiple and immensely resonant symbolic locations, as well as a consideration of the relation between the main Joycean characters and water and the sea or, in other words, of their attitudes to the sea. *Finnegans Wake*, with its "riverrun"...from swerve of shore to bend of bay" and its "commodius vicus of recirculation" (*Finnegans Wake*: 4), being in fact the supreme celebration of water, rivers and flux in Joyce's oeuvre – from a structural point of view as well – does not figure in the present discussion, since it would require an extensive consideration of its own.

A preliminary step suggests that the frequent images of the physical juncture between land and water may be analysed within a common or

---

[1] Katharina Hagena, 1996, *Developing Waterways. Das Meer als sprachbildendes Element in Ulysses von James Joyce*, Verlag Peter Lang, Frankfurt, Berlin, Bern.

contrasting framework, in relation to a possible confrontation between the Irish *insular* marine setting and a different, subsequent sea imagery that – for biographical reasons – had been influenced and enriched by Joyce's experience of the Mediterranean.

The representation of the sea and of watery surfaces and forms in general seems to be initially marked by overtly negative, threatening overtones in *Dubliners* and in the first chapters of the *Portrait*, insofar as it implies a traumatic severance from the land, or a missing bond with it. It is only with the end of the fourth chapter of the *Portrait* that the marine imagery in particular becomes potentially rich with fertile and reviving connotations in relation to a retrieved sense of connectedness, if not of oneness, with the earth and land. However, this aspect may read as paradoxical in the light of Stephen's flight from Ireland and frustrated return, and holds an ambivalence which becomes evident in the third chapter of *Ulysses*, where we find Stephen again ankle-deep in water, so to say, "walking into eternity along Sandymount strand" (Joyce 1986: 31).

Water, and especially sea-water, figures metonymically for passage, change, transition in Joyce. In *Dubliners*, "Eveline" is the episode that best exemplifies both this semantic trait and the initial adverse connotation of the sea in Joyce's work. One of the central structural oppositions in the story is precisely that between change and no-change (Hart 1969: 48), and the motif recurs in *Portrait* and *Ulysses* as well. Eveline proves unable to jump out of her stagnant living death, thus refusing the rite of passage symbolized by the sea journey that might liberate both her soul and her womanhood.

The new life that the protagonist dreams of – only to renounce – is conveyed and symbolized by an act of transgression, a stepping over from firm land to sea, by the unachieved crossing of a threshold that in its turn implies movement as detachment and departure. The two physical locations of the port quay and the mysterious sea are therefore the conflicting ends of a failed dialectic. "Passage" may be regarded as the key word in this case: the passage the two lovers seem to be longing for, the rite of passage from adolescence to adulthood which Eveline ultimately refuses to undergo.

> If she went, to-morrow she would be on the sea with Frank [...] their passage had been booked [...] (Joyce 1987: 35)

Frank, Eveline's fiancé, "had started as a deck boy" and is associated in the text with a life at sea; yet it is precisely the sea as an amorphous and engulfing mass of floods that becomes the crucial image which triggers this young Dubliner's paralysis and the petrifying terror which overcomes her:

> All the seas of the world tumbled about her heart. He was drawing her into them; he would drown her.
> Amid the seas she sent a cry of anguish. (Joyce 1987: 35, 36)

Eveline's fear and anguish, clearly charged with sexual anxiety, transfigure the Dublin quay, which functions as a metonymy of the mainland, into a step that will precipitate her into "all the seas of the world"; far from being a possible threshold of a new life, it proves a dark leap into the dark. Significantly, this sea that is so agonizing to her is the Irish sea, and then the Atlantic Ocean with its immense stretch of waters. Thus, a dichotomy emerges clearly between the city as a desiccating known and the open sea as the unknown; thus, "as for so many other characters in Dubliners, the banality of the known is for Eveline far preferable to the terror of the unknown" (Dettman 1996: 4).

The *Portrait* displays an impressive array of watery and marine imagery, which throughout the narrative shows a transition from a conventional and at times overwrought, hyperbolic use to a powerful, almost imagistic one. Further to W.Y. Tindall's seminal analysis of the importance of water and water imagery in Joyce, it is a common view that "in the first half of *A Portrait* water is commonly disagreeable, agreeable in the second" (Tindall 1970: 89). As one of the most important images of the novel, water serves to link the novel's books and is essentially ambivalent. This quality of ambivalence is also decisive to the enormous semantic resonance of the sea motif, which is, as remarked, coterminous with that of water and somewhat subordinate to it, particularly in the case of this text.

Most notably, the first chapters are punctuated by frequent references to the movement and the force of water and currents in relation to the protagonist's emotions and feelings, through words such as "wave/s", "tide", "flowing", "flowed", "breakwater", "outlet", "stream" and so on. Metaphors abound, together with similes, almost invariably linked to the ideas of sin, error, failure and strength in the perspective of the artist's quest for perfection.

> He had tried to build a breakwater of order and elegance against the sordid tide of life without him and to dam up, by rules of conduct [...]

the powerful recurrence of the tides within him. Useless. From without
as from within the water had flowed over his barriers: their tides began
once more to jostle fiercely above the crumbled mole". (Joyce 1989: 90)

In the hell-fire sermon Stephen is prompted to imagine eternity in
terms of uncountable "particles of sand" and "drops of waters of the mighty
ocean". After overcoming "the flood of temptation" of religious vocation,
for instance, "Pride after satisfaction" uplifts "him like long slow waves"
(Joyce 1989: 150).

Having "turned seaward from the road at Dollymount", Stephen sees
a group of Christian Brothers crossing the bridge: disquieted at their
sight, he tries "to hide his face from their eyes by gazing down sideways
into the shallow swirling water under the bridge but he saw a reflection
therein of their topheavy silkhats, and humble tapelike collars and loosely
hanging clerical clothes" (*Ibid.*: 151). The water surface reflects the suf-
focating reality of clerical Ireland he is trying to flee, it provides neither
shelter nor a way of escape, its function is not that of a nearby elsewhere
where Stephen can ideally turn to in his reverie. However, it foreshadows
as well the epiphanic potential of the watery and marine setting in the
following climactic scene at Dollymount beach, on the "day of dappled
seaborne clouds".

Progressively, the marine setting harbours and responds to Stephen's
creative imagination and mythopoeic skills; he loves "the rhythmic rise
and fall of words better than their association of legend and colour" (*Ibid.*:
152); now the rise and fall – though still evocative of waves and tides –
also implies a larger animistic sense of life. Stagnation is apparently slowly
giving way to throbbing life.

> He passed from the trembling bridge on to firm land again [...] A faint
> throb in his throat told him once more of how his flesh dreaded the cold
> infrahuman odour of the sea: yet he did not strike across the downs on
> his left but held straight on along the spine of rocks that pointed against
> the river's mouth. (*Ibid.*)

Here the synaesthetic representation of the sea points to a perception
of the marine stretch as something organic, if not deadly humanized, with
the rocks along the water forming a "spine", the river a "mouth" and "the
grey sheets of the water" a shroud.

When Stephen recognizes his mates who address him from their "div-

ingstone", "characterless" as they look, the sea water is once more oppressive, though apparently only to the body and its movements (the div-
ingstone rock is slippery, while "the towels with which they smacked
their bodies were heavy with cold seawater" (*Ibid.*: 153).

Approaching the encounter with his own new self as the fabulous ar-
tificer, Stephen wades the waters of the Irish sea, barefoot (*apparently* quite
unlike the tremulous Prufrock who will share his attire, though not his
elated spirit only a few years later). He clambers down the slope of the
breakwater, wondering at the endless drift of the affluent rivulets. The
Irish sea, the shroud-like stretch of grey is being infused with a new drift
of restless flow: and the deadly marine surface, human-like, is enlivened
with young "gayclad lightclad" figures of children. The girl who stands
"before him in midstream appears transfigured "into the likeness of a
strange and beautiful seabird", thus visually joining sea and air, "the emer-
ald trail of seaweed" "as a sign upon the flesh" hinting at the green of Ireland that still gets hold of the body and hampers it, a dominant image
in the previous passages. When the girl returns his gaze the water has become "gentle in sound", "low and faint and whispering": again, the water resounds with the human touch, as possibly evocative of the gentle
and low whispers of shy lovers.

Hearing is now the sense being focused on; Stephen strides "far out
over the sands, singing wildly to the sea, crying to greet the advent the
life that had cried to him". But Stephen's final move, further to his Pa-
ter-style epiphany, is now *landbound*:

> He turned landward and ran towards the shore and, running up the slop-
> ing beach, reckless of the sharp shingle, found a sandy nook amid a ring of
> tufted sandknolls and lay down there that the peace and silence of the evening
> might still the riot of his blood. [...] and the earth beneath him, the earth
> that had borne him, had taken him to her breast. (Joyce 1989: 156-57)

Water and the idea of a new world similar to sea depths figure once
more in the final description of his epiphany: "His soul was swooning
into some new world, fantastic, dim, uncertain as under sea, traversed by
cloudy shapes and beings" (157).

A merging between water and earth, between sea and land concludes
the chapter: "the tide was flowing in fast with a low whisper of her waves,
islanding a few last figures in distant pools" (*Ibid.*).

Once Stephen has envisaged his call and destiny, a reconciliation with the earth and the land which can shelter and welcome him again has been reached through the medium of the sea, on the liminal space of the shore. The spatial symbolism thereby becomes contradictory, to some extent, as that same newly re-appropriated land will be rejected in the fifth chapter as "fatherland", together with home and the Church. The sandy nook where Stephen finds shelter in fact foreshadows the recurrent image of the womb that occurs in the fifth chapter, in relation to his visionary experience of desire:

> a glow of desire kindled again his soul and fired and fulfilled all his body. Conscious of his desire she was waking from odorous sleep [...] enfolded him like a shining cloud, enfolded him like water with a liquid life; and like a cloud of vapour or like waters circumfluent in space the liquid letters of speech, symbols of the element of mystery, flowed forth over his brain". (Joyce 1989: 201)

Speech itself, like letters, becomes yet another watery element, liquefied as part of the mysterious world of flow, flux and transformation that Stephen envisages and longs for. What is more, it is the overt association of "liquefied" speech and words to desire itself that makes the passage particularly relevant to a coonsideration of the polysemy of water in the novel's texture, and in the whole of *Ulysses*.

(Sea) water is dispersed and multiplied, enfolding and pervasive. Together with the vision of birds, in the guise of birdlike and wavelike words, it is once more, as in the case of the transfigured girl on the beach, charged with sensuous and erotic overtones and explicitly evokes sexual desire through an allusion to masturbation and ejaculation conveyed by the repetition of "a soft liquid joy", which on the other hand may even hint at the pre-natal condition in the womb.

> A soft liquid joy like the noise of many waters flowed over his memory and he felt in his heart the soft peace of silent spaces of fading tenuous sky above the waters, of oceanic silence, of swallows flying through the seadusk over the flowing waters.
>
> A soft liquid joy flowed through the words where the soft long vowels hurtled noiselessly and fell away, lapping and flowing back and ever shaking the white bells of their waves in mute chime and mute peal and soft long swooning cry. (*Ibid.*: 204)

The ambivalence in the development of the water/sea symbolism identified by Tindall and previously mentioned is crucial to an analysis of the *Portrait* and, above all, of Stephen's attitude to the marine setting, as a character that prolongs its existence in *Ulysses*. Although in the final chapters the watery images have a generally positive connotation, Stephen never fully "enters" the sea, and his "affirmative contacts with the element are oddly tentative" (Day 1996: 10) so that his turning landward becomes emblematic of an unsevered and difficult bond that he is not yet willing to cut loose.

The act of wading the waters is indeed decisive, as it symbolizes the passage from land to sea as a kind of rite, a lay christening as well as a departure. In *Ulysses*, Stephen does not actually bathe – while Mulligan does, as Day remarks – instead he is merely a fascinated observer of the tide and floods who stays behind, and his diffidence towards water is associated with the motif of death, and particularly of death by water.

> A drowning man. His human eyes scream to me out of horror of his death. I […] With him together now… I could not save her. Waters: bitter death: lost. (Joyce 1986: 38)

After all, Stephen's name is Dedalus, and his soaring ambitions are those of an intellectual Icarus, who met his death by drowning in the sea.

The change and metamorphoses, as well as the dissolution of the human form brought by death by water, and – metonymically– by the sea itself, is an object of dread; Stephen is neither willing nor able to take on a protean attitude to his life experience ("I want his life still to be his, mine to be mine." *Ulysses*: 38); in other words, "he will not leave his self-enclosed identity or enter the lives of others for more than a moment. He is of the company of Joycean figures who hover at the verge of the water, safely dry, but who will not float, perhaps drown, in any case be changed" (Day 1996: 13). It is precisely this self-enclosed and self-absorbed identity, both in the *Portrait* and in *Ulysses*, that exposes Stephen's lack of empathy, when compared with the gentle, open, profoundly humane nature of Bloom.

The sea as a multivalent image and polysemic motif dominates throughout Ulysses, but is introduced in relation to the character of Stephen in the Telemachia, while in the second part, and especially in "Nostos", it becomes, so to speak, Bloom's cup of tea.

In *Ulysses*, the contextual and thematic importance of the strand and

sea, the cohesion and contiguity between water and land which are outlined in the final chapters of *Portrait* are revived and developed in the
third chapter, "Proteus" (the symbol of which is "tide") in terms of dynamic continuity and coexistence, eternal recurrence and transformation.
Hence the Protean image and symbolization of the watery flux as a natural foil and correspondence to the flux of the mind; hence the formal
correspondence of the subthoughts and (sub)motifs in the stream of consciousness as rivulets flowing into the mainstream.

"Am I walking into eternity along Sandymount strand?" (*Ibid.*: 31)
muses Stephen in the second paragraph: the shore as the paramount liminal space is thus, in ironic terms, the physical location of a reverie which
can turn the contingent into the timeless. The protean theme makes all
the senses coalesce: "Rhythm begins, you see? [...] I hear. [...] I will see
if I can see" (31).

Again, as repeatedly in *Portrait*, human passions are metaphorised in
terms of sea movements: "their blood is in me, their lusts my waves" (38).

Then the drowned man's leit-motif once again brings to the fore the
deathly, unfathomable power of the sea, which in this perspective, as a
"mighty mother" can also be figured in gendered terms as a devouring
mother; incidentally, the drowned man dies "off *Maiden*'s Rock".

The tone seems to change rather abruptly in the episode with the motif of death by water, which, now twofold, is associated with the drowned
man and with Stephen's mother, and is preceded by his wishing for the
safety of firm land. Here Dedalus acknowledges his unease with the idea
of the alterity of the sea and even his dread of its insidiousness:

> I am not a strong swimmer. Water cold soft [...] do you see the tide
> flowing quickly in on all sides, sheeting, the lows of sand quickly, shell-
> cocoacoloured? If I had land under my feet. I want his life still to be his,
> mine to be mine. A drowning man. His human eyes scream to me out
> of horror of his death. I [...] With him together now. [...] I could not
> save her. Waters: bitter death: lost. (*Ibid.*: 38)

So he connects to the sea change on an essentially intellectual plane,
without empathy; isolation at sea thus powerfully induces a kind of horror vacui that even Stephen, despite his bold defiance, is not yet ready to
face.

The rescued body of the drowned man and the quotation from *The
Tempest* ("full fathom five thy father lies") dramatise the sea change in be-

nign tones through images of life as literally persisting and spilling out of death ("the quiver of minnows", tiny fish, flashing through the slits of the dead man's button fly) and finally reassert the reconciling power of the sea: "Seadeath, mildest of all deaths known to a man" (*Ibid.*: 42). But it is worth noticing that, paradoxically, this restoration takes place out of the depths, once the body has been hauled onto the shore. Again, Stephen ideally turns *landward* and remains at a distance – a speculative and imaginative distance – from a sea that at times is to him like "a torpid Sargasso" (Kenner 1980: 60).

Ultimately, the closing image of the chapter is that of the three-master at sea, which resembling three crosses on the horizon conveys a double Christian allusion, in that it forecasts the "unholy trinity of which Stephen is a member with Bloom" (Tindall 1970: 147) and also alludes to crucifixion.

The open sea is then, once again, imbued with a sense of dismay quite explicitly symbolized. Albeit fascinated by the depths of the sea, Stephen seems to be unable to read "signatures of all things" *in* it and beyond the horizon. The open sea is still identified with the classic archetype of an unknown of that same darkness which Stephen knows we harbour in our souls.

While in the *Portrait* and especially in "Proteus" the sea can be said to figure and function as a chronotope that, "functioning as the primary means for materializing time in space", "emerges as a center for concretizing representation" (Bakhtin 1981: 250), the second and third part of the novel outline two basically opposite attitudes towards the watery and marine element that illuminate the sea as a cultural and historical motif. It is in "Ithaca" that we find their most eloquent characterisation in this regard. In the first part the main differences and similarities between the two men are outlined and the narrative is constructed by a sequence of questions and answers; here Stephen, the prim intellectual, confesses to his hatred of water, revealing "That he was hydrophobe [...] distrusting aquacities of thought and language" on account of "The incompatibility of aquacity with the erratic originality of genius" (Joyce 1986: 550). Stephen's avowed aversion is preceded by a page-long description of reasons substantiating the admiration that "Bloom, waterlover, drawer of water, watercarrier" (*Ibid.*: 549) has for the element. Among its infinite qualities he likes it for "Its universality", its physical characteristics in the most exotic seas of the globe, from the Pacific to the Dead Sea and, above all, "its infallibility as *paradigm* and *paragon*; its *metamorphoses* as vapour,, mist, cloud, rain, sleet, snow hail; [...] its *variety of forms* in loughs and bays and gulfs [...]"[2] (*Ibid.*: 549). His fondness for water and the sea, which is indeed recurrently evident in the book, is unmistakably a flair for its mutability and for its nature, comprehensive of the principle of life itself, beginning and end. Both Stephen and Bloom are fascinated by the flowing coexistence of life and death in water and the sea, but while Stephen is appalled by its annihilating power, Bloom is truly fascinated by the idea of a marine and acqueous *panta rei*, by the very formlessness of water as by the formlessness of life, as these lines from "Lestrygonians" epitomize: "How can you own water really? It's always flowing in a stream, never the same, which in the stream of life we trace. Because life is a stream" (*Ibid.*: 126).

Bloom often fantasizes about the sea and life at sea; in "Nausicaa", for instance, thinking of the "dreadful life sailors have" he wonders "how can they like the sea? Yet they do" (*Ibid.*: 309), but it is actually in the sixteenth chapter, "Eumaeus" that the sea becomes a prominent motif, if not a real topic of the narrative and that Bloom's "Mediterranean" idea of it unfolds. Navigation is in fact the art or science of the chapter, which counts numerous references to shipwrecks and the life at sea. As Katharina Hagena remarks,

> In no other chapter of *Ulysses* does the sea, to such an extent, ooze into the very texture of the language: even the letter combinations s-e-a, sea, or the sound combination [si:] are everywhere. [...] The whole chapter is overflowing with idioms and metaphors taken from the art of navigation: to fathom, to break the ice, to flounder, to sail under false colours, to hold water, to be all at sea, the coast was clear, to keep a sharp lookout, the superannuated old salt... (2002: no page indicated)

This thematic relevance of the sea is particularly interesting for the present discussion in that it discloses Bloom's ideal vocation as a seaman. It is associated with the two characters of Bloom and an old "*soi disant*" sailor named Murphy, who breaks into Stephen and Bloom's conversation and soon starts reminiscing about Simon Dedalus' presumed feats as a

---

[2] My italics.

marksman, thus betraying his true nature as an unreliable liar, since old Dedalus "is not the Ulyssean father of Telemachus Stephen" (Blamires 1996: 202). Within the main theme of the failed encounter between the father in search of a son and the son in search of a father, apart in their mindsets and sensibilities, the narrative in fact clearly outlines an antithesis between Blooms's marine *wanderlust*, his fondness for the Italian language and his interest in the spaces and confines of the Mediterranean, as well as for "the Spanish blood" and Stephen's deflating, indifferent attitude

Once again, while Stephen remains obsessed by Ireland in a self-centred, rebellious but "insular" perspective, Bloom is the true un-heroic seafarer, a Hibernian man of commerce of Mediterranean inventiveness who constantly devises new ingenious but improbable ways of making money and, above all, who conceives of Ireland as a country that should be open to all forms of social (and mercantile) contact, and who devises his own "Bloomusalem", an egalitarian society that embraces all creeds and classes. Furthermore, the Bloom-Stephen duality is also complicated by the figure of Murphy the seaman, "the bogus wanderer" (Blamires 1996: 211), who, as a yarn-spinner, is also a deflated version of the Ancient Mariner – along with the other fabulous figures of Sindbad the Sailor and the Flying Dutchman – as well as a sort of alias of the narrator's figure (the technique of the chapter being "relaxed prose"). But Murphy the sailor is also a kind of diminished, "bogus" foil to Bloom himself, as both are somehow wandering and telling stories, and fantasising about the wonders of the world, so that it could be argued with W.Y. Tindall that this "part of Ulysses seems divided between Bloom and the old sailor" (1970: 215).

The seas he claims to have sailed are many and exotic, but the most autobiographically Joycean spot he has touched is precisely the Adriatic shore in Trieste, as we learn when he refers to a knifing in a Triestine brothel. The episode, briefly sketched, is part of a discussion on the "passionate temperaments" of the peoples of the South, notably the Spaniards, that Bloom considers all "washed in the blood of the sun" (Joyce 1986: 521). Significantly, of all "the fair share of the world" seen by Murphy, what interests Bloom, married to the Mediterranean Jewess Molly, is the rock of Gibraltar; when prompted by Bloom's question, Murphy disappointingly answers in an elusive way and adds "I'm tired of all them rocks in the sea (…)" (*Ibid.*: 515).

It is in the midst of Bloom's "woolgathering" that we find one of the most significant statements of the whole novel for the construction of a discourse about the sea, which helps to define the positive, enabling vision that is associated with him. As often with Joyce, the resonant metaphor is paired with the platitude:

> Nevertheless, without going into the *minutiae* of the business, the eloquent fact remained that the sea was there in all its glory and in the natural course of things somebody or other had to sail on it and fly in the face of providence ... (*Ibid.*: 515).

Thus, in the Homeric parallel, Bloom is the true heir to Odysseus' natural ability and resourcefulness as a navigator, as well as to the archetypal Odyssean thirst for what Tennyson's Ulysses calls "the untravelled world", although his nature of quester spans a smaller world, the one a provincial man of commerce could long for in his mental meanderings.

Traditionally, however, the paramount Odyssean quality is cunning, and this should be firstly assigned to Stephen, according to the chronology of the narratives and the continuity of characters from the *Portrait* to *Ulysses*, and yet the whole point about the Homeric frame of references in this respect is that Bloom, and Bloom only, can be heir to the Greek hero's versatility of mind. Despite the often unsuccessful outcomes of his daily feats as a "cultured allround man" (*Ibid.*: 123), what really distinguishes him, rather than cunning and shrewdness, is the gift of a special kind of cautiousness. As R.J. Schork puts it, "In Joyce's epic universe, Leopold Bloom, like his Homeric analogue, is outstanding for his invariable 'circumspection'" (Schork 1998: 83).

It is precisely this quality, his capacity to traverse the whirlpools of the quotidian, the Scylla and Charybdis of modern life, and adapt to the changing circumstances and to an essentially hostile environment that could recall and well fit, in my opinion, the Greek notion of *metis*, that type of ready, alert intelligence applied to a practical context, as opposed to the purely speculative mind, which for the Greeks informed all aspects of life.

In a seminal study published in 1974, *Les Ruses de l'intelligence - la métis des Grecs*, the French scholars Marcel Detienne and Jean-Pierre Vernant explored the relevance of this "mental category" and "mode of knowledge" (1984: 34, 36) that comprises a complex mixture of mental attitudes and behaviours combining intuition, wisdom, cleverness and versatility, alertness and the capacity to overcome hostile forces in order to reach the goal. Originally, *metis* is a prerogative of hunting, war and, above all, of fishing, hence of the populations living by the sea. Needless to say, Odysseus is the "human incarnation" of *metis* in the Greek world. It would be tempting, therefore, albeit somewhat paradoxical, to consider Bloom, the somehow mediocre, middle class modern Everyman, so often clumsy and scarcely considered by his acquaintances and friends, as a champion of this peculiar brand of intelligence and as a counterpart to Odysseus, the slyest, shrewdest of the Greeks. And yet the parallel can be maintained as long as we hold to the Modernist paradigm of the deflated rewriting of the Homeric epic. Thus Odysseus' epithet of *polùtropos* (multiplex, versatile, shrewd), can find its equivalent in the "allround man" Bloom. The Greek gift of *metis* thus evolves and metamorphoses in Bloom into a flair for sharing in the "allroundness" of a human, social and natural microcosm that naturally expands into the macrocosm.

The association of the sea with a civilization that privileges intelligence and "cunning", as defined by Detienne and Vernant, and that would seem to identify Stephen as the seafaring hero in *Ulysses*, ultimately appears reversed. His "cunning", vaunted at the end of the *Portrait*, remains sterile since his intellectual framing of reality, his highly speculative mind which is perpetually at odds with his context of provenance, are not those of the true Homeric hero, curious, bold, eager to move on. While Stephen the hydrophobe leaves his father's house "to seek misfortune" (Joyce 1986: 506) and embarks on a round trip journey at sea, only to return to where he left from, Bloom ceaselessly sails the sea of life in his wanderings of the doubly watered Dublin, crossed by the Liffey, surrounded by the Irish sea, infused with the perfumes of the Orient.

Finally, the treatment of the watery imagery and of the sea motif in relation to an insular/Irish and to a universal/"Mediterranean" context can be further illuminated by the issue of gender, which revolves around two important links. The first connects water as a natural element to language, and is metaphorised in the *Portrait* by "the liquid letters of speech, symbols of the element of mystery" which flow forth over Stephen's brain (201). The second is the nexus between Ireland (traditionally represented as a woman) as a feminine island and its "masculine" sea.

Much has been written on the subject of the femininity of Joyce's writing, of Bloom and of Molly's *jouissance*; basically, however, the acknowledgement of the heterogeneity and multifariousness of experience is gendered as female, its refusal as male. To quote an authoritative Irish Joycean, Seamus Deane, in his introduction to *Finnegans Wake*,:

> Language, which is fissiparous, lends itself to the heterogeneous, prismatic nature of experience. To permit or to acknowledge such heterogeneity is to be feminine; to censor it is to be masculine. Women are forever the Greeks in an implacably male and Roman world. (Deane 1992: XXIX)

This gendered contrast between an acceptance of difference and chaos and its refusal, or at least a resistance to it, can also be regarded as an informing principle of *Ulysses*, as well as of Joyce's writing, and a crucial asset of his treatment of the question of history. In James Fairhill's words the "creative-deconstructive power of language as a source of freedom from history" (1995: 250), which dismantles the boundary between history and fiction (and myth), is in fact associated to the creative power of nature; the "creative mode" embodied in Ulysses is based on a "feminine generative principle", "incompatible with phallocentric mastery"(*Ibid.*: 252).

It is in "Proteus" that this nexus is fully developed, with Stephen's musings on the transformations of all things turning into linguistic virtuosity, so that, along with the imaginative sea change of the mother, the drowned man and Lycidas, language itself and linguistic creativeness metamorphose. Interestingly, the transformational power of the sea is gendered as feminine and maternal in the first chapter, "Telemachus", by Buck Mulligan:

> Isn't the sea what Algy calls it: a great sweet mother? The snotgreen sea. The scrotumtightening sea. *Epi oinopa ponton.* Ah, Dedalus, the Greeks! I must teach you. You must read them in the original. Thalatta! Thalatta! She is our great sweet mother. (Joyce 1986: 5)

Again, shortly after, Mulligan (ostensibly a sea-lover), calls it "our mighy mother" (*Ibid.*: 5), thus introducing the leitmotif of Stephen's dead mother. Both definitions are literary quotes, the former from A.C. Swinburne's "The Triumph of Time" which reads "I will go back to the great sweet mother,/mother and lover of men, the sea", the latter a favourite phrase for nature of the Irish poet and philosopher A.E. (Gifford 1988: 15).

In "Proteus", on the other hand, the gendering of the sea turns mas-

culine in Stephen's thoughts, as the correspondence with the Homeric figure of Proteus is made explicit in the epithet of "Old Father Ocean" (*Ibid.*:
42). It could be argued that it is not irrelevant that the cognitive
acknowledgement of the sea as a maternal, feminine and benign force
should come from a character, Mulligan, who is clearly at ease with the
marine element – he is a good swimmer and even bathes in "Telemachus"
while Stephen, the hydrophobe fascinated but also scared by the immense
metamorphic and creative power of the sea, addresses it as a fatherly archetype, referring to the mythological Proteus, while he is fighting against
the Proteus of the mind. Stephen also refers to the Irish analogue, Man-
anaan Mac Lir, the Irish god of the sea, who had Proteus' ability for self-
transformation using the metaphor "the steeds of Manaanan" for the waves.
If we proceed by associations, according to the gendering of the semantic
oppositions discussed above, Stephen, unlike his friend, seems to imagine
the sea as masculine, unfathomable like Proteus, and threatening, in accordance with the time-honoured stereotype of Ireland as woman – see the
old milkwoman in "Telemachus" – and of the island itself as feminine.

Thus, from the earlier to the later texts, and from the contrasting attitudes of Stephen and Mullingan first, and later of Stephen and Bloom,
a transition can be traced from an essentially insidious, negative connotation of the sea as a symbol of historical entrapment to a mythical, universal, and "natural" conception of the sea as the originating principle of
life: a conception that could be more easily associated with the calmer,
more benign and fecund Mediterranean, which is somehow epitomised
by Molly's final reveries of Gibraltar in "Penelope", than with the cold,
isolating, "wine-dark" Irish Sea. More specifically, these contrasting fig-
urations reveal a transition between the insular sea of Ireland, which somehow holds Stephen's aspirations hostage, and the imaginative sea "in all
its glory" that Bloom associates with his wife's Mediterranean origins and
which he navigates in his mental meanderings, as we saw in "Eumaeus".

The marine image that closes *Ulysses* is the sea Molly remembers
watching during the most sensual moments of her life, from the rock of
Gibraltar with her first love and from the hill of Howth the day she first
gave herself to Bloom. Thus Molly, whose overall fluidity, corporeal, mental, emotional, is thoroughly celebrated in "Penelope", achieves in her
stream the fusion of these two ideal seas in that of her youth: "O the sea
the sea crimson sometimes like fire" (*Ibid.*: 643). At last the opaque, unfathomable wine-dark has turned into a glorious crimson.

Bibliography

Bakhtin, Mikhail, 1981, *The Dialogic Imagination: Four Essays*, ed. by Michael Holquist, trans. Caryl Emerson & Michael Holquist, University of Texas Press, Austin.

Blamires, Harry, [1966] 1996, *The New Bloomsday Book* (Third Edition), Routledge, London.

Day, Robert Adams, 1996, "Joyce's AquaCities", in Morris Beja - David Norris (eds), *Joyce in the Hibernian Metropolis*, Ohio State University Press, Columbus, pp. 3-20.

Deane, Seamus, 1992, "Introduction" to James Joyce's *Finnegans Wake*, Penguin, London, pp. vii-xlix.

Detienne, Marcel - Jean Pierre Vernant, 1974, *Les ruses de l'intelligence – La mètis des Grecs, Flammarion*, Paris, Italian translation 1984, *Le astuzie dell'intelligenza nell'antica Grecia*, Laterza, Bari.

Dettman, Kevin J.H., 1996, *The Illicit Joyce of Postmodernism. Reading Against the Grain*, University of Wisconsin Press, Madison.

Fairhill, James, 1995, *James Joyce and the Question of History*, Cambridge University Press, Cambridge.

Gifford, Don - Robert J. Seidman, 1988, *Ulysses Annotated. Notes for James Joyce's Ulysses,* University of California Press, Berkeley.

Hagena, Katharina, 2002, "'Steamers from Mediterranean' – Seafaring in *Ulysses*". Unpublished paper presented at the Trieste 2002 International James Joyce Symposium, courtesy of the author.

Hart, Clive (ed.), 1969, *James Joyce's Dubliners. Critical Essays*, Faber, London.

Joyce, James, [1914] 1987, *Dubliners,* Grafton, Collins, London.

——, [1916] 1989, *A Portrait of the Artist as a Young Man*, Grafton, Collins, London.

——, [1922] 1986, *Ulysses* ed. by Hans Walter Gabler, Penguin, London.

——, [1939] 1992, *Finnegans Wake*, Penguin, London.

Kenner, Hugh, 1980, *Ulysses*, George Allen & Unwin, London.

McCourt, John, 2000, *The Years of Bloom. James Joyce in Trieste, 1904-1920*, The Lilliput Press, Dublin.

Schork, Robert J., 1998, *Greek and Hellenic Culture in Joyce*, University Press of Florida, Gainesville.

Tindall, William York, [1959] 1970, *A Reader's Guide to James Joyce*, Thames and Hudson, London.

Part III

SO CLOSE, SO DISTANT:
THE OTHERNESS OF THE MEDITERRANEAN

THE PRIMITIVES ARE UPON US:
JANE HARRISON'S DIONYSIAN MEDITERRANEAN

Giovanni Cianci

If I start my paper by mentioning John Ruskin, it is not because I wish to pay homage to the international conference that we held here in Milan on "Ruskin and Modernism" nine years ago[1]. It is simply because I cannot find, to start with, a better and a more fascinating distinction about the North and the South (or the North and the Mediterranean) than the one we find in the justly famous chapter "The Nature of Gothic" in the *Stones of Venice*. A distinction which appears to have become foundational for almost every Victorian critic or artist.

Notwithstanding this, there were intellectuals in the next generation, when Ruskin was still alive, who began to think of the South in totally different terms from those of Ruskin. One of these was Jane Harrison, a famous Cambridge classicist and anthropologist who came under Nietzsche's spell. In fact it is the massive presence of Nietzsche which accounts for her vision of the Mediterranean civilization represented by ancient Greece. Strangely enough, though she proclaimed herself "a disciple of Nietzsche"(Harrison 1925: 11-12), her name is never mentioned in the specialist studies devoted to the reception of the German philosopher. Harrison does not figure, for example, in Patrick Bridgwater's *Nietzsche in Anglosaxony* (1972), nor in John B. Forster's *Heirs to Dionysus* (1981). David S. Thatcher's *Nietzsche in England, 1890-1914*, confines himself to

---

[1] See the proceedings in Toni Cerutti (2000) and Giovanni Cianci and Peter Nicholls (2001).

citing Harrison and her colleague Francis Cornford among the first who "perceiv[ed] the originality of Nietzsche's work"(Thatcher 1970: 125), and that's all.

In fact, as was generally not the case with other Nietzschean enthusiasts, Harrison could read the philosopher directly in the original, as most classicists did.

From the volumes or essays devoted to the analysis of Harrison's personality and work[2], nothing much is gathered; or, if anything, it seems in any case inadequate to take stock of the impact Nietzsche had on her. As it comes out in one of her letters, by 1909 Harrison had "just been re-reading" Nietzsche's *Die Geburt der Tragödie* (Schlesier 1991: 171; 1990: 127-141). Indeed, without Nietzsche, it would be difficult to explain Harrison's receptiveness to modernity. And particularly to account for her assent to the revolution brought about by modernist continental movements and their imperative – in Pound's words – to "Make it New!". It was the revolution which was specially at work in England, on the London scene, in the first two decades of the twentieth century.

Harrison's pupils and colleagues (Cornford 1937: 408-9) confirm that her lessons on Greek drama were very far from showing formal, academic detachment. In her lecture on Orphism, she resorted to the use of special arrangements in a darkened room meant to be awe-inspiring: dramatic lighting, and sound effects – "bull roarers" and drums hidden in the background with a view to recreating the sound of pagan worship (Schlesier 1990: 133). This was done in order to create an experience of pathos, somewhat along the lines of what the *Birth of Tragedy* had described apropos of the audience of the primitive community, which was supposed to lose its identity under the ecstatic effect of the tragic event. The aim was to evoke the dark substratum from which the play sprang, trying to capture by way of sounds and images the sacrifical valence of the text, its archaic pathos.

[2] See for instance Robert Ackerman (1972); Sandra J.Peacock (1988); Hugh Lloyd-Jones (1996); Mary Beard (2000), Annabel Robinson (2002).

Nietzsche's vision of the South, which coincided to a large extent with his vision of ancient Greece, was totally dissimilar to the one we encounter in Ruskin's "The Nature of Gothic". To resort to Nietzsche's well-known polarity, if the radiant, luminous Apollonian, in terms of landscape, may partly coincide with Ruskin's famous description of

> ... great peacefulness of light, Syria and Greece, Italy and Spain, laid like pieces of a golden pavement into the sea-blue

the Dionysian is certainly not what the author of the *Stones* word-painted as a landscape:

> ... beaten by storm, and chilled by ice-drift, and tormented by furious pulses of contending tide (Ruskin 1905 X: 240)

and inhabited by

> ... plants that have known little kindness from earth or heaven, but season after season have had their best efforts palsied by frost, their brightest buds buried under snow, and their goodliest limbs lopped by tempest. (*Ibid.*: 241)

In Victorian and Edwardian culture – not actually very different from contemporary continental culture – the beautiful coincided, according to the idealized and aestheticized *fin de siècle* Hellenism, with the notion of proportion, equilibrium and harmony: the ideal, classical beautiful, had to be solar, harmonic, serene and rational. This idealistic, transfigured and sedate notion of the beautiful, went back notoriously to Winckelmann, about whom Walter Pater had written a penetrating essay in 1867. It was Winckelmann who famously lauded ancient masterpieces for being characterized by "noble simplicity" ("edle Einfalt") and "quiet greatness" ("stille Grösse").

Furthermore, let us remember that in England, when Harrison was at work, there still lingered on, either in criticism or in the emulative creative sphere, the "echoes" of Phidian marbles: the Parthenon's friezes brought to England by Lord Elgin and on show at the British Museum.[3]

---

[3] See Vincenzo Farinella and Silvia Panichi (2003).

Think also of the ancient world and its grand and most elaborate exotic marble settings as portrayed by Lord Leighton (he was described as being "marbellous") reduced, in most cases, to a conventional, ornamental, aesthetically pleasing, illustration (see Pl. 4).

Add to that the vogue, no less widespread, for classico-archeological sumptuous scenery inhabited by idealized figures of artistic grace, painted by the late Victorian Lawrence Alma-Tadema (see Pl. 5, 6, 7 and 8) and his imitators (like John William Godward: see Pl. 9 and 10) who in their works evoked a kind of Hollywood dream of ancient Greece and Rome: a prim and proper, respectable Victorian dream, of course, or in any way not breaking the expected standard of decorum. And again think – to adduce only one instance out of many – of the gracious iconography of the virginal white and demure classicism, perfectly controlled and promotional of propriety, represented by actresses in Greek dress set amid stately columns performing Antigone or Iphigenia: icons of Victorian staged Hellenism during the Edwardian years (Goldhill 2002: 108-177). See Fig. 1 and Fig. 2).

What it is essential to stress is that by the time of Nietzsche's *The Birth of Tragedy*, the secular canonization of the agreeable according to reason, of the harmonious and gracious beautiful (or Hellenism as "sweetness and light" in Matthew Arnold's famous slogan) had already reached exhaustion. For a long time it had been losing its original emotional thrust and discovery which had characterized Winckelmann's euphoric and enthralling undertaking ; in brief it had become a mere arid pedantic *kitsch*, the dead end of sheer vacuous forms, soon to provoke the strong reaction of the avant-garde aimed at deconstructing and rearticulating the beautiful.

What came to the fore was now Nietzsche's call for a renewed Dionysian tragic art, accompanied by a new sensibility which exalted the work of art as necessarily radical to the point of being violent and transgressive.

Here one is reminded of T.E. Hulme's and Ezra Pound's invectives against what they called the "caressable"[4] or what T.S. Eliot termed their

---

[4] "...an artist who has something to say will continually 'extract' from reality new methods of expression, and [...] these being personally felt will inevitably lack prettiness and will differ from traditional clichés" (T.E. Hulme, "Mr. Epstein and the Critics", *The New Age*, 24 Dec. 1913 now in Csengeri 1994 : 259).

Figure 1 – Lady Constance Stewart-Richardson, *The Tatler*, Vol. XXXV, No. 446, Jauary 12, 1910, London, p. 27.

Figure 2 – Maude Odell, *The Tatler*, Vol. XXXV, No. 453, March 2, 1910, London, p. 217.

"pleasantness": " the Georgians caress everything they touch" (quoted by Robert H. Ross 1965:160).

In his *Homage to Sextus Propertius* (1919) Pound (who had already accused Walter Pater of "sentimentalesque Hellenism" (Pound 1914: 109) under Nietzsche's spell, and disguised as the Latin poet Sextus, was eager to bring the "Grecian orgies into Italy (Pound 1984: 34)[5] and would decry the "Out-weariers of Apollo"(*Ibid.*), namely those who perpetuated a worn out classicism which by now turned out to be nothing but (in Eliot's words) "mummified stuff from a museum".

In the same vein, Wyndham Lewis in the pre-war years had been sarcastic against Bloomsbury's aesthetics (and, in particular, the Omega Workshops, the decorative arts company founded by Roger Fry in 1913) because, according to him, their "Idol" was still "Prettiness":

> As to its tendencies in Art, they alone would be sufficient to make it very difficult for any vigorous art-instinct to long remain under that roof. The Idol is still Prettiness, with its mid-Victorian languish of the neck, and its skin is 'greenery-yallery', despite the Post-What-Not fashionableness of its draperies. This family of strayed and Dissenting Aesthetes, however, were compelled to call in as much modern talent as they could find, to do the rough and masculine work without which they knew their efforts would not rise above the level of a pleasant tea-party, or command more attention. (Lewis 1913: 49)

Coming back to the North and South polarity, it is interesting to note that Harrison – starting from her book *Prolegomena to the Study of Greek Religion* (1903) and *The Religion of Ancient Greece* (1905) – locates the rough and brutal tribes of the Hellens in the cold North. From there they invade the warm southern countries of the "mild and peace-loving" Pelasgians, usurping their power. The long-term consequence is the meta-

---

"…you will never awaken a general or popular art sense so long as you rely solely on the pretty, that is, the 'caressable' " (Pound, 1915: 311). Even in the post-war climate of *rappel à l'ordre* T. S. Eliot stated that "In a sluggish society, as actual societies are, tradition is ever lapsing into superstition, and the violent stimulus of novelty is required" (Eliot, 1917: 82).

[5] "I who come first from the clear font/ Bringing the Grecian orgies into Italy".

morphosis of the local chthonic gods – the mystery gods, the gods of the earth, of which Dionysos was an outstanding example – into heavenly gods, expurgating "the cruder monstrosities", eliminating the "vague terror" (Harrison 1915: 203). Hence the orthodox and dominant Olympian hierarchy, bringing with it loss and separation from nature and emotional connectedness with the rest of the world (Peacock 1991: 171). As Harrison would write in her later book, *Themis* (1912):

> The Olympian gods – that is the anthropomorphic gods of Homer and Phidias and the mithographers – seemed to me like a bouquet of cut flowers whose bloom is brief, because they have been severed from their roots. To find those roots we must burrow deep into a lower stratum of thought, into those chthonic cults which underlay their life and from which sprang all their brilliant blossoming. (Harrison 1912: XI)

In the same book, following Nietzsche, she insisted on

> the difference between the Olympian and the mystery-god, between Apollo and Zeus on the one hand and Dionysos on the other: a difference, the real significance of which was long ago, with the instinct of genius, divined by Nietzsche. The Olympian has clear form, he is the *'principium individuationis'* incarnate; he can be thought, hence his calm his *sophrosyne*. The mystery-god is the life of the whole of things, he can only be felt – as soon as he is thought and individualized he passes, as Dionysos has to pass, into the thin, rare ether of the Olympian. (Harrison 1912: 476)

From books like *Themis* in particular, and *Ancient Art and Ritual* (1913), we gather that this meant the creation of a non-primitive, anthropomorphic and irreligious Northern ossified Olympus devoid of all but aesthetic content. To this Northern Olympus, desiccated in that it was cut off "from the very source of their life and being" – that is to say from the emotion of the *thiasos*, the emotion of the group of his attendant whorshippers – to this Northern Olympus, we were saying, which could not command her respect, Harrison attaches a series of strong negative qualities, so much so that one can easily perceive an anticipation of what would become the rationalist and oppressive Victorian establishment (she would call them "the Old Rats", to be later hated by the avant-garde). Whereas the positive, original *daimones* – namely the impersonal pre-deistic forces not yet crystallized into full-blown gods, the Chthonians – would "promote fertility and renew social unity by feasts of the annual cycle", "the all too perfect, magnified, official and canonical Olympians" who professed restraint (Schlesier 1991: 208, 210) were "a late and decadent development, the product of intellectual reflection, *objets d'art*, known, not felt and lived, desiccated and dead, chill and remote" (Robinson 2002: 228), deprived of that numinous dimension that could inspire awe. The Olympians "translated into stone" born "of pressing human needs and desires [...] pass into the upper air and dwell aloof, spectator-like and all but spectral"(Harrison 1913a: 203).

Not surprisingly for Harrison – a phenomenon deplored by her academic colleagues – the adjective "Olympian" came to have derogatory connotations.

In her diatribe against "the reverend Olympians"(Harrison 1915: 200) for whom Harrison felt an "intemperate antipathy"(Harrison 1912: VIII), Harrison's special targets are the luminous and rational Apollo[6] defined as "parvenu", "woman-hater"(Harrison 1991: 92, 394) and "ill-mannered prig"(Schlesier 1991: 216), and, at least for a certain period, Zeus whom Harrison characterizes as "the archpatriarcal *bourgeois*", "the prototypical usurper of older cults", and as an "impostor"[7]. According to Harrison, Zeus who "is always boasting that he is the Father and Councilor remains to the end an automatically explosive thunderstorm".[8]

With such abuses in mind, here one can't help recalling the would be dithyrambic Ezra Pound in his poem "Salutation the Second"[9] attacking the Establishment, when he invited his "impudent" songs to:

Go, little naked and impudent songs,/Go with a light foot!/ (Or with two light feet, if it please you!)/Go and dance shamelessly!/Go with an impertinent frolic!/Greet the grave and the stodgy/Salute them with your thumbs at your noses./ [...] Go! And make cat calls!/Dance and make people blush, /Dance the dance of the phallus/and tell anec-

[6] "Apollo has more in him of the Sun and the day, of order and light and reason" *Themis*, 443, (quoted by Schlesier 1991: 215).
[7] In *Prolegomena* (quoted by Schlesier 1991: 215-216).
[8] *Themis*, 455 (quoted by Gloria Cereda 1994-1995: 136).
[9] Composed in Oct.1912, collected in *Lustra* (1916).

dotes of Cybele!/Speak of the indecorous conduct of the Gods!/ (Tell
it to Mr.Strachey)/Ruffle the skirts of prudes... (Pound 1963:94-95)

Definitively positive and altogether different is another God to whom
Harrison felt very sympathetic indeed, namely wild Dionysus, the spirit
of "Life and of Life's ecstasy"(Harrison 1991: 654), the breaker of bonds
and limitations[10], the god of excess, who, in contrast to Apollo, "has more
of the Earth and the Moon, of the divinity of Night and Dreams" (Harrison 1912: 443).

> I had often wondered – Harrison wrote – why the Olympians –
> Apollo, Athena, even Zeus, always vaguely irritated me, and why the
> mystery gods, their shapes and ritual, Demeter, Dionysus, the cos-
> mic Eros, drew and drew me. I see it now. It is just that these mys-
> tery gods represent the supreme golden moment achieved by the
> Greek, and the Greek only, in his incomparable way [...] Dionysus
> is a human youth, lovely, with curled hair, but in a moment he is a
> Wild Bull and a Burning Flame. The beauty and the thrill of it! (Har-
> rison 1915: 204-5)

It is highly interesting to see the result of Harrison's updating of Nietzsche which occurs in her production in 1909-1912, at a time when
European literary and artistic avant-garde appropriated the German
philosopher on a grand scale (Pompili 2000). Harrison removes one of
the term of the Nietzschean formulation, that is the "Apollonian"component (the enlightened and sane dimension, which is on the contrary so
essential in Nietzsche's dichotomy) and exalts, as most modernist artistic and literary movements did, its opposite, the dark, disturbing and
frenzied Dionysian impulse.

But of course Harrison was not Filippo Tommaso Marinetti. She was
born in 1850 and though she was "eager to be in touch with all modern

---

[10] "...a breaking of bonds and limitations and crystallizations, a desire for life rather
than of the reason, a recrudescense it may be of animal passions" (Harrison 1991: 444).
On this aspect see Park McGinty (1978: 217, n.52): "interestingly, Harrison respect-
fully derived the conception of Dionysos as the breaker of barriers from Nietzsche's *The
Birth of Tragedy*."

movements" as she wrote in her *Reminiscences* (Harrison: 1925: 11-12) and a heretic – she actually promoted and founded the Society of "The Heretics"in Cambridge in 1909 (Robinson 2002: 233), – she could not help being a Victorian, although a rebellious one[11]. Thus, in her reading, despite what was perceived as her corybantic Hellenism, the wild and violent Dionysian impulse is rather downplayed since it has somehow to come to terms with the social dimension of community, thus losing the terror and horror under its veneer. The result is that in her interpretation Dionysos turns out to be a bit domesticated for it is seen not as the "patron of exultation" and "intoxication" but of "communion". (Schlesier 1991: 215)

"Communion". Here we find Jane Harrison drawing heavily on what hostile contemporary critics of her work called "French emotionalism", that is to say, the philosophy of Bergson's *élan vital*. But we have no space here to go into that[12]. What we can briefly point out is another French source of inspiration for Jane Harrison, in the movement Unanimism created by Jules Romains and with creative followers such as the poets René Arcos and Charles Vildrac. The most eloquent document in this respect is Harrison's paper on "Unanimism"[13], read at the Society of the Heretics in November 1912. This paper is persistently overlooked by the interpreters of her personality and vision, yet in my view it illuminates, to a far greater extent than her other writings, Harrison's early participation in the spirit of the avant-garde which informs so many other pages of her scholarly papers.

Harrison is fascinated by the Unanimists because "Like all living movements, Unanimism is positive, it affirms rather than denies"(Harrison 1913b: 5) and she likes them because "they protest against the undue sway of the traditional" (*Ibid.*: 6). "We as Heretics" – proclaims

---

[11] Bertrand Russell in his *Diary* (19 October 1903) among other Harrison's "faults" listed a "general restlessness and rebellion" and a "lack of self-control"(Robinson 2002: 131). Another aspect of her rebellion was her feminist militancy: on this see Arlen (1996).

[12] Bergson's concept of "durée" in *L'évolution créatrice* (1907) occurs frequently in *Themis*. See Schlesier (1991: 196, n.177).

[13] Published in the next year as *Unanimism – A Study of Conversion and Some Contemporary French Poets*, Cambridge, The Express Printing, 1913, collected in *Alpha and Omega*. Henceforth quotations are from the 1913 edition.

Harrison – "should, I think, expend our sympathies, not mainly on orthodoxies which stand stiff and secure in their own traditional buckram, but on young movements, like Unanimism, just trembling into life." (*Ibid.*: 6-7). She admires them because they reject academies of any kind and because they are very far from being averse to the primitives. In fact they proclaim "*il faut des barbares, des fauves*" (*Ibid.*: 7). She feels sympathetic to them not only for singing "of their own personal thoughts and emotion", for expressing "their own reactions, not those reactions handed down to us by others and labelled canonical and respectable." (*Ibid.*) but also for insisting on the shared emotions, interconnectedness and epic qualities of contemporary life (in the exciting scenarios of big cities[14], crowds, machines). She feels attracted to Unanimism since it expresses the joy of a collective experience in the breaking down of individual identities which become submerged in the spirit of the group in which "Individuality seems somehow submerged, partitions are broken down and there is a boundless sense of escape and emancipation from self " (*Ibid.*: 23) and "Oneness, the Individual Soul is lost in the All." (*Ibid.*), being released "from the prison house of self" (*Ibid.*: 26).

Again, Harrison is very fond of the Unanimists because they feel the need for emotion, unity with vital forces and life, "Union, Affirmation of Sympathy, Inclusion" (*Ibid.*: 5) and fusion: a vitalistic trend compounded of *élan vital* and Dionysian impulse. Harrison feels drawn to them because they seemed to herald the return of the dark gods, the return of the primordial. In this way Harrison reduces the distance between the atavistic, ritual Greeks and contemporary artistic movements eager to be released from the *fin de siècle* Ivory Tower and become one with the group emotion, with community.

It comes then as no surprise if Harrison agreed with the general tendency of what she names "Expressionism", i.e. the new movements in the arts after the Impressionists, "no matter by what name they call themselves, which believe in the expression "and communication of the artist's emotion" (Harrison 1913a: 232). Harrison welcomes in particular a rad-

---

[14] Harrison significantly felt no nostalgia for the country and the sentimentalized simple rural life as many contemporaries (and the Georgian poets in particular) did: "I can remember the time when life lived outside of London seemed to me scarcely life at all" (Harrison 1913b: 13).

ical movement under which both Nietzsche and Bergson loom large: Futurism. After taking stock that a new "movement towards or about art is all alive and astir among us. We have new developments of the theatre, problem plays, Reinhardt productions, Gordon Craig scenery, Russian ballets"(Harrison 1913b: 208), Harrison takes into consideration the vociferous Italian avant-garde. In fact Harrison was one of the few in 1912 who knew about Marinetti. In *Ancient Art and Ritual*, distinguishing herself from her friends of the Bloomsbury set (Clive Bell and Roger Fry) who were hostile to Futurism[15], and anticipating Wyndham Lewis's judgment that "Marinetti's services , in this home of aestheticism [England], crass snobbery and languors of distinguished phlegm, are great." (Lewis 1914: 29-32), she wrote:

> ... the Futurists are in the main right. The emotion to be expressed is the emotion of to-day, or still better to-morrow. [...] We are not prepared perhaps to go all lengths, to 'burn all museums' because of their contagious corruption, though we might be prepared to 'banish the nude for the space of ten years'. If there is to be any true living art, it must arise, not from the contemplation of Greek statues, not from the revival of folk-songs, not even from the re-enacting of Greek plays, but from a keen emotion felt towards things and people living to-day, in modern conditions, including, among other and deeper forms of life, the haste and hurry of the modern street, the whirr of motor cars and aeroplanes. (Harrison 1913b: 236-7)

We hear a lot about the connections between Jane Harrison and Bloomsbury[16]. Of course there are connections since Harrison liked to be in touch with intellectuals and artists. But there are significant dis-

---

[15] The Futurists were not on show in Roger Fry's second Post-Impressionist exhibition held in London in 1912. Both Clive Bell and Roger Fry – writes J.B. Bullen – "dismissed Futurist work as unworthy of inclusion in a modernist show". See his Introduction to Clive Bell's *Art* (Bell 1987: xxvi-vii).

[16] Harrison had a remarkable impact on the literary avant-garde. See Carpentier (1988). We are informed that "during the last two decades of Queen Victoria's reign, which Harrison spent in London beginning with her thirtieth year, she was one of a circle of artists and writers (Edward Burne-Jones and Walter Pater were also members) who devoted themselves to an 'aesthetic' enthusiasm for Greek classicism in the footsteps of Shelley and Keats" (Schlesier 1990: 129).

agreements, too. One of these, with which we end the essay, is very telling. Harrison was much more adventurous and radical than the Bloomsbury set. Understandably, having given the Dionysian, vitalistic, impulse a new lease of life, she did not agree with Clive Bell's influential notion of "significant form", according to which:

> to appreciate a work of art , we need bring with us nothing from life, no knowledge of its ideas and affairs, no familiarity with its emotions. Art transports us from the world of man's activity to a world of aesthetic exaltation. For a moment we are shut off from human interests; our anticipations and memories are arrested; we are lifted above the stream of life. (Bell 1963: 25)

In an essay of 1914, devoted to confuting Clive Bell's thesis, and anticipating D.H.Lawrence's famous piece on Cézanne[17], Harrison on the contrary stated that to her:

> Art [...]is very like a dead face or a sudden halt in a dance, but the noise of life and its flutter must be there if you are to feel the silence and the binding spell. (Harrison 1915: 217)

She concluded the argument underlining that, very far from being removed from emotions, art included:

> the noise of life and its flutter [...] Form without content is dead. It is the beat of the live bird's wing within the cage that makes form 'significant'. (*Ibid.*: 217, 220)

Bibliography

Ackerman, Robert, 1972, "Jane Ellen Harrison – The Early Work", *Greek, Roman and Byzantine Studies*, Vol.13, No. 2.

Arlen, Shelley, 1996, "For Love an Idea: Jane Ellen Harrison, Heretic and Humanist", *Women's History Review*, Vol. 5, No. 2.

---

[17] See his "Introduction to These Paintings" (1927) in Herbert (1998). On Lawrence and Cézanne see Cianci (2001: 213-29).

Beard, Mary, 2000, *The Invention of Jane Harrison*, Harvard University Press, Harvard.

Bell, Clive, [1914] 1987, *Art*, ed. by J.B. Bullen, Oxford University Press, Oxford.

Bridgwater, Patrick, 1972, *Nietzsche in Anglosaxony*, Leicester University Press, Leicester.

Carpentier, Martha, 1988, *Ritual, Myth and the Modernist Text: The Influence of Jane Ellen Harrison on Joyce, Eliot and Woolf*, Opa, Amsterdam.

Cereda, Gloria, 2004-2005, *Il Rito e la Danza in Jane Ellen Harrison*, Università degli Studi, Facoltà di Lettere e Filosofia, Milano [unpublished thesis].

Cianci, Giovanni, 2001, "D.H. Lawrence e la melità delle mele – O del recupero della corporeità", in Giovanni Cianci - Elio Franzini - Antonello Negri (a cura di), *Il Cézanne degli Scrittori, dei Poeti e dei Filosofi*, Bocca Editori, Milano.

Cianci, Giovanni - Peter Nicholls (eds), 2001, *Ruskin and Modernism*, Palgrave, London and New York.

Cornford, Francis M., 1937, "Jane Ellen Harrison", in *Dictionary of National Biography 1922-1930*, ed. by J.R.H. Weaver, Oxford University Press, London.

Csengeri, Karen (ed.), 1994, *The Collected Writings of T.E. Hulme*, Clarendon Press, Oxford.

Eliot, T.S., [1917] 1963, "Reflections on *Vers Libre*", in *Selected Prose*, ed. by John Hayward, Peregrine Books, Harmondsworth.

Farinella, Vincenzo - Silvia Panichi, 2003, *L'eco dei marmi – Il Partenone a Londra: Un nuovo canone della classicità*, Donzelli Editore, Roma.

Forster, John Burt, Jr., 1981, *Heirs to Dionysus*, Princeton University Press, Princeton.

Goldhill, Simon, 2002, *Who Needs Greek? Contexts in the Cultural History of Hellenism*, Cambridge University Press, Cambridge.

Harrison, Jane Ellen, [1903] 1991, *Prolegomena to the Study of Greek Religion*, Princeton University Press, Princeton.

——, 1905, *Religion of Ancient Greece*, Archibald Constable, London.

——, 1912, *Themis – A Study of the Social Origin of Greek Religion*, Cambridge University Press, Cambridge.

——, 1913a, *Ancient Art and Ritual*, Williams and Norgate, London.

——, 1913b, *Unanimism – A Study of Conversion and Some Contemporary French Poets*, The Express Printing, Cambridge.

——, 1915, *Alpha and Omega*, Sidgwick & Jackson, London.

——, 1925, *Reminiscences of A Student's Life*, Hogarth Press, London.

Herbert, Michael (ed.), 1998, *D.H. Lawrence – Selected Critical Writings*, Oxford University Press, Oxford.

Lewis, Wyndham, [1913] 1963, 'Round Robin', in *The Letters of Wyndham Lewis*, ed. by William Kent Rose, Methuen, London.

——, [1914] 1989, "A Man of the Week – Marinetti", in *Wyndham Lewis –*

Creatures of Habit and Creatures of Change: Essays on Art, Literature and Society, 1914-1956, ed. by Paul Edwards, Black Sparrow Press, Santa Rosa, California.

Lloyd-Jones, Hugh, 1996, "Jane Ellen Harrison, 1850-1928", in Edward Shils - Carmen Blacker (eds), *Cambridge Women – Twelve Portraits*, Cambridge University Press, Cambridge.

——, 2004, "Jane Ellen Harrison", in *Oxford Dictionary of National Biography – From the earliest times to the year 2000*, ed. by H.C.G. Matthews - Brian Harrison, Oxford University Press, Oxford.

McGinty, Park, 1978, *Interpretation and Dionysus: Method in the Study of a God*, The Hague Press, Mouton.

Peacock, Sandra J., 1988, *Jane Ellen Harrison – The Mask and the Self*, Yale University Press, New Haven and London.

——, 1989, "An Awful Warmth About Her Heart: The Personal in Jane Harrison's Ideas on Religion", in William M. Calder III (ed.), *The Cambridge Ritualists Reconsidered*, Scholar Press, Atlanta.

Pompili, Bruno (ed.), 2000, *Nietzsche e le Avanguardie,* Graphis, Bari.

Pound, Ezra, 1914, "The Exhibition at the Goupil Gallery", *The Egoist*, Vol. 1, No. 6.

——, 1915, "Affirmations – Jacob Epstein", *The New Age*.

——, [1928] 1963, *Selected Poems*, ed. by T.S. Eliot, Faber & Faber, London.

——, 1984, *Omaggio a Sesto Properzio*, ed. by Massimo Bacigalupo, Edizioni S. Marco dei Giustiniani, Genova.

Robinson, Annabel, 2002, *The Life and Work of Jane Ellen Harrison*, Oxford University Press, Oxford.

Ross, Robert H., 1965, *The Georgian Revolt, 1910-1922*, Carbondale, Southern Illinois Univ., Carbondale.

Ruskin, John, [1853] 1905, "The Nature of Gothic", in *The Complete Works*, ed. by Edward T. Cook - Alexander Wedderburn, 10 vols, George Allen, London, Vol. X.

Thatcher, David S., 1970, *Nietzsche in England, 1890-1914*, Toronto University Press, Toronto.

Schlesier, Renate, 1989, "Prolegomena to Jane Harrison's Interpretation of Ancient Greek Religion", in William M. Calder III (ed.), *The Cambridge Ritualists Reconsidered*, Urbana, Illinois.

——, 1990, "Jane Ellen Harrison", in Ward W. Briggs - William M. Calder III (eds), *Classical Scholarship: A Biographical Encyclopedia*, Garland, New York.

A FOOTNOTE TO CULTURAL HISTORY:
MODERNISM, IMPERIALISM, AND WILFRID SCAWEN BLUNT

Luisa Villa

As underscored by its title, the present paper has less to do with the aesthetic of Modernism than with the cultural history of Anglo-American modernity. It relates to research still in progress, and in its present form provides no more than a few hints as to how contextual material might be taken into account in any reassessment of the relationship between Modernism and empire. As a general premise on this overarching question, I will simply say that I belong to those who do think that direct experience of imperial peripheries was one of the factors which precipitated the British modernist literary mix. Whether we look at Schreiner's *Story of an African Farm*, or at Conrad's *Heart of Darkness*, or at Stevenson's *The Ebb Tide*, or even at some of Kipling's early tales (*The Strange Ride of Morrowbie Jukes*, *The Man Who Would Be King*), the feeling is that colonial encounters produced a perceptive shock, a de-centring of self, a questioning of the logics of imperial expansion, and its inherent gospel of "universal compulsory modernisation" (Parry 1997: 231) – all of which significantly contributed to the turmoil in representational strategies identified as "literary Modernism". This of course does not imply that a coherent anti-imperialist outlook was the result of such colonial encounters (in the case of Kipling, it was rather the opposite); nor does it imply that anti-imperialist arguments and feelings did not co-exist with sexist, classist and even racist attitudes which we today hardly associate with radical anti-hierarchical politics. What I believe is simply that the dramatic impact of exogenous modernisation on the colonial peripheries and their traditional cultures had a marked and perturbing vis-

ibility, which occasionally shook the self-confidence of the colonisers, forcing them – when literarily inclined – to experiment with new forms of
fictional or poetic (self) representation.

Within this conceptual context, it would be very naïve to assume that
modernist literature played a leading role in the dissemination of new
anti-imperialist awareness and attitudes, making the metropolitan readers *see* what was generally obscured by dominant discursive practices.
Though politics coherently inspired by colonial proclivities gained the
upper hand in late nineteenth century Britain, the public debate over the
issues involved was very intense in the 1880s and '90s, and in my experience (I have been working on colonial warfare in the Sudan; see Villa
2004) any sampling of material (drawn from newspapers, periodicals,
books on colonial issues) published around crucial dates and events will
show responses to imperialism ranging from downright opposition to
downright endorsement, with plenty of caveats and qualifications in-be-
tween. The expression of regret at the disappearance of traditional cultures is – for instance – a widespread *topos* in imperialist discourse: as anthropologist Renato Rosaldo (1989) has persuasively shown, such regret
is one of the most common emotional complications in dealing with "subject races" (as Evelyn Baring used to call them), and has little or no connection with actual political resistance to colonial practices. As to criticism of colonial malpractices and atrocities, it was very frequent, particularly among those who actively participated in the so-called civilizing
process, such as the members of the military engaged in operations on
colonial frontiers. Armed with Gatling guns and other state-of-the-art
weapons against swords, arrows, and assegais, when they happened to put
pen to paper, they rarely expressed unalloyed pleasure or conviction in
thus serving as (so to speak) the spearhead of modernisation.

Which is to say: modernist literature set in colonial contexts, subversive as it was of established literary practices, was produced – as a specific, 'elite' variant – in a much larger discursive context wherein imperial politics were variously and endlessly assessed, refracted and refash-
ioned. Of course, it remains a mooted question (which I do not propose
to enter into), whether the "critical edge against the real" of modernist
literature (that is, its potential for ideological critique and political resistance) (Williams 2000: 35) was any sharper than that of other, less sophisticated and less self-conscious, discursive productions that happened
to tread the same colonial ground.

It is as a follow up to this line of argument, then, that I propose here to approach briefly Wilfrid Scawen Blunt (1840-1922), who was surely no modernist, but who did contribute significantly to the connection of the metropolitan centre to its Mediterranean peripheries between the 1880s and 1910s. In so doing he seems to have struck chords generally left untouched by the prevailing Orientalist-paternalist discourse, so much so that postcolonial critics have pointed him out as one positive exception to the despicable and otherwise very stringent rule which governs the field of what can be said, in the West, about Islam and Islamic countries (Kabbani 1994: 95-103; Said 1993: 235-236). Blunt could well have orientalised his Egypt, had he been so inclined: he was a poet; with his wife he translated some fine specimens of ancient Arabian poetry;[1] he wrote essays thereon (Blunt 1896; 1904); and he was a man of taste who had bought himself a house (the beautiful Sheikh Obeyd walled garden) just outside Cairo. And indeed in their early approaches to the East, when they were still travelling "for pleasure", Wilfrid and Anne Blunt seem to have carried with them at least some of the Orientalist's customary luggage.[2] Besides, Blunt was not immune from the tendency to idealise the Arabs (the Bedouin tribes, but also the Mahdist warriors) as representatives of a chivalric tradition threatened by encroaching "progress", and had originally judged urban Egypt (at the time the avant-garde of modernisation in the Eastern world) to be far too westernised and cosmopolitan for his taste. But he welcomed the idea of a progressive Islamic revival as the basis for political and social reform in Arab countries, and soon came to consider Egypt as the cradle of a Moslem renaissance, with Arabi's nationalist movement as its harbinger. His Egypt, then, was not at all a picturesque place of sexual indulgence, sloth, cruelty, despotism, and perpetual political immaturity, nor a timeless repository of ancient

---

[1] Together with his wife, who was a more accomplished linguist than he ever managed to be, he published *The Celebrated Romance of the Stealing of the Mare* (1892), and *The Seven Golden Odes of Pagan Arabia Known as the Moallakat* (1903).

[2] "It is strange how gloomy thoughts vanish as one sets foot in Asia. Only yesterday we were still tossing on the sea of European thought, with its political anxieties, its social miseries and its restless aspirations, the heritage of the unquiet race of Japhet — and now we seem to have ridden into still water, where we can rest and forget and be thankful. The charm of the East is the absence of intellectual life there, the freedom one's mind gets from anxiety in looking forward or pain in looking back. Nobody here thinks of the past or the future, only of the present" (Blunt 1881: 1-2).

lore, but a site of austere simplicity and learning, where opinions were debated and, out of traditional religious concerns, new progressive politics could be shaped quite independently of European tutelage, British or otherwise.

Blunt was, indeed, a bit of an exception. At a time when the future for landed estates in Britain seemed increasingly uncertain, imperial politics, colonial administration and overseas investments provided opportunities for many ambitious or needy patricians, and it is no surprise that aristocrats generally shared in the Imperial ethos (Cannadine 1995: 109-129). To Blunt, the colonies provided neither a job abroad (which he did not need), nor a seat in Parliament (which he courted a couple of times, but did not get). Rather they offered him the sense of a personal "mission" – less benevolent critics would call it a hobby (Sheeran 1987: 153-160) – fuelled by traditional (pre-jingo) patriotic values[3] and by what he liked to imagine as a Byronian bequest of "sympathies in the cause of freedom of the East" (Blunt 1967: 6), a gallant and disinterested devotion to the cause of "liberty" to which his wife and himself were heirs. Admittedly, this was no great intellectual equipment to found a coherent anti-imperialist stance upon, but it did do the job, making Blunt something of a precursor, and much of a fellow traveller, of J.A. Hobson's.

Born into a family of the landed gentry of the South of England, Blunt served in the Diplomatic Service from 1858 to 1869, when he married Lady Anne Noel, Byron's accomplished and very rich grand-daughter and retired from the public service, henceforth leading the life of the leisurely, though restless, landowner – with plenty of travelling abroad, plenty of love-affairs, fox hunting and horse-breeding at home, some poetry writing, and from the early 1880s much vigorous anti-imperialist campaigning.[4] The latter consisted of writing controversial letters to the Press; publishing articles on colonial matters in leading journals as well as virulent polemical pamphlets at his own expense; helping Egyptian exiles and dissidents; shuttling between London and the southern shores of the Mediterranean (but also India and Ireland) and making himself so trou-

---

[3] "... the love of the land of one's birth, the passion of courage which prompted Thermopylae and Bannockburn and the thousand and one battles in defence of their homes against invading strangers" (Blunt 1911: IX).

[4] Blunt's life has been copiously written about. See Edith Finch (1938), The Earl of Lytton (1961), and Elizabeth Longford (1980).

blesome to the British establishment, at home and abroad, that for some years he was positively forbidden to set foot on Egyptian soil, besides being jailed for two months in Ireland; pestering – directly or through his network of influential friends – ministers and functionaries employed in the Foreign Office and in the colonial administration; and in later years editing and re-working into book form his very extensive private journals, which were given to the press as *Secret History of the English Occupation of Egypt* (1907), *India under Ripon* (1909), *Gordon at Khartoum* (1911), *The Land War in Ireland* (1912), and *My Diaries, Being a Personal Narrative of Events, 1888-1914* (1919-1920). They provided an insider's view of controversial episodes of recent history – a view, needless to say, often at odds with official versions.

The crucial decade for Blunt's militant activity were the 1880s, when he found himself drawn to two interlocking political concerns, the Egyptian situation and the Irish question. Decisive in the development of his anti-imperial awareness was the crisis in British foreign politics that culminated with the bombardment of Alexandria, the battle of Tel-el-Kebir and the military occupation of Egypt. The British government's siding with the Turk-Circassian elite gravitating around the khedival court against Egyptian nationalists in 1882 and particularly the role played by the British consular authorities, bent on defending the financial investors' interests in the region and ready to sacrifice to this objective any other consideration, deeply disgusted Blunt. To him, it amounted to a bitter disenchantment as to Britain's "providential mission in the East" (Blunt 1967: 7) – a veritable betrayal of the cause of liberty, whereby Britain's reputation among the Arab nations as "a redresser of wrongs and a friend of the oppressed" was lost (*Ibid.*: 9), Egypt's "great national hope was wrecked" (*Ibid.*: 143) and the work towards an Islamic liberal revival was discredited, to be replaced all over the Moslem world by a fanatical (today we would call it fundamentalist) reaction. Hence, in the very immediate future: the irresistible rise of the Mahdi in the southern Egyptian provinces and the wretched loss of life (British, Egyptian, Sudanese) due to the ensuing war, culminating (for the British) in the fall of Khartoum and the death of Gordon.

What enabled Blunt to experience fully the shock of Britain's first full-scale involvement in the "scramble for Africa" and to decipher its geopolitical consequences so perceptively was his own very peculiar involvement in two very different milieus: on the one hand, the charmed inner circle of Late Victorian high society and high politics, which made him privy to the ruses of diplomacy, the temperamental quirks of statesmen and the compromises of politics; on the other hand, the progressive Islamic intelligentsia in Cairo. The latter connection had developed through his travels in Arabia and his interest in Islamic religious reformation, which led him to study Arabic and to familiarise with members of the Al Azhar, the Islamic University of Cairo, such as Mohammend Khalil, and Mohammad 'Abdu – the future Judge of Appeal and Grand Mufti –, as well as with prominent Islamic intellectuals abroad, such as the journalist Louis Sabunji, who was for a while his interpreter and secretary, or Sayyid Jamal al-Din al-Afghani, the founder of revolutionary pan-Islamism.[5] Through them he gained easy access to the members of the nationalist faction in 1882, and through them, in later years, he kept in touch with current Egyptian events and opinions. This indeed accounts for the intellectual political slant of his view of Egypt – so remote, as I said, from the Orientalist *doxa*. Thus, as early as in *The Future of Islam* (1882),[6] and then in his Egyptian books or in his *Diaries* we would look in vain for the sort of aestheticised, learned approach to Egypt we find in Forster's guide to Alexandria. Blunt's Egypt is a public sphere caught in the early stages of its development, less free, because of Turkish despotism and oppressive British interference, but no less enticing than that

[5] Wilfrid Blunt deserves a chapter all to himself and his relationship with the Arab world in the authoritative Albert Hourani (1980: 87-103). See also Hourani (1983), which devotes lengthy chapters to Jamal al-Din al-Afghany, and Muhammad 'Abdu – the latter being Blunt's closest and most authoritative Egyptian acquaintance. Blunt's relationship with Louis Sabunji is discussed at length in Martin Kramer (1996: ch. 2). Blunt's contacts with the Moslem world are also dealt with by Riad Nourallah in his Introduction to *The Future of Islam* (Blunt 2002: 1-50).

[6] In his Preface to *The Future of Islam* (Blunt 2002: 58), Blunt acknowledged with satisfaction the recent political developments in Cairo, which had given rise to a lively public debate. This revival of political and religious thought, this fledgling public sphere, was what attracted him in the various Eastern regions he visited in the 1880s. Jeddah, the seaport of Mecca, for instance, which he visited in 1880 and which he described as a varied cosmopolitan place, where "even a European stranger [can feel] that he is no longer in a world of little thoughts and local aspirations. On every side the politics he hears discussed are those of the great world and the religion professed is that of a wider Islam…" (*Ibid.*: 62). Or the various Indian provinces which he visited in 1883-1884, where he got enthusiastically involved in local politics as he himself recorded in many a page of his *India under Ripon* (1909).

of the metropolis. It was, in other words, a place in which to live a rich intellectual life, to engage in debate, take sides, make statements, circulate radical newspapers;[7] not at all a place for aesthetic consumption by cultured Western tourists.

The specificity of Blunt's insertion in the imperial arena of discursive production is reflected in the diaristic quality of much of his writings, wherein political analysis and historical narration are intertwined in the minute record of his dense social life. The latter relates his ever frustrated attempts to have his role as mediator between the British establishment and Islamic politicians officially recognised, his sometimes really wild schemes to imagine political, even world-historical missions for himself (as a leader of the movement for an Arabian Khalifate in 1880-81, or as negotiator between Gladstone and the Mahdi over the Khartoum crisis), his frequent misjudgements of people and situations, his liability to overestimate his own role in the shaping of events. Perhaps it was vanity (Blunt – handsome, aristocratic, magnetically charming – was, according to all accounts, quite a peacock) that predisposed him to such merciless self-exposure; but to him it was a question of personal integrity, "in telling things disadvantageous to others" not to suppress his own private blunders (Blunt 1911: VII). And even today his books do make engrossing reading for their documentary value, for their layered textuality and for the insight they provide into the complexities of history – which intersect Blunt's discursive persona, pulling it one way and another, and ultimately baffling any rational attempt at mastering them. The more you know about history in the making, the less – it seems – you manage to grasp why on earth events come to pass.

Such epistemological doubts are, however, wholly imported (by myself), and have nothing to do with Blunt's conscious motives. As in his relentless letter-writing to the press and pamphleteering, in his historical books he simply wanted to expose the truth ("the *verité vraie* as it was known to me"), to rectify "misguided History" (Blunt 1967: XI). As such they must be read as further interventions – motivated by circumstantial

---

[7] In 1911 Blunt was the chairman of the committee that edited the monthly newspaper *Egypt*, where he himself published many articles under various pseudonyms. "After a career of little over a year the paper was forbidden circulation at Cairo, on the ground that it disturbed public order", but it was published and somehow disseminated till February 1913" (Finch 1938: 332).

pressure – in that enlarged, dialogical public sphere which included the metropolitan centre and its peripheries. This is particularly evident in the case of his books on Egyptian matters: *Secret History*, which Blunt had compiled in the 1890s and revised with the help of Mohammed 'Abdu in 1904, might have waited long for publication in Blunt's drawer, had it not been for the political earthquake of 1906-1907, when public indignation at the Denshawai incident brought about Lord Cromer's resignation.[8] And *Gordon at Khartoum* was, quite explicitly, Blunt's meditated answer to the (quasi-official) version of the events included in Cromer's monumental *Modern Egypt*, published in 1908. The withdrawal from the scene of Britain's most redoubtable "proconsul" opened up a new season for nationalist politics in Egypt, thus offering new opportunities for anti-colonial politics at home. Over the Denshawai incident Blunt had published the usual spate of vehement letters, and a pamphlet eloquently entitled *Atrocities of Justice under British Rule in Egypt* (T.F. Unwin, 1906). Indeed, he seems to have considered Cromer's debacle a sort of personal victory. The time seemed to have come at last for his views (corroborated by painful personal experience, and plenty of documentary evidence) to impact on public opinion, and especially on the younger generation.

But did they? Well, I have neither the will nor the necessary information to discuss the overall reception of Blunt's writings in Britain and in the Islamic world, and since Egypt does not seem to have figured very high in the Modernist literary agenda (Ireland might be a different matter, but unfortunately it does not lie in the Mediterranean)[9] I have very

----

[8] The dynamics of the incident are usefully summarised by Rana Kabbani (1994: 101) as follows: "Five English officers camped on a site in the Egyptian countryside, and commenced shooting pigeons at a village called Denshawai. They soon wounded four villagers who sought to refrain them from further shooting of the domesticated pigeons which constituted the village's main livelihood. A fight soon ensued when they remained adamant, and they were wounded by some villagers who surrounded them. One of the British officers, a Captain Bull, overcome at his attempt at flight, died from concussion of the brain and sunstroke. A villager attempted to save him but failed. English soldiers soon arrived on the scene, and seeing the same villager who was tending Captain Bull, they fell on him immediately and murdered him with the butt-ends of their rifles, assuming him to be the killer. Lord Cromer sent an English official advisor to the village, who arrested five bystanders and instantly requested the death-sentence", which was subsequently carried out after a mock-trial.

[9] The Blunt-Lady Gregory-Yeats-Pound connection to Modernism is the aspect of

little to go by, in winding up this short paper of mine.

To my knowledge the Denshawai incident and its consequences inspired only one significant literary effort: Hall Caine's enticing and, at the time, controversial *The White Prophet* (1908) – which, as the author himself remarked, owed something to Blunt's pro-Arabi poem *The Wind and the Whirlwind*. Of course Caine was no rising young modernist and his novel was a popular best-seller, which made it all the more attractive to Blunt, because "thousands of people will read it".[10] As for Egypt in general, the only prominent 'modernist' to spend a significant amount of time in Egypt and write thereon was Edward Morgan Forster, who reviewed Blunt's *Diaries* sympathetically as soon as they came out,[11] but does not seem to have made much of Blunt's *Secret History* (if he read it) while doing his extensive historical research for his Alexandria guide.[12] The ancient past of the city (the "immense ghost city" he had built for himself through his preparatory readings: the city of the Ptolemies and that of the early Christian times)[13] interested him much more than com-

Blunt's life which has been most thoroughly investigated by literary critics. See: Going (1971), Shasheen, (1983); Longford (1987), Faherty (2000); Carr (2000).

[10] See Hall Caine's Letter to Wilfrid Blunt: "I recall the fact that you kindly sent me 'The Wind and the Whirlwind' at the time of its publication. I read your poem again while writing *The White Prophet*, and it may possibly occur that certain of my own passages are coloured by yours ..." (Blunt 1923: 273). The book was a sort of political fantasy, set in the immediate future, where the autocratic decline and eventual resignation of the Consul-General was re-staged, within a plot redolent of memories of Gordon and clearly sympathetic with Egyptian nationalism and Islamic political-religious revival. Blunt was pleased with the novel which, though 'fantastic", "will do good as thousands of people will read it" (*Ibid.*). The Denshawai incident is prominent in Shaw's Preface to *John Bull's Other Island* – whose proof sheets Shaw sent to Blunt in January 1907 (Blunt 1923: 164).

[11] Edward M. Forster's reviews of "The Earlier Diaries (1888-1900)" (1919) and "The Later Diaries (1900-1914)" (1920) are now included in Forster (1996).

[12] Of course, the Alexandria guide-book does not exhaust Forster's engagement in Egyptian matters (his subsequent *Notes on Egypt*, for instance, reveal a much more politicised approach) nor does what I am writing here cover the whole subject of Forster's connection to Blunt, with whom he was corresponding in the late 1910s (Shasheen 1993; and 2004: ch. 4).

[13] "Very often I am happy and for good reasons. Ancient Alexandria – to mention one – is proving a most amusing companion. I'm constructing, by archeological and other means an immense ghost city". E.M. Forster, Letter to S. Sassoon, 3 August 1918 (as quoted in Forster 2004: XXX).

paratively recent events such as Arabi's revolt or the bombardment of Alexandria. Besides, what attracted Forster in Alexandria was the composite make-up of its population and its cultural tradition, its Greek (rather than Arab, or African) component which – in the 1910s – produced "such modern culture as is to be found in her". Alexandria was a city that had always belonged "not so much to Egypt as to the Mediterranean" (Forster 2004: 22), basically indifferent to "national susceptibilities" – and therefore, in Forster's opinion, the "natural foe" of Egyptian nationalists (*Ibid.*: 79, 6).[14] Predictably, these were the very attributes that had made Alexandria less attractive to Blunt. To him the Greeks, the Italians, the Maltese, the Spaniards, the French who had settled in such coastal cities as Tunis and Alexandria were "the mixed rascaldom of Europe" – money lenders, wine merchants, shopkeepers, who contrasted most unfavourably with the great tribes of the inner regions (with "their noble pastoral life [...], with their camel herds and horses, a life of high tradition filled with memories of heroic deeds") (Blunt 1967: 5), and even with the humbler agricultural communities of the Delta. Indeed, in Blunt's Egypt, the Greeks were prominent among such "needy adventurers of the Mediterranean seaboard" in their capacity as (hated) "usurers", involved – together with the Turkish tax-gatherers – in the merciless exploitation of the oppressed *fellahin* peasantry (*Ibid.*: 9). And much as he mourned the bombardment of Alexandria and its destruction due to the ensuing fire and bloody riots, he frankly felt it would have one positive consequence: it would "more completely get rid of the Greeks and Italians" (*Ibid.*: 282).[15]

It seems it was, then, their pronounced nationalist stance that made Blunt's Egyptian books somewhat old-fashioned and only moderately interesting to the younger anti-imperialist and pacifist intelligentsia, especially after the Great War had rendered late Victorian party politics,

---

[14] In later years Forster made the political point of his liking for Alexandria very clear: "The city symbolises for me a mixture, a bastardy, an idea which I find congenial and opposed to that sterile idea of 100% in something or other which has impressed the modern world and forms the backbone of its blustering nationalisms". E.M. Forster, *The Lost Guide* (1956) (rpt. in Forster 2004: 355-356).

[15] It is no surprise that an aristocratic couple such as the Blunts should value pure breeds, "good birth" and unmixed lineage (see for instance Blunt, Anne, 1881: 394, 405).

diplomatic intrigue and colonial warfare obsolete. The war had also cast a slur on competing national allegiances, and the ideal of a super-national government as an overcoming of localised greed and selfishness was foremost among the likes of Leonard Woolf and Bertrand Russell (Brantlinger 1996). And then, in the eyes of sophisticated metropolitan men and women of letters, nationalist resistance to the Empire – no matter how sympathetically regarded in theory – was bound to appear narrow and sectarian, as opposed to more large-minded humanitarian and existential, let alone artistic or cultural, issues.[16]

This generational gap is very evident when we come to consider Lytton Strachey, the one member of the Bloomsbury set who must have read Blunt's books attentively, since he included both *Private History* and *Gordon at Khartoum* in the sources of his chapter on General Gordon in *Eminent Victorians*. Strachey, a pacifist and a conscientious objector, qualified potentially as a sympathetic reader. He probably was, but his personal axe to grind did not lie in the field of anti-imperial politics, but rather in that of morality and religion, with a strong component of inter-generational animosity. The bitter political querelle over the responsibilities for Britain's long, expensive and bloody involvement in the Sudan, which Blunt had had so much at heart, was to him of little or no interest. Strachey represents himself in quasi-Jamesian terms, as a "curious examiner of the past", with a *penchant* for paradox and epigram – a far cry from the earnestness of Blunt's *engagé* political historian. Thus, Gordon's "tragic history, so famous, so bitterly debated, so often and so controversially described" is to him simply full of aesthetic "suggestion" (Strachey 2002: 221-222): *i.e.*, it allows free scope to his gift for elegant recounting, humorous vignettes, and merciless debunking of "eminent" ancestors.

Strachey's distance from Blunt is underscored by the different portraits they chose for illustrating their books. Strachey selected the famous photograph showing Gordon in full uniform as Governor-General of the Sudan. This was not at all representative of Gordon as he emerged from his own writings, or from the large bibliography on the subject which had been available to readers since the 1880s: it was quite simply Gordon at

---

[16] Said is probably right when he complains that third-world nationalism tended to be downrated as "the work of cliques, or of crazy millenarians" with little or no connection with the true wishes of the Oriental or African people (Said 1993: 239).

his most martial, the Victorian militarist icon whose ironical dismantling was Strachey's self-appointed task. Blunt's deliberate choice of a much more subdued, introverted image (a pencil drawing by Edward Clifford, 1889) signals an altogether different intention. Blunt, who knew Gordon personally and had written to him in January 1884 trying to avert his departure for the Sudan,[17] had been one of the very few in Britain who had genuinely rejoiced at the fall of Khartoum – much as he might have regretted the death of the General. Writing the true story of the Khartoum crisis was, to him, "a labour of love as well as of reparation" (Blunt 1911: VIII), an attempt to reclaim Gordon as, basically, a like-minded contemporary, and to place the blame for his misguided mission where it rightfully belonged, that is to say on Evelyn Baring and the British government.

Strachey's need of a target for irony (and not of a character to sympathise with) has, of course, much to do with the narrator's stance in *Eminent Victorians*, which is typical of the modernist avant-garde looking at the Victorian past with their customary patronising attitude. It is the irrepressible need to represent himself as less naïve (or more ironically disenchanted) than his predecessors that led him to overemphasise the general's temperamental eccentricities (for example, his highly conflictual yearning for influence over the public opinion) and alleged personal weaknesses (like his devotion to brandy and soda) at the expense of other, no less intriguing, aspects of his personality: the ideological dilemmas in which he was caught as a professional soldier who had become critical of war and of the so-called civilizing mission – which may appeal to us, as they appealed to Blunt –, or his irreverence towards grandees and politicians,[18] and his impatience with the rhetoric deployed by the press and his fans around his own public persona – which might have attracted Strachey, had his polemical verve been otherwise inclined. It was well known to the late Victorians that Gordon had been neither an uncritical tool of imperial politics, nor a "goody-goody saint" (Stables 1895: 106), nor an

[17] Blunt published his letter to Gordon in his *India under Ripon* (Blunt 1909: 168-69), since it was written during the months he spent in India between September 1883 and February 1884.

[18] As shown, for instance, by the humorous vignettes of prominent British diplomats which he inserted in his Khartoum diaries, a quasi-official document which he managed to smuggle out of the city towards the end of the siege and which was published posthumously (Gordon 1885a: 158, 237).

unreflecting dupe of media-driven "fame". To those who wrote to him praising his "noble work" with the "poor blacks" at the time when he was Governor of Equatoria, he had replied "acknowledging ourselves to be a pillaging horde of brigands"; and at the praise which accompanied military achievements such as his own he caustically commented with words that might have figured as an apt epigraph to *Eminent Victorians*: "Eminent services, etc., are eminent nonsense" (Gordon 1885b: 143, 362). Such stuff Strachey chose to gloss over, in order to create a suitable target for his avant-garde polemic.

Bibliography

Blunt, Anne, 1881, *A Pilgrimage to Nejd*, John Murray, London (facsimile reprint: Century Publishing, London 1985).

Blunt, Wilfrid Scawen, [1882] 2002, *The Future of Islam*, RoutledgeCurzon, London.

——, 1896, "Arabian Poetry of the Days of Ignorance", *The New Review*, XIV, pp. 626-35.

——, 1904 "The Odes of Hafiz", *Living Age* , 242, 2 July, pp. 52-56.

——, [1907] 1967, *Secret History of the English Occupation of Egypt*, Howard Fertig, New York.

——, 1909, *India under Ripon*, Fisher Unwin, London.

——, 1911, *Gordon at Khartoum*, Stephen Swift & Co., London.

——, 1923, *My Diaries: Being a Personal Narrative of Events 1888-1914*, 2 vols, Alfred Knopf, New York.

Brantlinger, Patrick, 1996, "'The Bloomsbury Fraction' versus War and Empire", in Carol M. Kaplan - Anne B. Simpson (eds), *Seeing Double: Revisioning Edwardian and Modernist Literature*, MacMillan, Basingstoke-London, pp. 149-167.

Cannadine, David, 1995, *Aspects of Aristocracy: Grandeur and Decline in Modern Britain*, Penguin, London.

Carr, Helen, 2000, "Imagism and Empire", in Howard J. Booth - Nigel Rigby (eds), *Modernism and Empire*, Manchester University Press, Manchester, pp. 64-92.

Earl of Lytton (The), 1961, *Wilfrid Scawen Blunt: A Memory by his Grandson*, Macdonald, London.

Faherty, Michael, 2000, "'Defeated but not destroyed': the Prison Poems of Ezra Pound and Wilfrid Scawen Blunt", in Helen Tennis (ed.), *Ezra Pound and Poetic Influence*, Rodopi, Amsterdam-Atlanta, pp. 212-223.

Finch, Edith, 1938, *Wilfrid Scawen Blunt 1840-1922*, Jonathan Cape, London.

Forster, Edward Morgan, 1996, *Abinger Harvest and England's Pleasant Land*, ed. by Elizabeth Heine, André Deutsch, London.

——, 2004, *Alexandria: A History and a Guide and Pharos and Pharillon*, André Deutsch, London.

Going, William T., 1971, "A Peacock Dinner: The Homage of Pound and Yeats to Wilfrid Scawen Blunt", *Journal of Modern Literature*, I: 3, pp. 303-310.

Gordon, Charles, 1885a, *The Journals of Major General Gordon at Khartoum*, ed. by A. Egmont Hake (facsimile reprint: Darf, London 1964).

——, 1885b, *Colonel Gordon in Central Africa 1874-1879: From Original Letters and Documents,* ed. by George Birkbeck Hill, Thomas de la Rue, London (facsimile reprint: Kraus, New York 1969).

Hourani, Albert, 1980, *Europe and the Middle East*, Macmillan, London and Basingstoke.

——, 1983, *Arabic Thought in the Liberal Age*, Cambridge University Press, Cambridge.

Kabbani, Rana, 1994, *Imperial Fictions: Europe's Myth of Orient*, Pandora/HarperCollins, London.

Kramer, Martin, 1996, *Arab Awakening and Islamic Revival: The Politics of Ideas in the Middle East*, Transaction Publishers, New Brunswick-London.

Longford, Elizabeth, 1980, *A Pilgrimage of Passion: The Life of Wilfrid Scawen Blunt*, Knopf, New York.

——, 1987, "Lady Gregory and Wilfrid Scawen Blunt", in Ann Saddlemyer - Colin Smythe (eds), *Lady Gregory, Fifty Years After*, Colin Smythe, Gerrads Cross (Bucks), pp. 85-97

Nourallah, Riad, [1882] 2002, "Introductory Essay", in W.S. Blunt, 2002.

Parry, Benita, 1997, "Narrating Imperialism: Nostromo's Dystopia", in Keith Ansell Pearson - Benita Parry-Judith Squires (eds), *Cultural Readings of Imperialism: Edward Said and the Gravity of History*, Lawrence and Wishart, London.

Rosaldo, Renato, 1989, "Imperialist Nostalgia", *Representations*, 26, pp. 107-122.

Said, Edward W., 1993, *Culture and Imperialism*, Chatto & Windus, London, pp. 235-236.

Shasheen, Mohammad Y., 1983, "Pound and Blunt: Homage for Apathy", *Paideuma*, 12 (Fall-Winter), pp. 281-87.

——, 1993, "Forster's Salute to Egypt", *Twentieth Century Literature*, 39: 1, pp. 32-46.

——, 2004, *E. M. Forster and the Politics of Imperialism*, Palgrave, Basingstoke.

Sheeran, Patrick F., 1987, "Wilfrid Scawen Blunt – a Tourist of the Revolutions", in Wolfgang Zach - Heinz Kosock (eds), *Literary Interrelations: Ireland, England and the World.* Vol. 3: *National Images and Stereotypes*, Gunter Narr Verlag, Tübingen, pp. 153-160.

Stables, Gordon, 1895, *For Honour, Not Honours: Being the Story of Gordon of Khartoum*, Shaw, London.

Strachey, Lytton, [1918] 2002, *Eminent Victorians*, Continuum, London-New York.

Villa, Luisa, 2004, "Omdurman 1898: The Representation of Colonial Warfare and Its Perplexities at the Fin de siècle", *Textus*, XVII, pp. 371-384.

Williams, Patrick, 2000, "'Simultaneous uncontemporaneities': theorising modernism and empire", in Howard J. Booth-Nigel Rigby (eds), *Modernism and Empire*, Manchester University Press, Manchester, pp. 13-38.

FORSTER, ALEXANDRIA, MODERNISM:
THE MEDITERANNEAN "AT A SLIGHT ANGLE"

Michael Hollington

First, Michael Haag's *Alexandria: City of Memory*, then Miriam Allott's mighty new edition of E.M. Forster's Alexandrian writings, both appearing in 2004, and now Peter Jeffreys's May 2005 publication *Eastern Questions: Hellenism and Orientalism in the Writings of E.M. Forster and C.P. Cavafy*. These books leave no room for remaining doubt that – to Dublin, Trieste, New York, St. Petersburg, Berlin, etc. – we must add Alexandria to the list of cities, in particular of Mediterranean cities, that figure centrally in the imagination of Modernist writers. It certainly holds a crucial place in E.M. Forster's imaginative development, though it is perhaps only with this recent work that the role in his life and work played by his wartime experience of the place is beginning to be fully understood. I shall try to show here that his two books on Alexandria approach it as one of the key centres of a Mediterranean civilisation felt now doomed to extinction as a result of the stupidity of war. The city's orientation being from the start towards the sea rather than towards the desert, he saw it as part of an international cultural system, its essential character cosmopolitan, its key activities dependent upon 'connectedness' in travel, trade, and the tolerant exchange of cultural influences and ideas. Founded, or rather refounded, by Alexander the Great as an 'universal city', and more or less governed in that spirit by the Ptolemies, right up until Cleopatra, its utopian origins had given way to recurrent episodes of violent intrigue and ideological strife opening the way to successive colonisations and, eventually, to catastrophic war.

Nonetheless, in active opposition to the idea of catastrophe we shall

find, in these texts, the central figure of the idea of passage. "Only connect" had been Forster's motto, of course, in *Howard's End*, and if the
Alexandrian texts tend to show that the ideal of 'connection' has become
virtually impossible to effect in any satisfying way as a result of the fatal
trends in its history, they continue to interrogate and worry the concept
in a profusion of metaphors and other devices. For if Alexandria is oriented towards Mediterranean sea passages, it is also, at its eastern end, a
pivotal hinge leading, particularly after the construction of the Suez canal,
towards the East. It forms, to put it in rough geographical terms, a kind
of half-way house between Forster's early ideal place, Italy, and its later
equivalent, India. But the relevance of the idea of passage to these texts
extends also to their position in Forster's *oeuvre*. I would claim that they,
more than any other work that Forster produced in the relatively fallow
years between *Howard's End* in 1910 and *A Passage to India* in 1924, bridge
the gap between Forster's two masterpieces. Though possibly Forster's
least known works, *Alexandria: A History and a Guide*, published in 1922,
and *Pharos and Pharillon*, dating from 1923, offer, I would argue, far more
convincing pointers to the scope and focus of his last great book than the
earlier *Maurice*, or the fragmentary *Arctic Summer*. They, I believe, constitute the essential passage to *A Passage to India*.

I would add that I see this passage, in crude terms, as a transition from
more-or-less full-blown realism, or at least realism with quirks, in the
early novels, to more-or-less full blown modernism, or modernism with
quirks, in *A Passage to India*. In literary terms, this transition can be traced,
I think, in the important exposure to Cavafy in Alexandria followed and
perhaps coupled with the yet more crucial reading of Proust in the later
stages of the writing of Forster's last novel. These influences help to account for the modernist 'quirks' in these later works, which are charac-
teristically oblique and often slyly hidden or 'byzantine.' Forster's love of
obliqueness is a feature of all his writings (Virginia Woolf described him
as "a vaguely rambling butterfly," and Furbank characterises him as "as
hard to catch or pin down as a butterfly" (Furbank 1979: 56-7); it is of
course most obviously manifest in his essentially ironic mode of writing
throughout his career. But it may be said to increase significantly after
Forster's encounter with Cavafy's (and later Proust's) sentences, dealing
with "the tricky behaviour of the Emperor Alexius Comnenus in 1096,
or with olives, their possibilities and price, or with the fortunes of friends,
or with George Eliot, or the dialects of the interior of Asia," regularly ef-

fecting in the syntactical frame a *passage* "to an end that is always more vivid and thrilling than one foresaw." He learnt from these mentors, in intensified fashion, to stand "at a slight angle to the universe," now that, having studied Cavafy and the Neo-Platonists, he realised that "the poet is more incapable than most people of seeing straight" ("The Poetry of C.P. Cavafy," Allott 2004: 245, 246).

But though we shall soon turn to oblique representations of the idea of passage in the Alexandrian texts, I want to say something about crucial changes that were taking place in Forster's personal and particularly sexual life at that time. It is obvious, of course, in the case of all three of these homosexual individuals, but most acutely in the case of the English writer, that oblique indirectness became a way of life, extending far beyond Proust's famous principle of referring to an 'Albert' through the mask of an 'Albertine.' Whilst writing about Alexandria, Forster was also writing at times in a kind of oblique code about his own personal experiences at that time. One example is the dedication of *Pharos and Pharillon* to "Hermes psychopomp," or 'Hermes the transporter of souls,' mediating between the earth and the underworld, meaning Mohammed el Adl, his Egyptian lover at the time of these texts. Whilst following in the footsteps of another homosexual writer, Whitman at the time of the American Civil War, in his devotion to the wounded soldiers he visited in Army hospitals as a Red Cross 'searcher' trying to gather evidence about men who had died or disappeared in the Dardanelles, he had fallen in love in a rather charming and appealing way with an Egyptian tram conductor active in the transportation of 'souls' about the city of Alexandria.

But Forster did not only mean that by his dedication. He meant primarily that it was Mohammed who had taken him across a vital threshold of adult life, comparable even to that between life and death. Like another great English modernist, the painter Stanley Spencer (himself heterosexual), Forster first experienced the joys of satisfied, reciprocated love and sexual fulfilment at a rather mature age – he was 38 when he wrote to his confidante Florence Barger in June 1917 that he had at last passed through to the other side: "I wish I could convey to you what I feel at this unique time – far from the greatest time in my life, but for me quite new. It isn't happiness: it's rather – offensive phrase – that I first feel a grown up man" (*Letters* I: 257). Having led hitherto a very sheltered suburban life, "full of aunts and tea-parties and little jokes and chocolate hares," in Furbank's entertaining and, as we shall see oddly 'Alexandrian' phrase

(Furbank 1979: 133), and thus far infantilised by his mother – "she is always wanting me to be 5 years old," he complained to Florence Barger in August 1915 – he now began to pursue happiness and self-fulfilment on his own terms, independently of his mother. He had given in to her in 1915 on his proposal to drive an ambulance at the Italian front, which she vehemently opposed, but at the end of that year he would set out on a three-year absence in Alexandria, where Mohammed would later become a major reason not to contemplate even temporary return to England to visit his mother. He had thus decisively moved beyond daring to write about homosexual love, as in *Maurice*, to life-altering experience of it, writing, again to Florence Barger in February 1918 that "it is certainly the most wonderful thing that's ever happened to me – has so outstripped my theories. The whole ending of Maurice and its handling of the social question now seems such timorous half hearted stuff" (*Letters* I: 287).

But it seems plausible, in fact, that Forster may have been at least partly conscious of what lay in store for him in Alexandria, which Murry may have discerned when he remarks half-jokingly that "being a dubious character he has gone off to a dubious city, to that portion of the inhabited world where there is obviously a bend in civilisation" (Haag 2004: 113). Fordonski and others writing on Forster from a 'gay' perspective stress the long-established tradition of "belief that there is a foreign place which would be more appropriate and sympathetic for gays [...] typical for gay literature since Ancient times," with Italy in particular a mythical 'homosexual heaven' (Fordonski 2001: 2, 3). Such myths of Italy figure in subterranean fashion in Forster's early work, and even its geographical position in the Mediterranean is discreetly eroticised in a letter of 1904 describing it as a "glorious limb sliding southward between warming seas" (Furbank I 1979: 126). After falling in love, unrequitedly, with his Indian pupil and friend Syed Ross, the centre shifted eastward, in the direction of India of course, and Mediterranean geography becomes mapped on to South Asian, with India as the new phallic appendage ("has it ever occurred to you," he writes to Malcolm Darling in October 1910, "that the south coasts of Europe and Asia are analogous – three peninsulas, with the chief mountain chain knotted up at the head of the middle one" (Furbank I 1979: 184). In the central image of *Arctic Summer* – an endless summer of midnight sun – utopia seems to have been pushed even further afield, but Alexandria in fact offered at least glimpses of a renewed vision of Mediterannean millennial bliss, most particularly at Montazah, the favourite convalescent home of recovering soldiers. From thence Forster writes almost ecstatically to Goldsworthy Lowes Dickinson of "hundreds of young men [...] and down by the sea many of them spend half their days naked and unrebuked and unashamed," and asks "Why not more of this? Why not? What would it injure? Why not a world like this?" (Haag 2004: 26, 31-2).

Despite, or perhaps even partly because of the severe disappointment Forster experienced in his dealings with Egypt and the vast majority of Egyptians, up to and beyond the meeting with Mohammed (he tended simply to contrast his 'cleanness' with the physical and moral 'mud' of Egyptians in general), we must retain this idea of Alexandria, both as a lost Mediterranean paradise – or one in the process of being lost – and as a potential future utopia, albeit of an reconstructed kind. For we shall uncover in these texts a dual movement – the passage to death, in the passing of a great civilisation, and the passage to life through 'impersonal' natural impulses that have to do with instinct rather than culture.

In turning now to the first of Forster's two Alexandrian books, *Alexandria: A History and a Guide*, we must first stress an essential quirkiness, even absurdity, about the undertaking (and fortuitously, about its publication history) that offers pointers to its oblique purposes and scope. "The only instance in recent years in which a distinguished man of letters has undertaken to present a work of this character," wrote Judge Brinton in the 1930s in promoting a new edition (Allott 2004: xiv), and indeed it is hard to think of many writers as prominent and established as Forster had become who have opted to tackle so prosaic and in Joyce's term 'kinetic' a genre as the guidebook – even though, at the end of the First World War, writerly hands were turning to many strange pumps (Lawrence, for instance, wrote a history textbook for schools). The sense of incongruity is heightened when we note, first, that Forster appears to have begun writing his guidebook in 1917, like *Maurice*, not with any clear idea of publication but "for my own amusement;" second, that no-one in England in wartime could possibly have contemplated a tourist excursion to Egypt, so that the book (described in a letter to Sassoon as "a superior sort of guide") could presumably only have been conceived of as addressed to the upper echelons amongst the recovering soldiers; and third, that most of the sights it describes were invisible. As history as much as guide, Forster insists, it requires a very active, even 'writerly' and thus modernist reader to decipher the ghosts of the Mediterranean past of Alexandria in the meagre remains of the present.

Thus there is a good deal of inherent paradoxicality in the texts, pointing to oblique layers of hidden meaning, the biographical one discernible
enough in the omnipresence, in the practical part of the book, of tram
routes, which ferry the 'souls' of visitors from sight to sight. Further contingent anomalies ensued; first, publication of the text was delayed till
long after its major potential clientele, the wounded officers, had left
Alexandria; most of the copies were supposedly accidentally burnt in a
warehouse fire, and insurance money paid out; then, after the embarrass-
ing discovery that they had not in fact all disappeared, they were *deliberately* burnt to avoid awkwardness with the insurance company! Forster's
humorous sense of life's ironies enjoyed this sequence of events, and in a
talk entitled "The Lost Guide" given at the Aldeburgh Festival in 1956
Forster related his favourite – how he got lost coming out of the railway
station during a return visit to Alexandria on the way back from South
Africa in 1932: "Can you imagine a more humiliating experience for the
author of a guide? Not only wasn't I a prophet in my own city, I couldn't
even find my way about in it" (Allott 2004: 358).

But Forster's devotion to so distinctly minor a literary genre as the
guidebook is in fact one of the keys to the book's strategies of oblique indirection. He is working (like Cavafy, with his sentences about the price
of olives or the dialects of Asia Minor) a vein of indirection that turns
standard notions of significance and insignificance on their heads. With
the concept of a subversive 'minor literature,' as outlined in Deleuze's important book on Kafka, very much in mind, we can readily highlight this
aspect of the book at many points, but perhaps most particularly at the
museum and in the section on the Alexandrian literary tradition. In the
museum, for instance, he exclaims with reference to an exhibit of Alexandrian glassware that "there is more beauty in this little case than in tons
of statues," or, of some terra cotta statuettes found in the tombs of children and women, that "they are the loveliest things in the Museum."
(100) The account of Alexandrian writing is similarly governed throughout by Callimachus's dictum "A big book is a big nuisance." Theocritus
is its characteristic ' big' writer, and his Fifteenth Idyll is quoted to represent the Alexandrian genius for the miniature and trivial. Here two Roman Alexandrian housewives – very much the equivalents in their time
of the Weybridge suburbanites with their chocolate hares – attend a Resurrection festival and hear a "beautiful hymn" to the dead Adonis (the
theme of male beauty again peeping through), but no sooner is it finished

than one of them remarks "all the same it's time to be getting home; my husband hasn't had his dinner and when he's kept waiting for his dinner he's as sour as vinegar." It is through such *madeleines* that the past comes truly alive for Forster, as he comments that "only through literature can the past be recovered and here Theocritus, wielding the double spell of realism and poetry, has evoked an entire city from the dead and filled its streets with men" (Allott 2004: 38).

Thus, those who conclude, or have in the past concluded, that these slim books (or indeed, Forster's slim corpus as a whole) are essentially slight and trivial have missed the dialectical aesthetic of the *multum in parvo*. The Alexandrian texts in fact carry quite large historical, political and philosophical ambitions, anticipating the much deeper, speculative scope of *A Passage to India*, where the example of Proust as a modernist philosophical novelist has clearly taken firm root. Somewhat fancifully, I shall group these around three instances of the syllable 'Med' – the 'med' of 'Mediterranean,' standing for the great cultural tradition of the past, the life of the spirit and the mind; the 'med' of 'Mohammed,' standing for the life of the senses, of friendship and sexual love; and the 'med' of 'mediation,' the peculiarly Alexandrian attempt to find a discourse to 'only connect' the two.

It is this third, major theme of *Alexandria: A History and a Guide*, that I shall briefly highlight here, acknowledging straight away a huge debt to Allott, whose masterly discussion leaves so little to add that above all I simply refer the reader to it. She focusses with particular acuteness on the major section 3 of Part One, 'The Spiritual City,' which finds in the various theological schools a common, Greek-inspired tendency of Alexandria's communities and religions, at least in the most living era of the city's history, to connect spirit and matter through some "intermediate being or beings who draw the universe together; and ensure that though God is far he shall also be near" (Allott 2004: xlvii, 56), and remarks on the 'unusual emotion' that informs it. That intensity of feeling, she rightly emphasises, has much to do with a kind of farewell tribute Forster is making here, under the influence of Cavafy, to the relatively naive Hellenism of his earlier work, but the first additional point we might also bring out – remembering the title of Edward Carpenter's book on homosexual love *The Intermediate Sex* – is how it may also be linked to the subterranean homage to Mohammed as an intermediate messenger from the gods that Forster's Alexandrian writings wish to express. She authoritatively draws

out the significance of Forster's praise of 'bastard' Alexandrian Greek-in-
fluenced Jewry in the figure of Philo – who "by the doctrine of the Mediating Logos, ensured that the deity should be at the same time acces-
sible and inacessible," (Allott 2004: xviii) – and of Alexandrian Neo-Platonism, exemplified by Plotinus and the Enneads, than which Alexandria "produced nothing greater" (Allott 2004: xlvii). But we may perhaps add one or two further points – how for instance Forster interestingly and portentously aligns Alexandrian Neo-Platonism with Indian
religion, even half-jokingly speculating whether Plotinus may have met
Hindu merchants on the Alexandrian waterfront, for "The Christian
promise is that a man shall see God, the Neo-Platonic – like the Indian
– that he shall be God." There is also perhaps a little more to say about
Forster's treatment of Alexandrian Gnosticism, which she finds "almost
equally moving" (Allott 2004: xlvii) – 'pessimistic, imaginative, esoteric
– three great obstacles to success,' he remarks ironically. (Allott 2004:
64). Valentinus is the great figure, once again a moderate and mediator,
seeing God as 'the centre of a divine harmony' who unfortunately produces a declining series of male and female creations ending with the thirtieth female, Sophia or 'Wisdom' who falls from spirit to matter, in a wonderfully imaginative double paradoxical design, not through excess of
hate or envy but through excess of love of the spiritual realm.

Turning now to his tiny second book about Alexandria, *Pharos and
Pharillon*, which occupies a mere fifty-five pages in Allott's edition, it is
not hard to discern its close relationship to the previous work. It too has
a binary structure, this time staying within the axis of historical time
rather than venturing into the spatial dimension essential in a guidebook,
whilst continuing the figure of conducting visitors on a tour of the sights,
its narrator once more clearly imagined as a kind of modern Virgil. It is
organised around two central Alexandrian symbols – the ancient lighthouse, the great Pharos, one of the wonders of the classical world, a triumph of the union of science and intellectual inquiry and power, and its
inferior modern descendant, the little Pharillon – in order more fully to
bring out the decline pattern in Alexandrian history. Forster calls it a
pageant, which is indeed an appropriate metaphor, but I shall treat it here
more as a series of epiphanies or scraps of secular illumination into the
"offence that drove a culture mad," to borrow Auden's phrase: in other
words, it lays bare how violence and strife have undermined a civilisation, and driven it to the point of extinction. One section is indeed enti-

tled simply "Epiphany," perhaps in acknowledgement of Joyce, though the specific reference is to Ptolemy V, surnamed Epiphanes, in 254 B.C.

Unlike the guide, however, which passed pretty much unnoticed, *Pharos and Pharillon*, published by Leonard and Virginia Woolf at the Hogarth Press, was enthusiastically reviewed by a number of contemporaries – John Middleton Murry and Goldsworthy Lowes Dickinson amongst them – and proclaimed as a small tragicomic masterpiece of direct representation in fragments that require active 'writerly' interpretation. Its style is remarkably accomplished, after the manner of Theocritus and Forster's conception of 'Alexandrian' minor literature, and equally justifies Lawrence Durrell's praise of the earlier guidebook as containing "some of Forster's best prose, as well as felicities of touch such as only a novelist of major talent could command" (Haag 2004: 81). Here Forster manages in 'Alexandrian' style to do what he felt he could not do in 1913 at the time of *Arctic Summer*, writing to Forrest Reid that "I must keep myself from trying to look round civilisation […] I haven't the experience or the power" (*Letters*: 200). Here he is able humorously "to look round civilsation" from the very first sentence with its sly 'Cleopatra's nose' litotes – "The career of Menelaus was a series of small mishaps;" huge historical consequences, it is being suggested, flow from trivial mistakes which are shaped and interpreted by later narratives. The rich vein of anachronistic muddling of ancient and modern that sustains many of the best jokes begins with Helen: she is bored at the mouth of the Nile, where Menelaus has brought her after the war of Troy, and Alexandria doesn't look to her like a promising place for a Club Med, for there is "nothing to see upon the island and nothing to eat […] its beaches […] infested with seals." So Menelaus has to try to amuse her, and begins by finding an old man to ask where they are and to whom the island belong, and the replies – it is 'Pharaoh's' or 'Prouti's' – are misheard as it is 'Pharos' the home of 'Proteus.'

We are thus in the realm of Beckett here – 'in the beginning was the pun' – and the same stress continues when Alexandria has been founded by Alexander, first as a Greek colony, but later and more crucially as the kind of harmonious universal city that befits a ruler who is hailed as the son of God. Only there is again a verbal misunderstanding, according to some, for the priest who is supposed to have "saluted the young tourist as Son of God" may have said, or may have meant to say, 'Paidion' (my child) rather than Paidios ('O Son of God'). Forster's version of history here, then, shows a high degree of postmodernist sophistication, being made up of delightfully inconsequential muddles and blurs worked into dubious texts that later carry spurious authority. His own historical 'epiphanies' are conceived of as equally spurious but this time strategically calculated tall tales designed to explode such authority and counter its ideologies.

Still, the Alexander/child/son of god muddle has felicitous consequences, for it establishes the Alexandrian tradition of tolerance and inclusiveness – a kind of ideal "human norm," perhaps, in the memorable phrase from *A Passage to India* that describes Mediterranean civilisation. Forster's 'heroes,' in the 'Pharos' section of *Pharos and Pharillon* very much typify this spirit. Ptolemy V, or 'Epiphany' is a suitable minor one – "the average later Ptolemy is soft," the Guide had informed its imaginary customers (Allott 2004: 29), and 'Epiphany,' in Forster's thoroughly mocked up history of him, is certainly that. He has the advantage of coming to the throne as a child, and his major 'contribution' to Alexandrian history is made at the age of five. He is surrounded by evil scheming usurpers who take on the role of regents, and hide him away. But the people long to see him, and when he is produced call out "Epiphany!", "Epiphany!," and ask "Shall we not punish your mother's murderers?" Forster has of course made up the question and its absurd suburban answer: "Oh yes – oh anything," in order to proffer an epiphany of your authentic soft and malleable Alexandrian: "This was the world, and he did not like it. He preferred his own little circle" (Allott 2004: 206). Yet out of Epiphany's cowardly 'childish' reply peace and harmony are restored for the time of his rule at least.

Philo the Jew and St. Clement the Christian are more major minors, contributing much to the tenor of this civilisation. In the marvellous piece on St. Clement of Alexandria, Forster begins with a great sly mock periphrasis for the Jewish and Christian dispensations, describing them as "assertions [...] made at one time or another in the uplands of Palestine," before introducing Clement as a Christian after his own heart, that is to say, as a well-mannnered and good-natured ironist: "when he attacks Paganism he seldom denounces: he mocks, knowing this to be the better way [...] shrines to the sneezing Apollo and to the gouty and to the coughing Artemis? Ha! Ha! Fancy believing in a goddess with the gout." Once again, in Forster's mocking of Clement's mockery, we hear the demotic of Weybridge and the world of chocolate hares – the accents of his mother, perhaps, a renowned waspish commentator on her entourage. But we learn too how Clement's moderation sustains and reflects the profoundest impulses of Alexandria's Mediterranean civilisation with his "hope that it may pass without catastrophe from Pagan to Christian" – a hope reen-acted most especially in Forster's prose in the superb sentence with which the sketch concludes: "And in that curious city, which had never been young and hoped never to grow old, conciliation must have seemed more possible than elsewhere, and the graciousness of Greece more compatible with the Grace of God" (Allott 2004: 213).

But with St. Athanasius the rot sets in. He wins the battle of the *fili-oque* against Arius at Nicaea in favour of what Stephen Dedalus terms "conjewbangsubstantiality," but after that the exemplary soft 'minor' Em-peror Constantine, who '"so easily got mixed," and wanted the Christians 'to imitate the Greek philosophers, who could differ without shedding one another's blood', asks him to calm down and patch up the theologi-cal quarrel, and he refuses. In Forster's fable, he is 'banished five times' refusing to wash henceforth (again like Stephen Dedalus), and generat-ing endless rounds of grimly comic theological guerrilla warfare – "ani-madversions against washing, accusations of sorcery, complaints that Athanasius had broken a chalice in a church in a village near Lake Mari-out, replies that there was no chalice to break, because there was no church, because there was no village" and the like. Athanasius too marks a pas-sage, but a distinctly negative one – "he weaned the Church from the tra-ditions of scholarship and tolerance, the tradition of Clement and Ori-gen" (Allott 2004: 218).

The "Pharos" half of *Pharos and Pharillon* ends with a section that of-fers a poem by Cavafy, 'The God Abandons Antony,' and since this poem has an important place in my argument, I quote it here in its entirety:

> When at the hour of midnight
> an invisible choir is suddenly heard passing
> with exquisite music, with voices –
> Do not lament your fortune that at last subsides,
> your life's work that has failed, your schemes that
>   have proved illusions.
> But like a man prepared, like a brave man,
> bid farewell to her, to Alexandria who is departing
> Above all, do not delude yourself, do not say that it
>   is a dream,
> that your ear was mistaken.
> Do not condescend to such empty hopes.

> Like a man prepared, like a brave man,
> like the man who was worthy of such a city
> go to the window firmly,
> and listen with emotion
> but not with the prayers and complaints of the coward
> (Ah! supreme rapture!)
> listen to the notes, to the exquisite instruments of
>   the mystic choir,
> and bid farewell to her, to Alexandria whom you
>   are losing.

It will be apparent, first, that the poem (described elsewhere by Forster as "exquisite") is offered to the alert 'writerly' reader absolutely without comment — that will come, on Cavafy as a whole at least, at the end of the second, minor section, "Pharillon." Cavafy thus provides closure to both parts of the texts, and here is the only modern allowed in the company of ancient 'Pharaonic' Alexandrians – a signal honour. And it is clear why – his poem is about the second vital passage in Forster's life during World War One, this one negative, whereas that given by Mohammed is entirely positive. It expresses the ideal of a stoic and dignified farewell to a doomed civilisation, saluting its glories, but moving on, beyond the Mediterranean, beyond civilisation.

Of that, a little more a little later, but for the moment the poem also signals an historical hinge, marking the moment of Alexandria's decline into the 'city of eggs and cotton and onions' as this is chronicled in the book's part two, "Pharillon." The Orient is in the process of being colonised yet again by the English and French in "Eliza in Egypt," which concerns the Fays and Alexandria "in the summer of 1779" when "that city was at her lowest ebb." Eliza is married to the deliciously named "Antony Fay," "an incompetent advocate, who was to make their fortunes in the East" like many colonisers before and since. Yet the sketch announces what will be the important 'third term' of *Pharos and Pharillon* – the natural world, which remains throughout from Menelaus to cotton merchants, as the "Conclusion" underlines: "only the climate, only the north wind and the sea remain as they were when Menelaus, the first visitor, landed on Ras el Tin, and exacted from Proteus the promise of life everlasting" (Allott 2004: 250). The Fays are hell bent on getting to the real Mecca, India, so that as they pass through Suez "her thoughts dwelt on the past with irritation, the future with hope, but on the scenery scarcely at all." It is Alexandria and Suez, the place of the passage itself, attracting once more Forster's favourite adjective of praise in this minor world – "that exquisite corridor of tinted mountains and radiant water" (Allott 2004: 230) – on which his eyes will increasingly be trained, realising the motto from Plotinus at the head of *Alexandria: A History and a Guide*: "to any vision must be brought an eye adapted to what is to be seen."

"Cotton from the Outside" is a brilliant epiphanic descent into hell offered by a Virgil who has inverted Dante's structure in his text. From the outside the shouts and screams from the Cotton Hall ("hall" and "hell" echo each other mockingly here) make it sound like a place of bloody slaughter like the Dardanelles front ("but they are murdering one another surely." "Not so. They merely gesticulate") in a deft epiphany of the link between capitalism and violence. Inside, Forster knowingly remarks, "it is usual to compare such visions to Dante's Inferno, but this really did resemble it, because it was marked out into the concentric circles of which the Florentine speaks" – perhaps with a side glance at the wonderful title of Cavafy's job as 'Clerk to the Third Circle of Irrigation' in Alexandria. Yet a hidden paradise remains a mile away in another building connected with cotton, a place where a more benign anarchy reins (hell is paradoxically orderly, paradise truly haphazard), cotton flying everywhere in "a snow storm, which hurtled through the air and lay upon the ground in drifts" – not unlike the pillow fight in Vigo's *Zero de Conduite*. This is where Arab labourers work to press the cotton "with the assistance of song. The chant rose and fell; it was better than the chant of the Bourse, being generic not personal, and of immemorial age – older than Hell at any events." Here in this text we already have in miniature the vast sweep of *A Passage to India*, with its depiction of the Indian palimpsest – successive colonisers, Moslem, English, successive religions, Hinduism, Islam, Christianity – as of the indifferent, impersonal Malabar caves from ages before all of them, that look "round civilisation" in a different sense from that Forster felt he could not do in 1913.

I shall deal briefly with aspects of two other epiphanies to further illustrate the extraordinary quality of this text. "The Solitary Place" is a beautiful prose poem about the flowers out in the desert around Lake Mariout near Alexandria – not unlike some of D.H. Lawrence's Mediterranean nature writings. The ubiquitous passage theme recurs in the depiction of spring, but it obeys its own quite different patterns – "there is nothing there of the ordered progress of the English spring [...] The flowers come all of a rush." They provide a beacon that outshines both Pharos and Pharillon – "a coloured ray of light playing on the earth" so brilliant that "they could seem afterwards like the growths of a dream." Yet there is nothing euphoric about the piece as it asks, whether "the extraordinary show" nature provides year after year "to a few Bedouins can continue when it is not sand she must cover but man's tins and barbed wire", and answers "Probably not [...] his old tins will be buried under new tins" (Allott 2004: 242).

"Between the Sun and the Moon" is about the central East-West oriented 'Corso' of Alexandria, the rue Rosette, which it contemplates through the lens of a lewd and of course minor Alexandrian novel written by a bishop about a hero with the suggestive name of Clitophon, searching in the city like many before and since for a lost lover, Leucippe. Once more at this stage of the text, Forster averts his gaze from any 'human interest' – it is the street itself that again interests him, and the jewel imagery implicit in the adjective "exquisite" recurs ironically in the phrase: "the passage gleams like a jewel among the amorous rubbish that surrounds it." The sketch again concludes by emphasising what still survives of Alexandria – neither Pagan nor Christian culture but nature and the place itself: "Paganism, even in the days of Clitophon and Leucippe, is dead. It is dead, yet the twin luminaries still reign over the street, and give it what it has of beauty. In the evening the western vista can blaze with orange and scarlet, and the eastern, having darkened, can shimmer with a mysterious radiance, out of which, incredibly large, rises the globe of the moon."

I want to conclude by seeing the Alexandrian books, above all else, as a means of passage through the unspeakable war of 1914-18. "All that I cared for in civilisation has gone forever, and I am trying to live without either hopes or fears ," Forster wrote at the end of 1915 to Syed Ross, the Indian man he adored unrequitedly at that time. He is writing despairingly but at least with some sort of programme, in the latter half of the sentence, as to how he might survive the catastrophe (*Letters*: 233). A similar consciousness haunted him in Alexandria, from whence he wrote for instance to Virginia Woolf in April 1916: "I imagine it is here that civilisation will expire" (*Letters*: 234), and similar feelings would dog him throughout his life. In July 1919 he wrote to Ludolf, the dedicatee of *Alexandria: A History and A Guide* expressing how "it's so puzzling to feel that one's the last little flower of a vanishing civilisation,' and in February 1940 to Hilton Young that 'the soil is being washed away" (Haag 2004: 107, *Letters*: 2, 172).

But if the God had abandoned, not only Anthony but half the seed of Abraham, one by one, Cavafy's poem could teach him a degree of stoic calm, the courage to go to the window and acknowledge the wonder of the passing music but move on. Mohammed could move him into the next phase of indifference to the "old bitch gone in the teeth" with her "two thousand battered books," realising in unexpected fashion – "generic not personal" – the Cambridge Apostle dictum that friendship was the supreme value : "When I am with him, smoking or talking quietly ahead, or whatever it may be, I see, beyond my own happiness and intimacy, occasional glimpses of the happiness of 1000s of others whose names I shall never hear, and know that there is a great unrecorded history. I have never had anything like this in my life – much friendliness and tolerance, but never this – and not till now was I capable of having it, for I hadn't attained the complete contempt and indifference for civilisation that provides the necessary calm" (*Letters*: I 269).

Bibliography

Allott, Miriam (ed.), 2004, *Alexandria: A History and a Guide and Pharos and Pharillon*, André Deutsch, London (Abinger Edition 16).

Doll, Rob, 2002, "Remembering Mohammed: E.M. Forster, Cavafy, and the Nexus of Memory", in http://www.emforster.info/pages/cavafy.html, 6 pp.

Fordonski, Krzystof, 2001, "The Homoerotic Function of Foreign Settings in the Early Fiction of E.M. Forster", in Humboldt Universität Centre for British Studies Working Paper Series, Berlin, http://www2.huberlin.de/gbz/index2.html?/gbz/publications/working%20papers.html, 17 pp.

Forster, Edward Morgan, 1983-1985, *Selected Letters*, ed. by Mary Lago - Peter Furbank, 2 vols, William Collins and Sons, London.

Furbank, P.N., [1977, 1978] 1979, *E.M. Forster: A Life*, 2 vols, Oxford University Press, Oxford.

Haag, Michael, 2004, *Alexandria: City of Memory*, Yale University Press, New Haven and London.

Jeffreys, Peter, 2005, *Hellenism and Orientalism in the Writings of E.M. Forster and C.P. Cavafy*, University of North Carolina, ELT Press, Greensboro.

"THE PALLID CHILDREN": AUDEN AND THE MEDITERRANEAN

Paola Marchetti

W.H. Auden's life was marked by different geographical landscapes and homes which corresponded to different stages in his evolution as a poet and as a man. Born a 'northerner' who was proud of his supposed Icelandic origins, he cherished Nordic sagas, the Anglo-Saxon literary tradition and, in his early poetry, combined the figures of mysterious secret agents and spies with the abandoned mining limestone landscape of Yorkshire. After leaving England he settled in New York, which for him became the symbol of the modern city where anonymity and existential isolation were the marks of a lonely existence. He came into physical contact with the Mediterranean world, and Italy in particular, in 1948. In that year he first visited Florence and the island of Ischia, where he was to spend all his subsequent summers until 1958. In 1958, after winning the Feltrinelli Prize, he bought a house in lower Austria – midway between the north and the south of Europe – and settled there until the end of his life in 1973.

The works published in the fifties, *Nones* (1951), *The Shield of Achilles*

The following abbreviations are used for bibliographical references to works by W. H. Auden

CSP   1966, *Collected Shorter Poems 1927-1957*, Faber & Faber, London.
F&A   1983, *Forewords and Afterwords*, selected by E. Mendelson, Random House, New York.
DH   1975, *The Dyer's Hand and other Essays*, Faber & Faber, London.
SW   1967, *Secondary Worlds*, Faber & Faber, London.

(1955) and *Homage to Clio* (1960), together with his many essays and fore-words, lay the ground for this inquiry into his relationship with the 'south', as does the perusal of both memorable texts like "In Praise of Limestone" (1948) and other, less memorable ones which nonetheless are all part of the political, philosophical, aesthetic and religious journey the poet undertook during his lifetime.

His encounter with Italy could be described in the terms he used for Goethe's *Italian Journey* in 1962 (*F&A*: 135,136), i.e. a quest undertaken in a moment of crisis ending up for both in a vision of eros. Auden in particular, who had already experienced love in the forties, transmuted his experience in the gradual awareness of the existential and religious meaning of freedom and choice through a revaluation and sanctification of the sensory life provided by the body.

As Mendelson points out, each geographical move in Auden's life "… coincided with fundamental changes in his work and outlook, and brought him to the landscape he thought most suitable to the kind of poetry he wanted to write." (1999: xviii)

The great philosophical (*New Year Letter*, 1941), religious (*For the Time Being*, 1941-42), aesthetic (*The Sea and the Mirror*, 1943-44) and existential (*The Age of Anxiety*, 1944-45) meditations of the forties in America had turned Auden's dualistic Freudian strife of the id and the ego into the struggle between eros and logos interpreted through the Kierkegaardian stages of the aesthetic and ethical. The third stage, the religious, represented the overcoming of such strife in an otherworldly dimension. He had thus adopted duplicity of form and content as the mirror of an existential dualism exemplified by the conflict between body and mind, the timeless and history, innocence and sin, fate and choice. This anti-romantic, ironic and detached viewpoint, (the 'gift of double focus'), allowed him to distinguish between two Blakean unsatisfactory extremes which cannot exclude one another without destroying one another (Blair 1995: 71).

In the Fifties in Italy, such dualism was to be included in the whole of his aesthetic vision, but his new, conversational style and a more relaxed, humorous approach to life emphasised acceptance of the opposites as a mark of our fallen existence. Therefore, when approaching the works directly dealing with his Mediterranean experience, the reader may detect three different series of opposites: first, the vision of a 'Catholic' community as the product of an 'erotic', archaic civilization opposed to the

'Protestant', abstract loneliness and 'ethical' mechanisation of the north; second, the rediscovery of his native landscape as both an aesthetic space suspended in time and as an earthly, material and mortal body which mirrors the inner character of its inhabitants; third, a landscape associated with Vincenzo Bellini and the language of Don Ottavio, i.e., with a world of music, opera and art as a stage of universal, mythical situations and characters. The operatic stage in particular mirrored for him the world of art as the gratuitous and innocent *lieu* of self-reflection through the 'artificiality' of form. Opera, the Italian *opera seria*, *opera buffa* and melodrama as opposed to *Musikdrama* were clear concepts in his mind when he wrote the libretti of *The Rake's Progress* (1948) for I. Stravinsky, *Delia* (1949 - unset), *Elegy for Young Lovers* (1959) and *The Bassarids* (1961) for H.W.Henze.

1. *North and South*

In poems like 'In Praise of Limestone' (1948), 'Ischia' (1948) and 'Goodbye to the Mezzogiorno' (1958), man's double nature is depicted as a conflict between two cultures represented by the erotic, fundamentally Catholic vision of the south, and the ethical, rational and basically Protestant *Weltanschauung* of the north. The link between the two is established through the limestone nature of the soil which, in Auden's words, combines the northern Protestant 'guilt culture' of his upbringing, with the Catholic 'shame culture' of the Mediterranean world.[1]

The two opposites are conceptually and imaginatively saturated in Auden's poetic and philosophical categorizing at a time when he was producing a composite neo-classical poetics which combined Goethe and Mozart with Greek mythology and tragedy and the Roman worlds of gentile Horace and Caesar with Christian Augustine.

Therefore, when the sophisticated, post-industrial traveller of "In

---

[1] In the Freud lecture of 1971 on a series of poems about the lead-mining world of his childhood, he mentioned 'In Praise of Limestone' and explained that the limestone landscape "was useful to me as a connecting link between two utterly different cultures, the northern *Protestant guilt culture* I grew up in, and the *shame culture* of the Mediterranean countries, which I was now experiencing for the first time." (Fuller 1998: 406, my italics)

Transit" arrived in Italy, he was magically projected into an exotic place where time seemed to have stopped and whose archaic nature – which was neither barbarous nor savage – [2] awoke half-forgotten images. These were connected with the Pennine landscape of his childhood[3] and brought him back echoes of the pre-classical world of myth. The picture of the land as a fertile, captivating mother who takes care of "her son, the flirtatious male," who never doubts, "That for all his faults he is loved" ('In Praise' *CSP*: 239) is typical of the archaic mind. In addition, the poet perceived a sort of beautiful balance between the works of nature and the works of man in the combination of rounded slopes, valleys, pools, ('In Praise' *CSP*: 238), boiling springs, twisted paths, shaded surfaces and sun-drenched spots ('Ischia' *CSP*: 242), with man-made fountains, squares, hill top temples, statues, vineyards.... and patron Saints ('In Praise' *CSP*: 239).

When compared with the inner and private experiences of loneliness transmitted by the open northern landscape, this land of external beauty and public light almost induced the poet to deny his 'antimythological myth' that art cannot change history ('In Praise' *CSP*: 240). There emerges then a contrast between a northern, often hostile landscape of magic won-

---

[2] Auden felt that Post-World-War-II Italy had not changed much since Goethe had visited it before the French Revolution: "Is there any other country in Europe where the character of the people seems to have been so little affected by political and technological change?" he observed. (*F&A*: 132).

In 1953, another eminent *émigré*, Hans Werner Henze, described Ischia as follows: "At that date Ischia did not even have its own supply of drinking water". "I found [...] a little farmhouse,[...] half in ruins, [...] with a little vineyard and an ancient lemon tree. It had a flat roof, which was reached by a flight of stairs, and on it figs had been laid out to dry on a basketwork. [...] An ancient *nonna* sat on these stairs. The first thing she asked me was whether the Germans were Christian. [...]. [...] Outside was a well fed by an underground spring and by rainwater and kept clean by an old eel that lived at the very bottom [...]. How silent it all was here!" (1998: 114, 118-119). Auden reports that Goethe was told by an Italian officer that Protestants were allowed to marry their sisters (F&A: 76). His experience on Ischia is well documented by Theckla Clark, in *Wystan and Chester. A Personal Memoir*, Faber & Faber, London 1995, Part I.

[3] "I hadn't realised till I came, how like Italy is to my 'Mutterland'". Letter to Elizabeth Mayer 8 May 1948 (quoted in Mendelson 1999: 290). For Mendelson his journey south woke up "his relations with the cold north of his childhood" (Mendelson 1999: 289-90). As a result the two landscapes are mixed in 'In Praise of Limestone' as they are in 'Prime' (*Horae Canonicae*) and in 'Not in Baedeker'.

der and awe[4] and an apparently Arcadian paradise where all appears fixed for ever as on Keats's urn. The former is a land of becoming while the latter is a space of being, the natural universe of repetition without change where innocent beasts 'repeat themselves' ('In Praise' *CSP*: 240). Here dwell Fate and the ancient amoral gods who enslave man by depriving him of choice. In so doing they compel him to chase off fear either by means of noise, such as the fireworks exploded in honour of a patron Saint, the radios played up to full volume, the "silencers off the 'Vespas", or by physical proximity which makes him feel "immune/ To all metaphysical threats" ('Goodbye' *CSP*: 340).

On the other hand, the pallid child "Of a potato, beer-or-whiskey / Guilt culture" ('Goodbye' *CSP*: 338) is the son of a man fallen from Arcady who knows that this beckoning and apparently unchanging paradise is a lie and that it is itself a historical place once raided by crusaders and the homeland of both dictators and saints like Mussolini and St Francis of Assisi ('Ischia' *CSP*: 241). The northern traveller of 'Goodbye' is thus confirmed in his vision of life as a dynamic journey of self-knowledge on an open road – a *Bildungsroman* ('Goodbye' *CSP*: 340) – where choices change one's life. However, Auden observes that the encounter with the south may become a chance to consider things from a distance, to stop thinking about one's journey and to accept life for what it has to offer in a momentary suspension of pain and anxiety.[5]

The poet makes it clear that, by being both a real world and a projection of the mind, the lie this erotic Eden exhibits is ambiguous. In reality, in the erotic, impersonal world of Homer, there exist no lies.[6] Its

---

[4] References to the north may be found in many poems of the period starting from 'In Praise' where people "had to veil their faces in awe / Of a crater whose blazing fury could not be fixed."(*CSP*: 239). In 'Ischia' the Pennine landscape is recalled as the green valleys 'where mushrooms fatten in the summer nights / and silvered willows copy/ the circumflexions of the stream."(*CSP*: 241). His 'Mutterland' is an often humid, hostile place where "Man still [..] / Exists [..] and [...]/ is not uncheerful" ('Not in Baedeker', *CSP*: 250). Furthermore, the north had historically generated the barbarous hordes which invaded the Roman Empire and came from a land where "herds of reindeer move across / Miles and miles of golden moss, / Silently and very fast." ('The Fall of Rome', *CSP*: 219).

[5] The peace of the place 'is a cure for, ceasing to think / of a way to get on, we / learn simply to wander about' ('Ischia' *CSP*: 242).

[6] Such a land does not 'lie about pain or pretend[s] that a time / of darkness and outcry will not come back" we all know that 'all is never well' and 'what is the case' ('Ischia' *CSP*: 243).

apparent simplicity is that of art where "nothing remains hidden but all is manifest" (*SW*: 58). As a consequence, when the traveller is deceived by its external and public appearance, it is only his fault in that he has not been able to read the duplicity of its hard and soft primitivism (Johnson 1973: 126-171).

Auden's emphasis on the impossibility of reconciling the two opposites of north and south automatically excludes self-deception and enhances self-awareness when it portrays the southerner as the innocent Arcadian and the northerner as the guilty, introverted Utopian.

In his view, the former may be associated with either the Homeric pre-classical hero or the tragic hero of classical drama. The latter corresponds to the wise man of classical Greek philosophy, whom he connects with the rational and scientific drives towards civilization and power represented by the Roman Empire or by our scientific, mechanised culture.

The southern man shares with the exceptional Homeric warrior a life in an unchanging and indifferent nature and a lack of any moral sense of good and evil. His actions are gratuitous and determined by the passions which guide him at the moment and which are sent by the gods. He has no sense of fear (like Siegfried in Wagner's *Ring*) because he displays his heroism in situations which are given to him. With respect to this, the poet observes that

> The world of Homer is unbearably sad because it never transcends the immediate moment; one is happy, one is unhappy, one wins, one loses, finally one dies. That is all. Joy and suffering are simply what one feels at the moment; they have no meaning beyond that; they pass away as they came; they point in no direction; they change nothing. It is a tragic world but without guilt for its tragic flaw is not in human nature [...] but a flaw in the nature of existence. (*F&A*: 18; see also *DH*: 172-73.)

The tragic hero of Greek tragedy represents a sort of evolution in that he is a moral example not to be followed. He is the lucky, strong and gifted man who has unwittingly made the wrong choices by following his *hubris*, an excessive confidence which "makes him believe that he, with his *Arete*, is a god who cannot be made suffer" (*F&A*:19). When his guilt is discovered he suffers but, by accepting his suffering, he is reconciled with the law of the gods. Auden believes the Catholic practice of confession to be a sort of archaic heritage as it frees man of his burden in a public and ritual practice which reunites him with God. This is a ritual, similar in scope, to the public performance of Greek tragedy in which the catharsis produced by the moral example of the suffering hero, purifies the chorus of average men. He therefore justifies the emphasis on ritual in Catholic shame culture as a collective act of the community designed to attract divine favour (*DH*: 173).

As to the ethical man of modern drama, the poet suggests that he originated in the Greek philosopher, whose God is reason (logos), and whose laws are universal and just. There is thus a shift from the amoral Homeric hero to the philosopher who is able to transcend his passions through reason and by appealing to universal ideas such as the law or the good. However, like the erotic man he is not tempted and does not choose although, like the Freudian ego, he instructs the id in the good or imposes it on him (*DH*: 175). The coming of Christianity and of revealed religion transformed him into the fallen, self-conscious individual who has entered time and history and knows that sin is his rebellion against God rather than ignorance of God or lack of relation with him. Therefore, his sin is not *hubris* but pride, i.e. the refusal to accept the "limitations and weaknesses which he knows he has" (*F&A*: 21): hence, in opposition to the tragic hero, he is not reconciled with the law.

Christianity thus changed the philosopher into the ethical man of modern drama and placed an emphasis on the difference between Socrates and Faustus when it pointed out that the latter will not automatically choose the good when he knows it because, by exercising his free will, he has the option of choosing evil. Auden's statement that 'The personal I is by necessity Protestant' (*F&A*: 54) is thus reinforced by his assertion that

> Protestantism set out to replace the collective, external voice of tradition, by the internal voice of the individual conscience [...]. In religion, it shifts the emphasis from the human reason, which is a faculty we share with our neighbour, and the human body which is capable of partaking with other human bodies in the same liturgical acts, to the human will which is unique and private to every individual. (*F&A*: 83)

If choice is the mark of the ethical man, and his choices are made in history, the danger inherent in them is related to the way this man approaches nature: he is often the scientist who pursues knowledge at any cost, transforming his object of study into a lifeless cipher and himself into a god.

With respect to this and to the two irreconcilable natures of the erotic and ethical man, Auden argues that they are ingrained in the Christian world as it emerged from the combination of two conflicting spiritual worlds, namely the intransigent historical and religious experience of the Jews, and the Gentile speculation upon and organization of that experience. In such a precarious balance, the Gentile in us teeters between two fundamentally pessimistic visions i.e., a superficial worldly frivolity and a false spiritual other-worldliness; while the Jew in us is arrogantly tempted to show no tolerance towards the dissenter, the different or the stupid. Auden thought that the Inquisition was the example of both a Gentile interest in rationality and a Jewish interest in truth (*F&A*: 14) and, implicitly, read the Holocaust as the product of protestant political and social intolerance.[7]

The danger for the Christian of relapsing into the aesthetic or gentile world, seems to annul the difference existing between shame culture and guilt culture, namely the fact that man is or is not free, (as Christians, the Catholic and the Protestant, both are) and appears rather to produce the impression that they still confine their choices to either/ors, i.e. between strength or weakness, ignorance or knowledge, which exclude a more comprehensive 'and', which the poet finds in the true religious dimension of otherworldly love.

A vision of love capable of transcending the limitations imposed by choice is unknown to both the aesthetic and the ethical man because they experience it as either eros (the force of procreation of Caliban and Papageno) or as Dame Kind (the absolute romantic love of Tristan).

Dame Kind as the sublime and mystical encounter with the holy existence of mountains, rivers, seas or flowers and beasts is typical of northern people with a Protestant upbringing (*F&A*: 58, 60), while the vision of eros belongs to the Mediterranean countries. Here "the individual experience of nature as sacred is absorbed and transformed into a social experience expressed by the institutional cults [...] the local Madonna and the local saint" (*Ibid.*: 60).

By offering two more conflicting opposites, Auden stresses once again duplicity, but insists that the vision of eros is more important than the

---

[7] Mendelson interprets Auden's writing about the flesh and the ordinary human body as a response to the war and a refusal "to believe all myths of power and exclusion." (2004: 56)

vision of Dame Kind. The latter is not really sensual but mental in that it is a revelation of the general beauty and multiplicity of life. On the contrary, the vision of eros appeals to the individual and to his erotic body, thus enabling man to transcend Dame Kind and to reach a personal correspondence with God.

2. *The body as a Mediterranean landscape*

In the dialectical relationship between the north and the south, the ethical and the aesthetic man, the body is metaphorically associated with the rock, the mind with a desert plain.

In the fifties the plain became the subject of many of Auden's poems, of other *paysages moralisés* that metaphorically displayed the reduction of rock to soft earth by erosion. Out of metaphor, such landscapes showed how the Protestant man of the north had subdued the body/stone in his will to power. As in *Paysage Moralisé* in 1933, the ethical man who, in lines 43-60 of 'In Praise', metaphorically abandons an Eden without choice to move to other landscapes, embodies different choices made in history: the desert of clay and gravel chosen by the saint (the self-inflicted suffering of the tortured soul); the plain to be drilled for oil (by a mercantile civilization); the river to be curbed for energy (by the scientist); the plain turned into the habitat of an unintelligible multitude (by politicians) where also armies march (led by Caesars and Hitlers).

The dismal plain of soft earth, the blood-stained, impersonal, democratic and mechanised reign of Clio, is confronted with the hilly landscape of the moderately high Apennines and Pennines where the "rock creates the only and truly landscape".[8] The body or the rock, whose voice was first heard in *The Sea and the Mirror* as Caliban's, is the expression of a sound, archaic wisdom of acceptance rather than of the denial of one's condition. Whereas the egotistical, romantic imagination of the northerner approaches life as a mythical mountain to be conquered in a quest

---

[8] This quotation is taken from a letter Auden wrote to his friend Elizabeth Mayer when working on 'In Praise'. The whole passage runs as follows: "... that rock creates the only truly human landscape, i.e. when politicians, art etc. remain on a modest, grandiose scale. What awful ideas have been suggested to the human mind by huge plains and gigantic mountains!" (in Fuller 1998: 407).

(see also Auden's own Ransom in *The Ascent of F6* or Byron's Manfred or Shelley`s 'Mont Blanc'), the Arcadian man of the south accepts it as a mystery, i.e., a frivolous, gratuitous, manifestation of life. Hence, if for the mind life is an incessant tragic struggle, for the body it becomes, in perspective, a comic experience. And comic detachment can only be the result of acceptance.[9]

Auden thus overcomes the distinction between mind and body already expressed by Plato as the co-existence of the immortal, rational soul (the 'scintilla of the divine archetype' – capable of knowing and recognizing truth) with the finite, mortal, unredeemed body, whose freedom the mind limits (*F&A*: 34). In the forties, Auden had described the mind/body relation as that between master and servant (Prospero and Caliban) and had observed that the mind's aspiration to resembling timeless God (the unique pure being of order and love – logos, ratio and spirit) made it oppress the body which exists in time and is becoming. In the fifties he finally realised that the relation did not imply a choice between an 'either' and an 'or' but an acceptance of both as part of a creaturely condition already expressed in the doctrine of creation. He observed that if matter is not intrinsically evil, then:

> the order of nature is inherent in its substance, individuality and motion have meaning, and history is not an unfortunate failure of necessity to master chance, but a dialectic human choice. (*F&A*: 36)

The rehabilitation of the body as the true human voice expressing one's limits was further developed in Auden's comments on Augustine's idea of the Fall. He maintained that the Fall did not mark a contrast between body and mind, but between "flesh, i.e., all man's physical and mental faculties as they exist in his enslaved self-loving state, and spirit, which witnesses within him to all that his existence was and still is meant to be." (*F&A*: 36). The proof is that orthodox Platonism and orthodox

---

[9] Therefore, when at the end of 'Ischia' he describes the beautiful days spent in Italy as a 'marble / milepost in an alluvial land', he combines the two visions of eros and art with logos, further developing the dialectics existing between north and south. Like Goethe, Auden was interested in geology and was well aware that marble is similar in chemical composition to limestone, the rock that dissolves in water, just like the body dissolves in the earth.

Christianity saw the act of creation as good: hence the acceptance of the Aristotelian cosmos and the joy and inspiration the Middle Ages found in it. Such humble acceptance allowed St Francis of Assisi to draw a parallel between our creatureliness and that of the animals and likewise produced the vision of eros which Dante celebrated as the necessary stage to attain a vision of God. (*F&A*: 42) Radically dualistic theories which spread in the years of the Antonine peace brought forth a change in perspective which saw the cosmos as created by an evil spirit and the body as a curse. As a consequence, the latter was looked at with disgust and, for many, the sin of fornication came to substitute the archetypal sin of pride. The saints in the desert of sand and gravel described in 'Islands' or in 'In Praise' thus give voice to the logic of the mortification of the flesh to attain salvation, a logic Auden finally rejected in the late forties and fifties. At this time he came to see the body as the sacred receptacle of the spirit, just as in the sixties, his house in lower Austria was the many-sided container of his bodily and spiritual life ('Thanksgiving for a Habitat', 1964).

Accordingly, Mendelson's assertion that when Auden first arrived in Italy "...he began to write about the inarticulate human body [...] as if he had found the missing link in Don Giovanni's list' (Mendelson 1999: 277) describes the poet's new-found sense of harmony between his 'Protestant' mind and his 'Catholic', natural, and innocently curious body.

The vision of eros mediated by Dante made him realise that, in the passage from erotic love to the personal love of God, 'Eros is transfigured but not annihilated" (*F&A*: 68), and taught him to read the doctrine of the resurrection of the body as a declaration of the body's sacred importance. Both Plato and Dante agree that the revelation of eros is a vision prefiguring "a kind of love in which the sexual element is transfigured and transcended," (*F&A*: 65) thus pointing to the vision of the Uncreated. Plato's vision is that of the gentile aesthetic man, i.e. a general one, while Dante's is that of the Christian, ethical man, i.e., personal.

Thus, when at the end of 'In Praise', the poet associates the forgiven and risen body of man with the marble statues of the athletes and the fountains of the Italian landscape, he makes two points: he goes back to the metaphor of the stone and the mother as a female body; yet, by asserting that these were made 'solely for pleasure' (innocent, gratuitous pleasure), he reveals that they are the erotic products of art. And art in the poem is the masculine energy of an evidently false Priapus presented in the first version of the poem as the statuary image of the 'nude young male who lounges / Against a rock displaying his dildo.'(lines 12-13 of the *Nones* version). It has been observed that the dildo is the symbol of the Freudian sublimation of sex into art and that mother earth, the rock, is not charmed by sex but by art (Fuller 1998: 407). That the young nude male should be Kallman is not a coincidence as, in Auden's life, he represented the Beatrice who made him aware of both a resurrected body and a resurrected spirit and who introduced him to the world of opera.

3. *Art and opera*

Kallman, as a recalcitrant and unfaithful erotic 'object', was therefore responsible for both Auden's theorisation of art as free play in the aesthetic circle and for his association of words with music and the human voice in the complex relationships established in operatic dramaturgy. The poet's approach to the genre seems thus to reveal another aspect of the conflict between two cultures. These are exemplified on the one hand by the music of the north possessed of an abstract tension towards the absolute in the perfect, rational structure of Bach's oratorios; on the other, by the Italian tradition of the south where music and the voice, in Palestrina's or Scarlatti's oratorios, blend in a less ascetic and more sensual fashion.[10] Furthermore, the implication that Italian music is primarily associated with the voice and the stage fascinated the poet in its double aspect of gratuitous pleasure in the beauty of the produced sounds, and of delight in a theatricality which requires a fully conscious suspension of disbelief on the part of the spectator.

Auden's encounter with opera apparently brought together different issues which had constantly engaged his mind since the thirties: first, the erotic nature of music which is the most gratuitous of all arts; second, the dramatic or semi-dramatic quality of all his works that display the static confrontation of abstract concepts; third, his vision of man as an actor and of art as a fictional stage where parables and myths are acted out. During and after his collaboration with Igor Stravinsky on *The Rake's Progress*, Au-

---

[10] 'I am eternally grateful [...] to the musical fashion of my youth which prevented me from listening to Opera until I was over thirty, by which age I was capable of appreciating a world so beautiful and so challenging to my own cultural heritage' (*DH*: 40).

den set out to reorganise his views about music and the poetic word and finally produced the most articulated reflections on opera ever to have been formulated by a librettist in the twentieth century.

His vision of music as a privileged art dated back to his collaboration with Benjamin Britten in the thirties. He then felt it to be less limited by the laws of denotation than language so that it could more freely and self-consciously play with its own formal configurations and could openly declare itself as 'pure contraption' ('The Composer', 1938). The views expressed in 'In Praise' that music is his only comfort, the one art which 'can be made anywhere, is invisible, / And does not smell' (*CSP*: 240) refer to a poetics of music already stated in 'Anthem for St. Cecilia's Day' (1941) namely, a neo-classical vision, going back to Dryden, which considers it as the expression of a worldly heaven - the music of the spheres. Auden imagines it as a pure disembodied singing voice which can neither grow nor lie and can only play (*CSP*: 174).

As to the intrinsic erotic nature of music associated with the human voice, and the importance of the erotic experience for a balanced life, the poet argued that the nude in painting aroused in him the same sensual delight as the voice in opera. He observed that

> In both [painting and music] there is an essential erotic element which is always in danger of being corrupted for sexual ends but need not be, and without this element of the erotic which the human voice and the nude have contributed, both arts would be a little lifeless. (*DH*: 505, 506)

In his theoretical reflections on the different arts, Auden came to see all aesthetic realms as secondary universes of signs governed by technical and aesthetic conventions which 'recreate', with a lesser or major degree of 'realism', the primary world of nature/history. In their feigned simplicity and in their use of artifice they are closer to the real essence of life in that they speak through universal abstractions, images, metaphors and symbolic or mythical characters ('The truest Poetry is the most Feigning' 1954, 'Words' 1957).

This form of proclaimed self-reflexivity affects music and language differently, although both occur in time. Thus words, depending on the limits set by meaning and communication, enhance reflection and choice, whereas music, as the art which manipulates time as repetition and becoming, becomes a paradigm of man's existence as both an erotic and a rational creature on a more general and abstract level than language (*DH*: 465).[11] Moreover, Auden finds that the combination of music and the human voice in singing represents the most gratuitous of all acts, as singing does not involve communication but is an expression of pure joy. In such genres as the song, the lyric or the aria, music and words interplay with different degrees of freedom. Paradoxically, in the aria or the ensemble, which are purely musical forms, repetition and becoming magically suspend the time of the plot and reveal the feelings of the characters as individuals (aria) or as a group (ensemble) (*DH*: 470).

Auden's shift in interest from spoken drama to music drama suggests that the latter's more formalised and self-conscious conventions contributed to his new awareness about the real nature of his own writing which, since the thirties, tended to stage the dramatic confrontation of differing ideas through the polyphonic voices of abstract and larger-than-life characters. It has often been observed that his longer poems may be described as 'mini-dramas' and his short ones as 'implied dramas,' in that they display dramatic qualities which have been associated with the morality play (Blair 1995: 96-98). His works do not exhibit psychological conflicts, or a dramatic action following a cause-and-effect logic, but rather proceed by way of juxtapositions, tableaux or fragments which have a life of their own. Hence his poems may be said to represent states of being suspended in time where the lyrical voice, the emotion or conflicts of emotions are distilled and heightened as happens in the operatic aria or the ensemble.

Already in the thirties, when working on the plays with Isherwood, he had realised that poetic drama differed from prose drama in that prose tended to be more realistic, while poetry dealt with a mental world of abstractions. In the fifties, when writing opera libretti, he came to the formulation of art, not so much as the northern, romantic self-expression but rather as the erotic southern 'act', the love of self-conscious performance, heightened rhetoric, exaggeration, theatre (Blair 1995: 103).

In *The Sea and the Mirror* Auden had already shown such a duplicity in Prospero's personality. He was the artist divided between the two fac-

---

[11] If words, which belong to the ethical man, can express love for a single person, thus indicating choice, music can just express a general idea of love as the aesthetic and innocent impulse of the erotic world (*SW*: 91).

ulties of intuition, ie. Caliban's body, and conscience and will, i.e. Ariel's mind. Significantly, he had given Caliban the task of spelling out the nature and function of the aesthetic world, equating man to an actor. He had thus conjured up the vision of a creature who could simulate or imitate other realities and consciously live in both. If the good actor should not mix up the fictional and the real worlds, the poet should likewise alert his reader that his art is contraption, an intellectual game reproducing his double nature of creature bound to necessity and of man capable of conscious moral choices. The aesthetic point of view thus produces a tension between the amorality of poetic artifice and moral seriousness by presenting both as an interplay of interesting possibilities beyond good and evil. From the aesthetic point of view art is ritual, magic, a formal game the ethical artist has to play within a closed aesthetic circle, without confusing the must of nature and the wishful ought of art.[12]

In the Fifties he thus blended his vision of opera with his aesthetics of gratuitous play when he transferred the double vision of words versus music to the double vision of actor versus singer. For the singer or ballet dancer there is no

> question of simulation, of singing the composer's notes "naturally", his behaviour is unabashedly and triumphantly art from beginning to end. The paradox implicit in all drama, namely that emotions and situations which in real life would be sad or painful are on the stage a pleasure, becomes in opera quite explicit. [...] in a sense there can be no tragic opera because whatever errors the characters make and whatever they suffer, they are doing exactly what they wish. [...] On the other hand, its pure artifice renders opera the ideal dramatic medium for a tragic myth.' (*DH*: 468,469)

With respect to the use of myth in opera, the poet seems careful to distinguish between myth and parable as fictional correlatives of worlds where choice and freedom are or are not contemplated and which may produce tragedy and/or comedy.

---

[12] Echoing Goethe, Auden suggested that "In der Beschränkung zeigt sich erst der Meister, / Und das Gesetz nur kann uns Freiheit geben." (*F&A*: 126 " ... only in limitation is mastery revealed, and law alone can give us freedom." Translation by Auden).

In opera Auden clearly distinguishes between Wagner's *Musikdrama* – or continuous opera – in which everything is expressed by way of the music, and Italian opera, made up of arias and recitatives. The two operatic forms also reveal different cultural approaches: *Musikdrama*, the secular elaboration of the Bachian oratorio, follows the principles of the romantic ethical man and becomes a complete secondary reality through the application of an exclusively musical logic; *opera seria* and *opera buffa*, on the other hand, in their interplay of sung and spoken parts, produce a constant dialogue between the primary world of words where one is sensible (rational) and the secondary world of music where one is not.

The poet was likely to formulate these notions when working on *The Rake's Progress* which, in Stravinsky's intentions, was to be a meta-opera engaged in playing with, echoing and alluding to the Italian tradition as a whole. *The Rake's Progress* was also a homage to Mozart, who had himself blended the operatic world of the south with the symphonic tradition of the north.

In Auden's new-found neoclassical vision, opera thus perfectly mirrored an innocent and erotic world of art where freedom and choice are reduced to pure essences and actions become exemplary deeds. As a consequence, the operatic genre turned the central issues of the erotic and ethical man concerning freedom and choice, sin and repentance into the gratuitous acts of the aesthetic hero. This latter thus stands as the correlative of the mythical hero in a Christian world.

Tristan and Don Giovanni, who for him exemplify such a figure, are possessed by just one passion and 'feign' choices because they passionately cultivate that one passion (*DH*: 470). They embody a modern, Christian way of approaching the myth of love which is neither classical madness nor violent passion but rather a human attempt towards a love of beauty and justice in the first, and that of Good in the second instance. Moreover, as the two heroes represent *Musikdrama* and Opera respectively, they again typify the vision of Dame Kind and the vision of eros surviving in a Christian world.

In the myth of Tristan and Isolde two noble characters possess *Arete,* fall in love and see the other person as an ultimate good or as an absolute. However, since they cannot physically merge "their ultimate goal is to die in each other's arms" (*F&A*: 24). In the myth of Don Juan, the hero is no real hero but a villain dressed up as a hero who, like a scientist, has no interest in his victims as persons but as members of a species, as numbers he can study and add to his list. This is a further indication that the two characters represent diseases of the Christian imagination in the form of monism and dualism, again mirroring the ethical and the erotic man. However, as operatic characters, they are perfect because opera

> is an imitation of human willfulness: it is rooted in the fact that we not only have feelings but insist upon having them at whatever cost to ourselves. Opera, therefore, cannot present characters in the novelist's sense of the word, namely people who are potentially good *and* bad, active *and* passive, for music is immediate actuality and neither potentiality nor passivity can live in its presence. (*DH*: 470)

Both Tristan and Don Giovanni are tragic heroes in that they refuse the "and" of real love and choose to remain in the either/or of the erotic and ethical condition. In particular, when discussing Wagner's heroes, Auden, like Nietzsche, emphasises the tragic self-inflicted misery of these neurotic characters and finds in the erotic desire of the south represented by *Carmen* a similarly tragic, but more positive stance (*F&A*: 254, 255) .

The importance attached by Auden to reconciliation, understanding and forgiveness points in two directions: to Metastasian *opera seria*, which always ended with reconciliation, and to *opera buffa* which stressed the parabolic acceptance of division and limit. He found this comic acceptance in Verdi's *Falstaff* and in Mozart's works. In *opera buffa* suffering and joy, the erotic and the ethical are combined, thus leading to self-knowledge, repentance, forgiveness and love. This mirrors the other-worldly love where duality is overcome and where north and south are reconciled. Mozart's vision is truly Christian because "Feelings of joy, tenderness and nobility are not confined to 'noble' characters but are experienced by everybody, by the most conventional, the most stupid, the most depraved" (*DH*: 469).

In the libretti Auden wrote with Kallman, these issues are explored in ways which would require more space than allowed by these pages. In *The Rake's Progress*, Tom Rakewell is half Faustus, half Don Juan, and saved by Anne who represents a gratuitous, quasi-religious love. However, being a twentieth-century hero, he will not marry her but remain insane and die.

Mittenhofer, in *Elegy for Young Lovers,* is the self-possessed god-like artist who, by following his one passion – art, bends reality to his own aesthetic needs and thus becomes a murderer. In *The Bassarids*, which rewrites the myth of Euripides' *Bacchae*, barbarism and violence, the darker side of the erotic man, are depicted after the Holocaust in a manner which ironically, on the part of Auden and Henze, echoes Wagner's *Musikdrama*.

In his later life Utopian Auden became a self-conscious Arcadian by choice. He had lost his innocence and did not believe – as the romantics did – in Art with a capital "A", but conceived it as the limited expression of a body and mind which had to learn and live together.

Bibliography

Blair, John G., 1995, *The Poetic Art of W.H. Auden*, Princeton University Press, Princeton, New Jersey.

Fuller, John, 1998, *W.H. Auden. A Commentary*, Princeton University Press, Princeton, New Jersey.

Johnson, Richard, 1973, *Man's Place, An Essay on Auden*, Cornell University Press, Ithaca and London.

Henze, Hans Werner, 1998, *Bohemian Fifths: An Autobiography*, Faber & Faber, London.

Clark, Theckla, 1995, *Wystan and Chester. A Personal Memoir*, Faber & Faber, London.

Mendelson, Edward, 1999, *Later Auden*, Farrar, Straus and Giroux, New York.

——, 2004, "The European Auden", in Smith Stan (ed.), *The Cambridge Companion to W.H. Auden*, Cambridge University Press, Cambridge, pp. 55-67.

WYNDHAM LEWIS IN MOROCCO:
SPATIAL PHILOSOPHY AND THE POLITICS OF RACE

Brett Neilson

Published by Black Sparrow Press in 1983, *Journey into Barbary* is ostensibly a work of travel literature that recounts Wyndham Lewis's journey to Morocco and its surrounding territories in the spring and summer of 1931. Divided into materials initially published as *Filibusters in Barbary* (1932) and a series of unpublished chapters collected under the title *Kasbahs and Souks*, the text supplies a compelling, if problematic, register of the increasingly complex cultural relations between Europe and this particular part of the Mediterranean basin. As one of the principal figures of the English avant-garde, both in the fields of writing and painting, Lewis constructs a discontinuous narrative of his journey, interspersed with materials from his sketchbook and complicating the familiar linear structure of the travelogue. Apart from drawing on academic studies and tourist guidebooks, the text incorporates fictional episodes, making it as much a work of fantasy as a serious ethnographic or cultural study. Indeed, Lewis's embellishment of what he presents as an authoritatively researched and factual text works to distinct ideological ends, reflecting not only European anxieties about the rising tide of decolonization in North Africa but also a deep cultural phobia toward the Islamic religion.

Although Morocco at the time of Lewis's visit was a French protectorate, there is a long history of British colonial ambitions in the area dating from the sixteenth century to the French colonization in 1912. An agreement signed between France and England in 1904 gave England a free hand in Egypt in exchange for its colonial interests in Morocco. And in 1906, the signing of the Algeciras agreement between the European powers opened the way for the French occupation, which would last until 1956. The French hold over Morocco, however, was always precarious, mainly due to uprisings by the Berbers, the majority ethnic group in the territory. Indeed, in 1930, the year before Lewis's visit, the French introduced a measure known as the Berber Dahir to overcome this nationalist resistance, dividing the Berbers from their Arab counterparts and seeking, through the banning of Sharia law in the Berber areas, to remove them from the Islamic influence that was perceived to fuel their opposition to the colonial presence. Lewis's efforts to distinguish the Berbers from the Arabs as well as the military language and observations that appear occasionally in the text have led at least one commentator to suggest that he made the journey under the sponsorship of some guild or governmental organization (Dellal 2004). But, whatever the truth of this matter, Lewis's interests in Morocco far exceed a merely reactive disapproval of decolonizing movements. He projects European concerns in the encounter with distant people and places, laying the blame for the fragility of empire not only on the competition between the imperial powers but also on progressive social tendencies within Europe, which, to his mind, encourage white self-hatred and the erosion of cultural identity.

Lewis's trip to Morocco was motivated not only by the familiar modernist impulse to visit and artistically represent extra-European lands but also by one of the most unfortunate episodes of his literary career. In 1931 he published *Hitler*, a text that reacted favourably to the emerging phenomenon of National Socialism in Germany, especially its attempts to control the apparently increasing influence of mass culture on political and social life. *Journey to Barbary* promotes the fiction that its author's journey to North Africa was prompted by the desire to escape the hostile reception of this work. At one point, Lewis describes himself as the "Luther of Ossington St.", forced to flee "dying European society" by the "dust of moralist and immoralist England" (24). His language here is not entirely free of the Orientalist fantasies of escape and rejuvenation that he had earlier criticized in *Paleface* (1929). *Journey into Barbary*, however, aims not to establish the incommensurable otherness of the Berber people but rather their similarity to the European. Lewis discredits familiar stereotypes of Oriental exoticism and untrustworthiness by turning these accusations back on the Europeans themselves. "Is not the European capable of plots?" he asks, suggesting that his attribution of conspiratorial schemes to British "Chamberlains" (much like his praise of Hitler) might "dislocate the pattern of personalities in [... his] neighbourhood" (128). For Lewis, the Berbers (or at least some idealized vision of the Berber past) provide an example of what Europe might become under a strong system of centralized rule. Thus while he freely expresses disapproval for the political forms of parliamentary democracy, he frequently compares these indigenous people to his fellow Europeans, organizing his arguments about the claim *"Capitalism and Barbary breed the same forms"* (71).

As a central feature of the book's rhetorical operations involves the slippage between the words Berber, Barbary, and barbarism, it is appropriate to note that these are not words with which the indigenous people of North Africa identify themselves. The term Berber, which derives from the Greek *barbaros*, was used to describe the heterogeneous indigenous groups of North Africa by conquerors of Greco-Roman times (Phoenicians, Greeks, Romans, Vandals, and Byzantines) and its use continued during the period of Arab domination from the seventh century onwards. Later French and British imperialists adopted the term, adapting it to describe the geographical area where these people live as Barbary (a word that circulates in the English language at least as early as Shakespeare's *The Merchant of Venice*). By contrast, these people call themselves Amazigh (plural Imazighen) and reject the term Berber as an imposition. While Lewis's text predates the ascent of Amazigh national consciousness in the late twentieth century (particularly following the uprisings in the Kabylie region of Algeria in 1980), his repeated play on the words Berber, Barbary, and barbarism reveals an important element of his attitude toward these people. *Journey into Barbary* avoids relegating the Berbers to a primitive otherness, seeking at once to establish their similarity to the European and to separate them from Arab influence, a move that, in the context of the Berber Dahir, cannot be innocent of an ongoing interest in the French presence and a deep suspicion of Islam.

Berbers, Semites and the authority of the eye

Lewis's efforts to distinguish the Berbers from the Arabs are most evident in the section of the text entitled *Kasbahs and Souks*. The strategy here is to repeatedly draw parallels between the Berbers and the Europeans, while identifying the Arab invaders of North Africa, who he claims imposed Islam upon the Berbers, as members of the Semitic race. His use of this later term, given the racial situation in Europe itself, would also seem to carry implicit reference to the Jews. In many ways, Lewis conceives the supposed divide between Berber and Arab as analagous to that between Europeans and Jews. He thus makes sense of the colonial situation in Morocco through implicit reference to the more familiar racial politics of Europe itself. That he could consider these very different historical and geographical situations analogous is clear from a passage in *Paleface*, where he compares colonial race prejudice with "another racial superstition, the most intense and inveterate that the world has ever known-namely that of the *inferiority* of the Jew" (18). Such parallels are not unfamiliar to contemporary readers of postcolonial criticism. For instance, in his classic text *Black Skin, White Masks*, Frantz Fanon calls upon Sartre's *Anti-Semite and Jew* to develop his critique of colonial racism. But unlike these thinkers, Lewis's repudiation of anti-Semitic racial superstition is ambiguous and vexed.

To be sure, *Hitler* dismisses the phenomenon of anti-Semitism as incidental to Nazism, a German national characteristic that need not accompany fascist modes of social, cultural, or political organization. And, in unfortunately titled *The Jews, Are They Human?* (1939), Lewis distances himself from all accusations of anti-Semitism, claiming that he never endorsed such race prejudice. Some critics accept these claims at face value. For instance, Reed Way Dasenbrock affirms that "Lewis was not at all anti-Semitic at any point" (1992: 94). By contrast, David Ayers argues that Lewis's work displays a consistent if understated anti-Semitism. Concentrating on the depiction of fictional characters such as Jan Pochinsky in the 1928 *Tarr* and Julius Ratner in *The Apes of God* (1930), Ayers concludes that Lewis's writing functions according to a dynamic by which an "unstable and incoherent self projects anxieties about its own deficiencies on to the Jew" (1992: 30). My purpose here is not to settle this question once and for all. Rather, I want only to note how anti-Arab (or more precisely anti-Islamic) prejudice holds intimate relations to European anti-Semitism. This connection was certainly not lost on Edward Said, who, in *Orientalism*, his celebrated account of the European subjectification and subordination of Islamic cultures, notes that he found himself "writing the history of a strange, secret sharer of Western anti-Semitism" (1979: 27). Lewis's identification of the Berbers as Occidentals pitted against Semitic forces confirms and strengthens this parallel, revealing the extent to which his fellow-travelling Nazism, explicitly announced in *Hitler*, shaped his experiences in Morocco.

Kasbahs and Souks presents a series of parallels between the Berbers and the Europeans, and in particular between the Berbers and the Celts, a comparison Lewis draws from Budgett Meakin's 1902 book *The Moors*. This contrast is relevant in terms of Lewis's earlier assessment of Irish nationalism in *The Lion and the Fox* (1927), a work that attacks the Celtic literary revival and the corresponding struggle for Irish political independence on the basis that such nationalist claims institute unnecessary divisions between people who are racially alike. In a chapter entitled "The Berber as 'European,'" Lewis translates these claims into the North African context, arguing not for the invalidity of Berber nationalism, which in any case he understands as an unrealistic proposition in the current geopolitical climate, but for the racial affinity of the Berbers and the Europeans. "The Berber, " he writes, "inclines far more, in the matter of the more intimate springs of conduct, to the Occidental than the Oriental" (211). What remains striking is the force with which Lewis makes this claim, declaring that his experience as a visual artist grants him an indubitable authority to pronounce on such matters:

> I shall be so bold as to place before you the results of my own unaided investigation with the naked eye – of far more value in such a case than the microscope – and I believe, where the evidence is so conflicting, of first-rate importance, supplementary to the gropings of the often almost eyeless historian: provided of course that the eye brought into action in this informal field-work is in the head of a trained observer. In my own case the organ of sight can be said to answer to that description I suppose: for the kind of art I have always practiced, the art of design, founded as it is – like the severe linear art of the Renaissance Masters or the Greeks – upon a constant, in the truest sense scientific, study of Nature, qualifies me far better than many professional ethnologists to pronounce myself in a matter of this nature [...] It is perfectly clear that the Berber people – the Riffs, Hahas, or Chleuhs – do not belong to the Semitic race, like their Arab overlords. This much is established by mere eyesight at once. (190-91)

Here Lewis contrasts his investigations "with the naked eye" with "the gropings of the often almost eyeless historian". His assertion of the primacy of vision in ethnographic investigation entails a rejection of historical analysis oriented toward an understanding of modernity in a temporal frame. *Journey into Barbary* is full of passages that declare the power of vision above "historically-minded" analysis. For instance, when discussing the Koutoubia (a mosque tower in Marrakech), Lewis contrasts "what a thing signifies *historically*" with the evidence of sight, complaining that a positive appraisal of this monument would constitute "a triumph of history over the eyes in the head" (53). These claims for the primacy of vision confirm the earlier arguments of *Time and Western Man* (1928), which anchors its critique of Henri Bergson's time-philosophy in the attempt to develop a '*spatial philosophy*' that will also be a '*philosophy of the eye*' (392).

For Lewis, Bergsonism is at once the cause and symptom of a variety of social and aesthetic ills. These include the proliferation of interior monologue in narrative fiction, the psychoanalytic emphasis on sexuality and the body, the uptake of revolutionary socialist ideals among the European artistic avant-garde, and most importantly, the worldwide dissemination of modern Western (Enlightenment) values and modes of social organization. What he objects to most vehemently in Bergsonism is its association of all forms of qualitative difference with an uninterrupted temporal flux (or *durée*) and its consequent understanding of space as the realm of quantitative homogeneity, an empty and undifferentiated domain given to simple forms of calculability and stasis. Thus, rails he against what he calls a "bastard universalism" or "mercurial spreading-out in time" that results in an absorption of "otherness", institution of global "Oneness" and "overriding of place" (241). What he hopes to achieve by identifying the Berbers as Europeans while establishing the otherness of "their Arab overlords" is not only a justification of the Berber Dahir and all it entails but also the establishment of wider cultural boundaries that might limit this homogenizing flow and, through the isolation of an irreducible Islamic otherness, legitimate the claim for the uniqueness of European civilization.

Land, Land-Land

Lewis's hopes for the superiority of European civilization are seriously compromised by the perception that it has entered into a state of decadence or degeneration; a condition that, as mentioned above, he diagnoses primarily through the spread of Bergson's philosophy. One way he copes with this is by inverting his claims for the Berber as European by describing the European as barbarian: "We are, of course, *nous autres Européens*, complete barbarians, and those of us who affect to believe they are not so, are the most barbarous of all" (213). At stake in this inversion, which takes the rhetorical structure of the chiasmus, is a quasi-Nietzschean claim for barbarism as the reinvigorating force of an enervated Western civilization, an argument that finds confirmation earlier in the text where Lewis notes the similarity between the dispossession of Western workers under the collapse of finance capitalism and the effects of imperialism upon Berber culture.

In *Filibusters in Barbary*, Lewis compares the mobility of capitalist workers, particularly US workers during the Great Depression, and the nomadism of traditional Berber society. To be more accurate, he identifies the Berbers as *transhumants* or "semi-nomads" (70) who alternate between pastoral tent dwelling and highland village life. But for the purposes of the comparison with Western capitalist societies, he emphasizes the nomadic aspects of Berber culture. "[O]ur civilization", he writes, "with the impetus given it by machines, is turning from the settled to the restless ideal" (75). This prompts a contrast between Bidonville, a makeshift town of petrol-tin dwellings on the outskirts of Casablanca, and a similar Hobo-city of unemployed workers outside Chicago. More importantly, Lewis applies the language of nomadism to the French colonialsts in Morocco, claiming they are "as unfixed, restless and incalculable" (76) as the people over whom they rule. French Morocco, he writes, is a protectorate "built upon sand", "transitory" and "the last thing of that sort our society will witness" (76). Here, Lewis prefigures the onset of decolonization by noting the increased mobility of culture on a global scale. His likening of the Berbers to Europeans licenses an equation of barbarism and capitalism that announces the demise of the European imperialism.

These sections of *Journey into Barbary* emphasize the tendency of capitalism to break down established national units and generate profit along increasingly global lines of flow. In particular, Lewis registers the way in which imperial expansion fractures the territorial integrity of the nation-state or what Benedict Anderson describes as "the inner incompatibility of empire and nation" (1991: 93). But Lewis's identification of capitalism with barbarism also registers the territorializing aspects of imperialism. This becomes clear when he explores the political complexities surrounding the Rio de Oro, the vast stretch of desert to the south of Morocco. Lewis characterizes this area variously as "one of the most mysterious countries in the world" (160), "an almost complete *terra incognita*" (161) and a "kind of geographical *blank*" (166). He goes on to insist that it "is the only territory of dimension in the world where, definitely, it is quite impossible for the White Man to go" (164). At one point, he even relates a story, supposedly told to him by the French author Antoine de Saint-Exupéry, of a French corporal who narrowly escaped being "boiled alive" in this stretch of desert (182). In short, Lewis saves for his description of the Rio de Oro all the most stereotypical clichés of native treachery and ruthlessness. At the same time, he explains that the zone's inaccessibility is due not to the fierceness of the Berbers or "the feebleness of European arms" but to the "competitive susceptibility of European nationalism" (164). The French, he claims, could quickly assert control over the Rio de Oro, but because Spain ostensibly owns the area, international complications would result.

Earlier in *Journey into Barbary*, when he describes the circumstances surrounding Hubert Lyautey's resignation as Resident-General of the Protectorate of Morocco, Lewis examines in detail the land transactions in the colony. As commander of the French forces in Morocco, Lyautey sought to exercise control through an agreement with the "Great Lords of the Atlas", Berber leaders "who refused to allow their subjects to sell land to the foreigner" (120). But in 1925, under pressure from European entrepreneurs who had no patience for this strategy of cooperation, he was forced to stand down. Lewis laments this event and pictures Lyautey as a forgotten hero. "Land, *Land-Land* is the cry" (121), he writes, describing the many crooked dealings (vanished title deeds and false documents) that surrounded the French commander's demise. This declaration registers the seizure of land that accompanies the nomadism of imperial venturing. In so doing, it confirms the mutual implication of territorial seizure and global nomadic wondering in the workings of modern imperialism.

Lewis captures something of this split logic when he identifies the Berbers as *transhumants*. He follows E.F. Gautier by writing of "*the dual soul of the Maghreb* – the *nomadic* and the *sedentary*" (208). While this is not intended to describe the interplay of nomadism and territorialism in imperialism, it is relevant for his understanding of barbarism and its significance with regard to the contemporary state of European civilization. Lewis's discussion of this dual soul provides him with an opportunity to reflect on the question of nationalism. The Berbers, he contends, have never had a *nationality* due to the Arab imposition of Islam:

> So all the great things the Berbers have done, they have done in the name of the Arab God. And from that it was only a step to doing it *as Arabs*. And that must after all be the main reason why the Arab has got the kudos for anything done by the Berber; that seems patent enough. [...] It is because of *the dual soul of the Maghreb* – the *nomadic* and the *sedentary*. That is why Maghreb was never one. The imported religion – and such a religion too, one of the most ferocious and fantastic engines of mass-bigotry that the world has to show, anywhere in its well-stocked Chamber of Horrors – *that* gave them one "dual soul". And the fact that they have from earliest times been half nomads, half "sedentaries", did not make things any better – that bestowed upon them a *second* "dual soul" as it were [...] Indeed (to show how the "double soul", twice-over, worked) their great *national* upheavals have always been in the name of religion rather than of race. These spasms have been a sort of civil war, about the Arab Allah. They have always taken the form of a tribal *ruée*. Maghreb having been split up into these two parts – *one* that wandered outside the sedentary fold, and the *other* that sat down like a good boy inside it, and cultivated its garden – these "national awakenings" have invariably taken the painful and destructive form of the former flinging themselves upon the latter – on the ground that they were not religious enough. (208-209)

Lewis attributes the disunity of the Berbers to the fact that they are Occidentals under the influence of Semitic religion. Indeed he altogether sidelines the impact of European imperialism upon the Berbers to attribute their tribal conflicts to the earlier penetration of Islam, a religion for which he saves his most exquisite words of condemnation: "one of the most ferocious and fantastic engines of mass-bigotry that the world has to show, anywhere in its well-stocked Chamber of Horrors". In so doing, he implies that only a consciousness of race can give the Berbers a collective identity sufficient to overcome this internal dissension.

Remembering the parallel Lewis draws between the Berbers and his fellow Europeans, it is possible to read this as an indirect comment concerning the conflicts of European nationalism. In *Paleface*, Lewis proposes that such internal friction between the European powers might be avoided by an extensive nationalism that aims to unify Europe as a bastion of white civilization. The anti-Semitism of *Paleface* is indirect, obscured by Lewis's analysis of whiteness, decolonization, and modernist primitivism. Even so, the text advocates the exclusion from its proposed European union of "the asiatic elements in Southern Spain, Italy and Russia" (279). David Ayers understands this as "a plain statement of antisemitism in the broad sense of the term: it is a sentiment that stems from a fear of the dissolution of identity" (1992: 193). *Journey into Barbary* displays a more conspicuous racial phobia, directed against the supposed Arab domination of the Berbers. While clearly at a remove from Nazi anti-Semitism, Lewis's identification of the Berbers as Europeans (a categorization that acquires added force due to the Islamic signification of the term Maghreb, meaning west) and the Arabs as Semites betrays an anxiety regarding the overrunning of Europe by Semitic peoples.

At stake here is a racial discourse that legitimates European fears of Semitic conspiracy. As Edward Said explains in *Orientalism*, "what has not been sufficiently stressed in histories of modern anti-Semitism has been the legitimation of such atavistic designations by Orientalism, and [...] the way this academic and intellectual legitimation has persisted right through the modern age in discussions of Islam, the Arabs, or the Near Orient" (1979: 262). Like *Paleface* before it, *Journey into Barbary* indirectly validates an expansive nationalism based on principles of racial identification. This is a distinctive element of reactionary ideology, which also finds confirmation in Lewis's support of the *"Action Française"* (117). In this sense, the book's registration of the territorial logic of imperialism is inseparable from a celebration of the territorializing impulses of the fascist state.

Time and again

The racial discourse in *Journey into Barbary* constructs a cultural continuum upon which the Berbers are placed in proximity to a white European standard from which the Arabs are held to depart. Lewis explains that "there is nothing *abstract* about" the "Berber nature", which supposedly displays *"an indigence of ideas, and a passionate attachment to persons"* (121). For him, the Berber is a "Barbarian – if it is to be a Barbarian to be [...] attached to things of the physical world, repelled by the abstract" (212). He also reports that the Berber "respects the law" (210) and is capable of "detachment and objectivity" (215). But while stressing the

Berbers' supposed disdain for the abstract, Lewis also highlights aspects of their culture that imply a respect for abstraction and transcendence. For instance, he commends their building of *Kasbah* (or fortresses), remarking that these castles display "all the resources of a monumental aesthetic developed for some far more abstract and lofty purpose" (214). He goes on to claim that the "structural repertoire" of these buildings "is suggestive of a civilization of the first order" (217). Such a civilization, he imagines, to be regulated by strong leadership and a written legal code. While he is aware that these are not qualities possessed by the Berber tribes of the time, they provide the model for his construction of an ideal Berber society of the past.

Lewis's understanding of Barbary is fashioned by various aspects of the groups he encounters (or more often just reads about), gathered into an idealized projection of a strong society that resists Arab domination. He expresses great interest in the Ikounka tribe of the Anti-Atlas who, according to the French ethnologist Robert Montagne, possessed "a well-preserved, written legal code for the management of their communal fortresses" (137). He also celebrates the power of the "Blue Sultans" of the Rio de Oro, who led what he calls the "Berber Renaissance", a successful revolt against Arab dominance that lasted from the sixteenth to the twentieth century (178-79). Furthermore, he praises the "empire building" of the so-called Almoravides, seagoing Berber pirates of the fourteenth century, whose descendants supposedly form the Touareg tribe, "the purest Berbers of the desert" (201-02). Behind this enthusiasm for the Berber past lies Lewis's identification of the Berbers with the Europeans and his projection of an idealized society, ruled by strong leaders and pitted against Semitic forces. But Lewis faces a difficulty with this parallel, since at the time of his visit the Berbers are subjugated by European imperialism, their powerful Lords laid low by foreign invasion. He attempts to overcome this problem by comparing the "brigandage" of the Almoravides with the "empire building" of "white colonization" (201-06). But while he employs the metaphors of nomadism and territorial conquest to describe this empire building, he finds the linear form of the travel narrative insufficient to describe the territorial politics of imperial conquest.

At one point in the narrative, when he declines to describe the topographical layout of the city of Agadir, Lewis declares, "language is not suited for [...] abstract mapmaking" (122). Elsewhere in the book, there is further evidence that language introduces to intercultural relations a differentiating force that resists containment within spatial parameters. When Lewis attempts to justify his spelling of the word *souk* (meaning a market or bazaar), his discussion registers the power relations that obtain within imperial systems of translation and transliteration. Noting that the Berber pronunciation has "a complicated guttural tail to it", he explains "these sounds often have no equivalent in a European tongue" (54). Lewis's conventionalized French spelling (he rejects several English variants) submits the foreignness of language to a process of overcoding, resolving the original Arabic into merely phonetic elements. But within this process, which has everything to do with imperial domination, it is possible to detect the disruptive movement of linguistic/cultural difference, which emerges to interrupt the linear temporal form of Lewis's travel narrative. Earlier in the text, when describing the Algerian city of Oran, Lewis asserts that his entire experience on the southern side of the Mediterranean partakes of a "radical temporal displacement":

> Oran is more interesting than anything upon the European side of the Latin Sea (without setting up Carthage against Rome, because the former is so deliciously "oriental" or any such exotic shallowness of the marvel-loving savage of the West). With us it is *the time* that is out of joint, not the place. [...] Oran is of course not the best example to take — Morocco carries all this out far more fully. The farther south you go, again, to the Atlas and the Sous, the more complete is the illusion of a radical temporal displacement. (42)

There is something more at stake here than the standard anthropological denial of coevalness that imputes the non-contemporaneousness of geographically diverse but chronologically simultaneous zones. Lewis's emphasis on "radical temporal displacement" and disjointed time complicates the Enlightenment ideal of rational human progress by interjecting the iterative disturbance of linguistic difference. In these moments, the narrative exceeds its chronological form and the disruptive force of linguistic/cultural difference questions the projection of barbarism within the two-dimensional plane of cartographic representation ("language is not suited for [...] abstract mapmaking"). From a narratological point of view, this anomalous temporal movement can be identified with the book's dual obsession with "lethargy and incessant movement"

(54). This implies a movement of deferral that questions the linear/episodic time-scheme of the travel narrative.

Understood from this perspective, Lewis's identification of capitalism and barbarism mobilizes the disruptive energy of repetition, derailing the narrative of modernity in much the same way as postcolonial critics like Bhabha (1994) claim the irruption of linguistic/cultural difference in modernity shifts its location to the postcolonial site. This prospect remains undeveloped at the thematic level of *Journey into Barbary*, but Lewis's concern with temporal displacement and the process of linguistic translation/transcription cannot simply be discounted, put aside in favor of an analysis that respects the limits of his own *spatial-philosophy* and implicitly endorses his claims to ethnographic authority. In stressing the spatial dimension of capitalist expansion, Lewis also registers the way in which the territorializing activities of the fascist state aim at modernity's disruption, seeking a path beyond modernity that, as Andrew Hewitt writes, "would not reinscribe itself in the transgressive logic of modernism" (1992: 55). It is not sufficient to think the end of modernity with exclusive reference to either fascism or imperialism. To understand the complexity of this plural event it is necessary to consider the mutual implication of these forms, examining the intertwining of space and time in the traumatic and materially violent processes of capitalist globalization.

Too often in contemporary cultural and social theory there is a split between deconstructive approaches that emphasize the temporal difference of writing and geographical models that stress the production of space. The racial politics of *Paleface* and *Journey into Barbary* attest the danger of a spatial approach that forgets the disruptive temporality of writing. Only by remaining attentive to the material processes by which the earth is carved into territories, is it possible to mobilize the powers of difference without losing sight of the everyday spaces so rapidly inscribed and reinscribed by the forces of capital. My analysis of *Journey into Barbary* suggests that Lewis's rejection of the qualitative flux of the Bergsonian *durée* is insufficient or incomplete. As much as he seeks to dispel this differential movement from his work, it makes a surreptitious return in the "radical temporal displacement" that marks his narrative technique. More is at stake here than a simple recovery of the *durée* from the clutches of Lewis's *spatial-philosophy*. Try as he might to escape the temporal complexities of narrative, to make writing approximate painting,

Lewis cannot but write a text that is discontinuous and open to disruption. To note this complexity is not to underestimate the force of Lewis's cultural fear. But it is to signal that a proper theoretical attention to both space and time must be central to any project that seeks to reconstruct a geopolitical formation such as the Mediterranean through its representation in literature, art, and other cultural forms.

Bibliography

Anderson, Benedict, 1991, *Imagined Communities: Reflections on the Origins and Spread of Nationalism*, 2nd ed., Verso, London.

Ayers, David, 1992, *Wyndham Lewis and Western Man*, Macmillan, London.

Bhabha, Homi K., 1994, *The Location of Culture*, Routledge, London.

Dasenbrock, Reed Way, 1992, "Wyndham Lewis's Fascist Imagination and the Fiction of Paranoia", in Richard J. Goslan (ed.), *Fascism, Aesthetics, and Culture*, University Press of New England, Hanover, pp.81-97.

Dellal, Mohammed, 2004, "Re-presenting Minorities: Wyndham Lewis' Journey into Berber Athena", *Working Papers on the Web*, http://www.shu.ac.uk/wpw/morocco/Dellal/Dellal.htm.

Fanon, Frantz, 1967, *Black Skin, White Masks*, trans. Charles Lam Markmann, Grove Press, New York.

Hewitt, Andrew, 1992, "Fascist Modernism, Futurism, and Post-Modernity", in Richard J. Goslan (ed.), *Fascism, Aesthetics, and Culture*, University Press of New England, Hanover, pp.38-55.

Lewis, Wyndham, 1927, *The Lion and the Fox: The Role of the Hero in the Plays of Shakespeare*, Grant Richards, London.

___, [1927] 1993, *Time and Western Man*, Black Sparrow Press, Santa Rosa.

___, [1928] 1982, *Tarr*, Penguin, Harmondsworth.

___, 1929, *Paleface: A Philosophy of the 'Melting Pot'*, Chatto & Windus, London.

___, [1930] 1981, *The Apes of God*, Black Sparrow Press, Santa Barbara.

___, 1931, *Hitler*, Chatto & Windus, London.

___, 1939, *The Jews, Are They Human?* Allen & Unwin, London.

___, 1983, *Journey into Barbary*, Black Sparrow Press, Santa Barbara.

Meakin, Budgett, 1902, *The Moors*, S. Sonnenschein, London.

Said, Edward, 1979, *Orientalism*, Vintage Books, New York.

Sartre, Jean-Paul, [1948] 1965, *Anti-Semite and Jew*, George J. Becker (trans.), Schocken Books, New York.

WALKING THE TIGHTROPE:
SACHEVERELL SITWELL'S REWRITING OF THE MEDITERRANEAN IN *SOUTHERN BAROQUE ART*

Laura Scuriatti

In an article entitled *Settecentismo* published in 1922, Roger Fry writes:

> It is not uninteresting to note that a generation of Italians that has been brought up on Futurism should turn with such zest to their own painters of the seventeenth century. In many ways Caravaggio was an expression of a turbulence and an impatience of tradition similar to that which Futurism displays. Like the Futurists he appealed to the love of violent sensations and uncontrolled passions. Like them he loved what was brutal and excessive. Like them he mocked at tradition. Like them he was fundamentally conventional and journalistic. The strange thing is that the aspect of the Italian character which creates Futurism and Fascism should have taken so long to find expression in art. (158)

Here Fry sets out to condemn the contemporary fashion for the Italian "Settecento"[1] amongst young British and Italian scholars. Not only has admiration for the works of art of the seventeenth century become indiscriminate and unjustified, Fry argues, but this neglected period in art history has turned in contemporary art criticism into a tool of self-promotion for numerous ambitious young men. Fry does not reject Italian

---

[1] It is rather confusing that in this article "Settecento" defines at times not only the eighteenth century, but also the seventeenth, in spite of Fry's knowledge of the meaning of the Italian word, as witnessed by his correct use of it in previous articles.

Baroque art *tout court*, indeed he does not fail to emphasise his own plea
to do justice to the "Settecento" painters, expressed in 1905 in his introduction to Sir Joshua Reynolds's *Discourses*. He objects, however, to the
lack of discernment of the admirers of these painters and architects, who
fail to grasp that vulgarity in art was born in precisely in that century,
ushered in by none other than Caravaggio and passed on to other painters
and architects:

> It is not merely a question of violence of character so much as of at-
> tributing an aesthetic value to violent sensations. In fact the Italian
> artists of the seventeenth century invented the modern popular picture,
> invented the view of art which culminates in the drama of the cine-
> matograph. In fact, they may be said to have invented vulgarity, and
> more particularly vulgar originality in art. (158)

In the reference to young scholars and to Futurism we may recognise
in *Settecentismo* a polemic echo of the essays on Futurism published by
Roberto Longhi in 1913. Longhi compared the capacity of the Baroque
to communicate movement to the masses of the Renaissance with the relationship between Futurism and Cubism (Raimondi, 2003: 83).[2] More
importantly perhaps, the dubious genealogy allying Baroque painting to
the "drama of the cinematograph" and the depreciation of Caravaggio's
"essentially journalistic talent", which would make him a suitable "impresario for the cinema" (Fry 1922: 163), express Fry's fear of the spreading of mass culture through the new media of cinema, but represent simultaneously a specific inflection of the Baroque. This appropriation
serves as a critique of contemporary aesthetics, which aligns Fry's essay,
but not his stance, with a host of other texts published in Europe since
the end of the nineteenth century, reassessing and rediscovering Mannerist and Baroque architecture and painting.

From Heinrich Wölfflin's influential *Renaissance und Barock* (1888)
and Alois Riegl's *Die Entstehung der Barockkunst in Rom* (published posthumously in 1908) to Walter Friedländer's studies on the painter Barocci

---

[2] Fry may have known about Longhi's article through Bernard Berenson, whom
Longhi approached in 1912, and with whom he corresponded at irregular intervals until 1957. For the correspondence between Berenson and Longhi see: Berenson *et al.*
(1993).

(1912) and on Mannerism (1925), Nikolaus Pevsner's *Gegenreformation und Mannerismus* (1925) and Erwin Panofsky's *What is Baroque?* (1934), to name but a few, Baroque, Mannerism and Rococo started to occupy the stage of European art theory and became in many cases the vehicles for the redefinition of the epistemological categories of the discipline, as well as the markers of the modernity of the art historians and authors involved in the "rediscovery". Indeed, a few years after Fry's article, Walter Benjamin quoted the prophetic observations of a literature historian in 1904, who had equated the search for a new style of eighteenth century baroque literature with fin-de-siècle literature, and suggested that by 1925 the Baroque "Kunstgefühl" had become mainstream, at least in the literature of German-speaking Europe (1978: 37).

Sacheverell Sitwell's *Southern Baroque Art. A study of Painting, Architecture and Music in Italy and Spain of the Seventeenth and Eighteenth Centuries* (1924) may be considered as part of this renewed interest for Baroque art and architecture. Sitwell wrote it after a long trip in Italy and Spain with his brother Osbert, who published in the following year a travel book entitled *Discursions on Travel, Art and Life*, focused on the Baroque cities in Puglia, Campania and Sicily. Of the two brothers, Sacheverell continued to write on Baroque art, music and literature for the rest of his life, collaborating also, amongst others, with art and architecture historian Nikolaus Pevsner for a volume of essays on German Baroque sculpture, published in 1938.

In spite of Sitwell's consistency, his writings on the Baroque are less scholarly pieces than very diverse collections of impressions and appreciations, often seemingly aimed at communicating more about the author's taste than about the artworks discussed. This attitude, emerging at different levels in both brothers' writings, has gained the Sitwells an aura of eccentricity and *dilettantismo*, which they even publicised, to the point that their friend, the art historian Kenneth Clark, stated that they used the Baroque "to express something of their own, slightly frivolous, un-English attitude to life" (Ziegler 1998: 203).

The introduction of *Southern Baroque Art* is aligned with the works of the majority of British authors writing on Baroque art at the beginning of the twentieth century, as Sitwell feels he needs to justify his choice of subject matter and simultaneously constructs his own authority by proclaiming the novelty and originality of his own approach. In the case of *Southern Baroque Art*, certainly most of the cities and monuments in Puglia and Sicily, even the least known, had already been "discovered" by Northern European learned travellers long before the Sitwells decided to publish their books; the Portuguese and Southern American baroque buildings, however, were probably less familiar. Goethe, for instance, had written in his *Italian Journals* about the grotesque statues at the Villa Palagonia in Bagheria (entry of Monday, 9th April 1787), whereas Ferdinand Gregorovius and Paul Bourget had already described in the second half of the nineteenth century the cities and monuments of Puglia, just to mention some of the most obvious instances. A prominent and more recent British traveller to Puglia, indeed not one of the most frequented regions of Italy, was George Gissing, who published in 1905 his travel account *By the Ionian Sea.* Moreover, in 1910 the British architect Martin Shaw Briggs published in London his study of the art and architecture of Puglia, concentrated mainly on the city of Lecce and its peculiar inflection of the Baroque, entitled *In the Heel of Italy. A Study of an Unknown City.*[3]

However, unlike most of his predecessors writing about Southern Italy, Sacheverell Sitwell is not interested in the Greek and Roman origins of these places; he does not lament the loss of the culture and civilization of antiquity, nor is he fascinated by the local mores and folkloristic details that fill the pages of his brother's book and of Briggs' *In the Heel of Italy.* If in his studies of German Baroque art and sculpture Sitwell's discussion of the artworks adopts the form of the scholarly essay, in *Southern Baroque Art* the characteristics of numerous genres and art forms are fused into a phantasmagorical whole, in which geographical boundaries and temporal markers are constantly blurred. The aim of the narrator is to:

> soak myself in the emanations of the period, that I can produce, so far as my pen can aid me, the spirit and atmosphere of the time and place, without exposing too much the creaking joints of the machinery, the iron screws and pins of which are the birth dates and death dates of the figures discussed. (10)

This passage expresses the wish to recreate the conditions and forms of Baroque art in his own writing, to merge form and content into an or-

---

[3] For selections from Briggs' book in Italian translation see Cecere (2000).

ganic architecture that should not openly declare the means of its construction – a theme to which the text returns in the tale about the relationship between the melancholic Charles III of Spain and Farinelli, appointed by the king to cure him:

> What the emperor advised was no less than this: that Farinelli should throw away all the fireworks and technical tricks of his art and should adopt instead a style of absolute simplicity, under which his technique should lie hidden. (181)

It is perhaps too far-fetched to read Sitwell's "machinery" as a reference to the stage machinery of the courtly masques and of operas that feature copiously in the book, but certainly this passage is preoccupied with the position of the reader as a spectator, who should suspend his or her beliefs as in a theatre. It also introduces a preoccupation, typical of Baroque and Rococo art and architecture, with the status of artifice and with the viewers' modes of perception. But this desire to conceal the skills and effort necessary to produce a work of art (which, incidentally, is almost a version of the high Renaissance concept of "sprezzatura") should also alert us to consider them in relation to the position of the narrator as observer and producer of narratives. It is this position or, rather, *these positions*, that I will analyse in this article, focusing particularly on the function and failures of the textual narrative strategies aimed at providing an account of the art and architecture of Southern Italy.

The book consists of four chapters, entitled respectively "The Serenade at Caserta", "Les Indes Galantes", "The King and the Nightingale" and "Mexico". Neither the titles nor the promises of the narrator should be understood literally, since, except for the last essay, none of the first three focuses on a specific geographical area. In "The Serenade at Caserta" the reader is taken on a sweeping tour of a few Baroque churches and monuments in Naples and in the Palace of Caserta, to the baroque villas of Bagheria and Palermo, with long digressions on Venice, the *commedia dell'arte*, Sicily and Puglia, to finish again at the Royal Palace of Caserta. The second and the third chapters are connected by the story of the melancholy of Philip V of Spain, cured by the heavenly singing of the castrato Farinelli. It is especially in these chapters that the narrator's position continuously shifts, diving into a painting by El Greco, then digressing on the life of an Eastern Sultan in order to explain Venetian painting, going back to look at the genius of Luca Giordano and turning a museum visit into a daydream inside an imagined baroque spectacle, in which past and present are blurred.

Kenneth Clark was not alone in his scepticism towards Sitwell's version of the Baroque. Another art historian, Michael Kitson, observed in an article on the relationship between Nikolaus Pevsner and Denis Mahon in the 1930s that "the predominant approach to Baroque art" in the first decades of the twentieth century in Britain did not offer anything beyond its "frivolous late manifestations developed on the fringes of Europe and in Latin America, popularised by Sacheverell Sitwell" (1998).

And indeed, *Southern Baroque Art* cannot be compared in any way to the ground-breaking (if now less acclaimed) books of Pevsner, even though, interestingly, Pevsner decided to collaborate with Sitwell in 1934, as I mentioned above. *Southern Baroque Art* does not propose any interpretation of the term Baroque, the artworks and their ideology, nor does it seem to differentiate systematically between the different schools and geographical areas; fundamental terms such as Baroque, Rococo, Renaissance, Realism and Classicism are often used incoherently, whereas Mannerism, then a relatively new concept dominating the debates on the art of the sixteenth and seventeenth centuries, features only once in the text.

There are many complex issues which emerge from this text, *in primis* the meaning and usefulness of the term "Baroque" as an all-encompassing definition of painting, architecture, music, theatre and literature. However, in this article I will concentrate on the following questions: first of all, on the possible reasons for and the significance of the choice of Baroque Puglia, Campania and Sicily as objects of enquiry, and secondly on the perspective from which Southern Italy is construed in the text. My contention is that if, on the one hand, the book is partially structured as a voyage of discovery in unknown countries – a narrative strategy which is typical of travel books and tends to construct the foreign countries as typically exotic and "other" – on the other hand the textual narrative strategies undermine the proposed task not only of presenting a "study" of southern Baroque art, as promised by the title of the book, but also of representing the cultures and countries in which that art has been produced.

Why southern Baroque art? I have briefly mentioned how throughout Europe art historians and critics had been debating, and would continue to debate, the value and definition of the Baroque. From the very first few lines the text distances itself from the terms of the renewed interest for Baroque art in British culture: in the teens and twenties, for example, the *Burlington Magazine for Connoisseurs* published numerous articles on Italian Baroque art, including a few by Roger Fry and the architecture historian Martin S. Briggs, whose book *In the Heel of Italy* was a source for both Sitwell brothers (Briggs 1915a and 1915b; Constable *et al*. 1923; Fry 1913, 1921, 1922; Gamba 1922; MacLagan 1922). Most of these pieces sound remarkably similar, in that they begin with either a justification for the choice of subject, or the plea for a re-evaluation of long neglected artists such as Bernini, Borromini, Caravaggio and El Greco, seen as the representatives of the period.

In the introduction, the narrator positions himself outside the circle of academics and connoisseurs of the Baroque, but, like numerous other authors, he does so by proclaiming his own difference and originality and by justifying himself:

> It will be remarked by anyone who has the patience to read these four Essays that in their range of subject they have but little contact with the accepted or famous names of the period: Bernini and Borromini are not examined, and they play no more part in these pages than do Venice and Rome. (9)

When, towards the end of the introduction, he imagines writing a book on Renaissance painting in Venice, Sitwell polemically laments that "such a work could hardly find a place, so huge is the plethora of facts, and so fierce and dogmatic are the critics" (11). Thus, the exclusion of famous painters and places creates a niche of freedom:

> Thus the subject eventually chosen, concerning these painters and architects once famous and now considered obsolete, gave a free rein to the search into forgotten volumes, through which it was undertaken. (11)

The tone of the text at this point is reminiscent of the many travel writers' narratives which claim the privilege of an authentic and original experience in a foreign country, especially in the cases of popular destinations; it performs the creation of a kind of virgin territory, which guarantees both the authenticity and authoritativeness of the writer: "I have chosen virgin ground, where nothing of the kind was in existence before, for the scene of my experiment" (131).

On the one hand the text affirms from the start the author's innocence of the trappings of scholarship and connoisseurship, and, on the other hand, it never ceases to remind the readers of their presence. This is achieved not only through the assertion of the importance of artistic knowledge by its repeated and careful refusal, but also through hints at and critiques of other contemporary aesthetic positions:

> Too many people, looking like each other, and all talking in one and the same voice, may be heard at this time loud in the praise of Matisse and André Derain, while they have already returned to Raphael, and will soon come back to admire Guido Reni, falling victim by this to a strong and complete mental boomerang. The negro sculptors, obscured in a black anonymity, are now extravagantly praised, so it is surely time for someone to set up again the crumbling statues to Gòngora and to Luca Giordano. (10)

Sitwell here is tracing a panorama of the new developments in art history and theory, especially in Britain. More specifically, this passage is clearly an attack directed against the supporters of Post-Impressionism, and of Roger Fry in particular, whose *Vision and Design*, published in 1920, happened to discuss and praise Matisse, Derain, negro art, and even to "return to Raphael" in "Retrospect", the last essay in the book (1990) . But this vague map of contemporary art theories and scholarship is there also to enable the author to call himself out of it and to build a space of authorial freedom. Part of this strategy is also the creation of a version of the Baroque which may be suitable to the project. From the start, Sitwell's Baroque is defined by its theatricality and artificiality, in opposition to more realistic forms of representation:

> I know that the subjects of my choice are arraigned, in the earnest tones of the learned, because they possessed an entirely scenic conception of life, but the camera should have taught us by now how elusive a quarry is realism. Life, in its human aspect, is very ugly, and has always been so, it being the duty of Art to improve and select, transmuting for our own eyes that which we know to have been sordid into what we can be persuaded was beautiful. (11)

This also explains the reference to Guido Reni in the previous quotation – a reference which, once again, leads us back to Roger Fry, whose article "Pictures Lent to the National Gallery" (1917) attempts to justify his appreciation of a painting by Guido Reni at the National Gallery at a moment of crisis during the war, while all the most important paintings have been safely stored away:

> One may, even, after exhausting the Italian rooms, wander among the despised masters of the seicento [*sic*] to recognize that in spite of the superficial sentimentality of a picture like Guido Reni's *Ecce Homo* there is a unity of rhythm and a beauty of tone and colour which repay the effort to overcome one's first movements of revulsion. (198)

It is difficult to determine if Sacheverell Sitwell knew about the article; but whether the reference to Guido Reni was aimed against Fry or not, the Italian painter stands in Sitwell's critique in opposition to the Post-Impressionists, as a representative of a sort of realistic *rappel à l'ordre*. And indeed the "elusive quarry" of realism is one of Sitwell's targets. In *Southern Baroque Art* "realism" is not associated to a specific period in art history, but is presented as a recurring anti-vital force, it means ugliness, grotesqueness, and is most of all anti-artistic: it is therefore ascribed to the eighteenth century sculptures by Antonio Corradini, Francesco Queirolo and Giuseppe Sammartino in the San Severo Chapel (Pietatella) in Naples, that have a "far Eastern deadliness of realism" (26); to the late fifteenth century "Lamentation Group" of Guido Mazzoni, displaying a "horrible Tussaud-like realism" (27); and to Caravaggio, whose style is defined as "melodramatic realism" (90).

The programmatic refusal of realism is reflected also in the narrative style and in the positions occupied by the narrator. If compared to one of its main sources, Martin S. Briggs' *In the Heel of Italy*, Sacheverell Sitwell's text appears to be lacking well organized historical information about (in the specific case) Puglia and its monuments, as well as folkloristic details on its inhabitants, and constant references to antiquity and *Magna Grecia*. On the contrary, Osbert Sitwell's *Discursion on Travel, Art and Life* follows Briggs' format, fusing accounts of travel experiences, local folklore, and learned appreciation of artworks.

Southern Baroque Art disrupts this form and deploys Baroque visual techniques, proposing new formal solutions for the genres of the travel book and of the essay on art. We are presented with contradictory forms of narration. The beginning of the first essay has a tone recalling contemporary learned travel books. The narrator is present on the scene and tries to convey the atmosphere of the foreign place by deploying the commonplace attributes associated by Northern travellers to Southern Europe – the heat, the blinding light and a constant surrounding music:

> Six o'clock in the morning, and already the heat of Naples was such that it required confidence to believe in any hours of darkness. Most of the houses were still latticing the light with their barred shutters. They were skimming a soft music off the stillness, and as, one by one, the windows were thrown back, so that the shutters threw their shadow on the wall, the very strings of this fluttering music were visible, lying there as plain as anything for skilled fingers. (15)

But the perspective of the narrator undergoes rapid changes. Upon entering the church of Gesù Nuovo in Naples and sensing the effect that Solimena's fresco, *The Expulsion of Heliodorus from the Temple* (1725) has on himself and on his perception not only of the interior of the church, but also of its immediate external surroundings, the narrator feels suddenly forced to reconsider his own task: the theatricality of these spaces – "as if a stronger hand dragged you on to the stage of a theatre" (16) – forces the narrator to abandon his temporal anchorage, blurring past and present. He seems first to enter into the picture and then inhabit a vaguely recognisable seventeenth century Venice, conjured up by the comparison with Palladian waterside buildings. Merging with the artworks or spectacles observed is a recurrent narrative strategy of Sitwell's narrator. The description of Philip V's palace, for example, prompts a long digression in which the narrator becomes part of a courtly masque on the conquest of India.

Diving into the foreign landscape, being overwhelmed and getting lost in art is a topos of travel literature; it is a fantasy of pure visibility, the expression of the wish to annul the distance between the observer and the observed. But Sitwell's narrator is definitely not overwhelmed by art. The numerous instances of the blurring of spatial and temporal boundaries are orchestrated as epistemological strategies: the evocation of the masque on the conquest of India is defined as both an "experiment" and a "dream" (131); Naples's artistic wealth demands the adoption of changeable and artificial points of view:

> So great is the profusion of work of our period that I am forced to conduct a bird's eye promenade over the roofs of this quarter of the city, making it understood that by some X-ray process the insides of all the

churches may be seen without moving far away from our post in the air.
(48-49)

The tension between a subjective point of view and a seemingly neutral, ordering gaze is extreme here and is reminiscent of a similar tension
which architecture historian Robert Harbison sees in the Baroque recurrence of bird's eye views and perspectives from below. These are, according to Harbison, the expressions of the Baroque preoccupation with the
value of subjective response (2000: viii): if the perspective from below vi-
sualises the feeling of awe and surprise of a spectator before a work of art
or a building, the bird's eye view, manifesting an "enthusiasm for scenery,
that is of heightened dramatic presentation of the need for order, [...]
makes a plan look like a view and subjectivises it as if from a real though
unobtainable vantage point, not just thinking about it with some quasi-
mathematical parts of the brain" (10). However, the wish for control on
the processes and objects of one's own perception is not fully realized in
Sitwell's text, where the narrator's efforts are often disrupted by a reality
that takes at different times the shapes of labyrinths, webs, tangles, folds
and optical illusions: "The illusive architecture was, however, peopled by
a whole army who knew their way through this maze of building and
seemed to know by instinct what was painted and what was true" (1924:
170). In these moments artifice takes over, and the narrator willingly or
unwillingly partakes of the confusion he ascribes to the protagonists of his
fantasies – a confusion that he embraces already in the introduction, and
that expresses itself also through images of tightrope walking and vertigo:
in the first pages, the book is defined as a stage with "perilous boards" and
clouded in smoke, which Solimena, Churriguerra, Sanfelice and others are
to tread; the Sicilian town of Bagheria looks suspended in the air, "like a
lot of swings and trembling platforms" (20); the effect of marching music "was like walking a tight-rope, from which you dare not look down"
(110). Domenico Tiepolo's pulcinellas are painted as if suspended in mid-
air; Sanfelice's staircases are beautiful because maze-like (54).

That Sitwell should dedicate so much attention to the Neapolitan
architect Ferdinando Sanfelice (1675-1748) attests to his sophisticated
knowledge and understanding of Baroque architecture, but is also symptomatic of his interest in the reflections on aesthetic experience in the
art of the period. Harbison sees the ceremonial stair halls built in Naples
and South Germany as architectural elements, which "soften or
strengthen the building's relation to its context, making it more active

and dynamic by stretching out the moment of transition between in-side and outside": they "call into being a double or triple space through which we move at once laterally and vertically", and "are instruments of delay as of progress" (2000: 17-18). In Sitwell's text Sanfelice's stair halls are not so much remarkable because they blur the boundaries be-tween inside and outside, but rather because they are maze-like, because they confuse the narrator, forcing him to reconsider his position and to accept that his wish for full visibility and understanding of the foreign country is illusory (1924: 53-54).

The vertiginous experiences, the visual tricks and numerous meta-morphoses of the texts are not only meant to recreate an epoch and its spectacles for the reader – and in this sense they may be read as projec-tions of the author's exoticising gaze on the Mediterranean countries – but they also function as representations of modernist aesthetic percep-tion. In *Southern Baroque Art* scenery and landscape are often indistin-guishable. Water, music, and air are subject to constant metamorphosis, and so are trees and animals:

> The whole of one of those vast encampments is, as if by a miracle, con-verted from its temporary appearance into permanence; all that was wood becomes stone. Through long waiting the great wooden wagons change into stone houses […]; for it is a process of petrification. (138)

Water is "transmuted" by the golden sun rays (159), Philip V's melan-choly transforms the world into a living nightmare in constant change: "The too voluble nightingales who disturbed him after dark were so many venomous serpents; and the sunlight was like a heavy funeral pall that his enemies were waiting to throw over him" (160).

As Giorgio Melchiori noted, the mannerist images of tightrope walk-ing and vertigo were characteristic descriptions of the epistemological ex-perience of the inter-war period. In 1924, the year of the publication of Sitwell's essay, Virginia Woolf in her essay *Mr Bennett and Mrs Brown* fa-mously compared the experience of reading Eliot's poetry to the dizziness and the dangers of the acrobat's movements (Melchiori 1963: 154; Woolf 1992: 84). Perhaps this is not enough for reading Sitwell's text as the product of an age of disquiet and instability, rather than as a "frivolous" account of newly rediscovered artists and locations in the Mediterranean. But it is to be observed that the confused visual experiences created by the author leave very little room for an essentialising picture of the Southern Mediterranean countries. To him, art must be the expression of race and nation, but he cannot find any confirmation for his theory in Southern Europe:

> Whereas the towns of Northern Europe, and the great monasteries over the countrysides, had developed a style that perfectly interpreted their feelings, it can be said that the great towns of Southern Europe remained without any means of expression peculiar to themselves. (1924: 90)

This is not the case when Sitwell talks about Mexico. There, everything seems to be a transparent expression of the genius loci: the names of the Mexican cities are "skilful imitations of the sounds that nature makes" (225); the extreme ornamentation of the Baroque churches is not "a test of the disconcerting energy from the Indians" but, "on the contrary, a tribute to their slow and contented laziness" (236). In this sense, the impossibility to ascribe a specific identity to Southern European art should not be read as a disparaging judgment (i.e. as an assertion that there is no such identity), but rather as an admission of the problems in the narrator's own position when perceiving the complexity of the much admired objects of his investigation. Sacheverell Sitwell's racism does not allow him to see the problems of his own stance in relation to the cultures of Central America, but his Southern Mediterranean and its artworks are conceived as unstable entities requiring daring stylistic leaps and constantly fostering uncertainty – they are, in other words, remarkably modernist.

Bibliography

Benjamin, Walter, [1925] 1978, *Ursprung des deutschen Trauerspiels*, Suhrkamp, Frankfurt.

Berenson, Bernard - Roberto Longhi, 1993, *Lettere e Scartafacci 1912-1957*, ed. by Cesare Garboli - Cristina Montagnani, Adelphi, Milano.

Briggs, Martin S., 1915a, "The Genius of Bernini", *The Burlington Magazine for Connoisseurs*, Vol. 26, No. 143, pp. 197-202.

——, 1915b, "The Genius of Bernini (Conclusion)", *The Burlington Magazine for Connoisseurs*, Vol. 26, No. 144, pp. 223-228.

Cecere, Angela, 2000, *Viaggiatori inglesi in Puglia nel Novecento*, Schena, Fasano.

Constable, W.G. - C.H.S. John, 1923, "Italian Rococo at Cambridge", *The Burlington Magazine for Connoisseurs,* Vol. 42, No. 238, pp. 46-53.

Fry, Roger, 1913, "Some Pictures by El Greco", *The Burlington Magazine for Connoisseurs*, Vol. 24, No. 127, pp. 2-5.

——, 1917, "Pictures Lent to the National Gallery", *The Burlington Magazine for Connoisseurs*, Vol. 30, No. 170, pp. 198-202.

——, [1920] 1990, *Vision and Design*, ed. by J.B. Bullen, Oxford University Press, London.

——, 1921, "The Baroque", *The Burlington Magazine for Connoisseurs*, Vol. 39, No. 222, pp. 145-148.

——, 1922, "Settecentismo", in *The Burlington Magazine for Connoisseurs*, Vol. 41, No. 235, pp. 158-169.

Gamba, Carlo, 1922, "The Seicento and Settecento Exhibition in Florence", *The Burlington Magazine for Connoisseurs*, Vol. 41, No. 233, pp. 64-75.

Harbison, Robert, 2000, *Reflections on Baroque*, Reaktion Books, London.

Kitson, Michael, 1998, "A Memorable Meeting of Minds: Nikolaus Pevsner and Denis Mahon in the 1930s", *Courtauld Institute of Art, News Archive* 5. Available online at: http://www.courtauld.ac.uk/newsletter/spring_1998/04memorablemeetingSP98.html.

MacLagan, Eric, 1922, "Sculptures by Bernini in England", *The Burlington Magazine for Connoisseurs*, Vol. 40, No. 227, pp. 56-63.

Melchiori, Giorgio, [1956] 1963, *I funamboli. Il manierismo nella letteratura inglese da Joyce ai giovani arrabbiati*, Einaudi, Torino.

Pevsner, Nikolaus, [1925] 1968, "The Counter-Reformation and Mannerism", in *Studies in Art, Architecture and Design,* Vol. 1, Thames and Hudson, London, pp. 11-33.

——, [1932] 1968, "The Crisis of 1650 in Italian Painting", in *Studies in Art, Architecture and Design*, Vol. 1, Thames and Hudson, London, pp. 57-76.

Raimondi, Ezio, 2003, *Barocco moderno. Roberto Longhi e Carlo Emilio Gadda*, Bruno Mondadori, Milano.

Sitwell, Osbert, 1925, *Discursions on Travel, Art and Life*, Duckworth, London.

Sitwell, Sacheverell, 1924, *Southern Baroque Art. A Study of Painting, Architecture and Music in Italy and Spain of the Seventeenth and Eighteenth Centuries*, Grant Richards, London.

——, 1938, *German Baroque Sculpture*, Duckworth, London.

Woolf, Virginia, [1924] 1992, "Mr Bennett and Mrs Brown", in V. Woolf, *A Woman's Essays*, Vol. 1, ed. by Rachel Bowlby, Penguin, Harmondsworth, pp. 69-87.

Ziegler, Philip, 1998, *Osbert Sitwell*, Chatto & Windus, London.

Plates

Plate 1 – Einar Forseth, *Queen of the Mälaren*, 1923. Mosaic, Golden Hall Stockholm.

Plate 2 – Roger Fry, *View of Cassis*, 1925. Oil on canvas, 71.5 × 105 cm. Inscribed 'Roger Fry 1925'. Musée d'Orsay, Paris. Gift of Mrs Pamela Diamand, daughter of the artist, 1959.

Plate 3 – Roger Fry, *Excavations at St Rémy*, c. 1931-3. Oil on canvas, 60.5 × 81.5 cm. Inscribed 'Roger Fry'. Manchester City Art Galleries.

Plate 4 – Frederic, Lord Leighton, *The Bath of Psyche*, exhibited 1890. Oil on canvas, 189.2 × 62.2 cm. Tate Gallery, London.

Plate 5 – Lawrence Alma-Tadema, *A Favourite Custom*, 1909. Oil on canvas, 45 × 66.1 cm. Tate Gallery, London.

Plate 6 – Lawrence Alma-Tadema, *Silver Favourites*, 1903. Oil on panel, 69.1 × 42.2 cm. City of Manchester Art Galleries.

Plate 7 – Lawrence Alma-Tadema, *A Difference of Opinion*, 1896.
Oil on panel, 38 × 22.9 cm. Private collection.

Plate 8 – Lawrence Alma-Tadema, *The Favourite Poet*, 1888. Oil on panel, 36.9 × 49.6 cm.
Lady Lever Art Gallery, Port Sunlight.

Plate 9 – John William Godward, *Song Without Words*
(*Sweet Sounds*), 1918. Oil on
canvas, 59 × 80 cm. Private
collection.

Plate 10 – John William Godward, *Idleness*,
1900. Oil on canvas, 99.1 × 58.4 cm. Private collection.

Plate 11 – Johann Martin von Rhoden, *Am Rand der Campagna in der Valle Dell'Empiglione* (*On the Outskirts of the Campagna in the Empiglione Valley*), c. 1806. Oil on cardboard, glued on canvas, 36 × 63 cm. Kassel, Staatliche Museen (detail).

Plate 12 – Wilhelm Ahlborn, *Die Bucht von Pozzuoli bei Neapel* (*The Bay of Pozzuoli by Naples*), 1832. Oil on canvas, 29 × 83 cm. Berlin Staatliche Museen, Alte National Galerie.

Plate 13 – Johann Wilhelm Schirmer, *Landschaft bei Civitella mit Blick auf Rocca Canterano und Rocca Santo Stefano*, (*Landscape by Civitella with a View of Rocca Canterano and Rocca Santo Stefano*), 1839. Oil on paper, 37 × 53 cm. Karlsruhe, Staatliche Kunsthalle.

LIST OF ILLUSTRATIONS

Figures

Plates

INDEXES

Index of Names

Index of Places

List of Contributors:

Massimo Bacigalupo is Professor of American Literature at the University of Genoa, and Head of the Department of Sciences of Linguistic and Cultural Communication (DISCLIC). He is the author of *The Formed Trace: The Later Poetry of Ezra Pound* (1980), and *Grotta Byron: Luoghi e libri* (2001), and the editor of many volumes of essays and translations, most recently *America and the Mediterranean* (2003) and Ezra Pound, *Canti postumi* (2002). He has contributed to *The Modern Language Review, American Literary Scholarship, Journal of Modern Literature, Paideuma, The Paris Review*, etc.
massimo.bacigalupo@unige.it

J.B. Bullen is Professor of English at the University of Reading. He has had a long-standing interest in interdisciplinary studies. He has published articles on Coleridge, Ruskin, Dickens, George Eliot, Browning, and Pater; he has edited Roger Fry's *Vision and Design* (1981), Clive Bell's *Art* (1987) and has compiled *Post-Impressionists in England* (1988). In 2003 he published a history of the Byzantine Revival entitled *Byzantium Rediscovered*, and his latest book is *European Crosscurrents: British Criticism and Continental Art, 1810-1910* (2005). He is currently writing a critical biography of Dante Gabriel Rossetti.
j.b.bullen@reading.ac.uk

Giovanni Cianci is Professor of English Literature at the State University of Milan. He is the author of *La Scuola di Cambridge* (on the literary criticism of I.A.Richards, W. Empson and F. R. Leavis, 1970) and *La Fortuna di Joyce in Italia* (on the reputation of Joyce in Italy, 1974). He has promoted, edited or co-edited: *Quaderno 9* (a special issue on *Futurism /Vorticism,* 1979), the international collection of essays *Wyndham Lewis: Letteratura/Pittura* (1982), *La Città, 1870-1930* (1991) and *Modernismo/Modernismi* (1991). He has promoted and coedited *Ruskin and Modernism* (2001), *Il Cézanne degli scrittori, dei poeti e dei filosofi* (2001) and *T.S. Eliot and The Concept of Tradition* (2007).
giovanni.cianci@unimi.it

Laura Colombino is Lecturer in English Literature at the University of Genoa. She has written the book *Ford Madox Ford. Visione/visualità e scrittura* (2003), as well as articles on Thomas Hardy, Ford Madox Ford, J.G. Ballard and Geoff Ryman. She is currently working on a book about London's architecture in relation to contemporary British writers.
laura.colombino@lingue.unige.it

Renzo S. Crivelli is Professor of English Literature at the Faculty of Fine Arts of the University of Trieste and Head of the Department of Foreign Literatures, Comparative and Cultural Studies. He is the author of many essays on English, American and Canadian literature. Among his books figure *James Joyce: Triestine Itineraries* (1996, 2nd ed. 2001), *Lo sguardo narrato: Letteratura ed arti visive* (2003), *A Rose for Joyce* (2004). He is the director of the literary review *Prospero* and contributes to the cultural sections of numerous Italian newspapers.
crivelli@units.it

Francesca Cuojati is Research Grantee in English Literature at the State University of Milan, where she obtained her PhD in 2003. She has articles published and forthcoming on Thomas De Quincey, John Soane, John Clare, Alfred Tennyson and James Joyce. Her research interests include mental illness and poetic writing in the Victorian Age, cultural and literary cityscapes, Modernism and the Adriatic.
francesca.cuojati@unimi.it

Gabriella Ferruggia teaches American Literature at the University of Genoa. She has written on African-American theatre, radical writers of the Thirties, and authors such as John Dos Passos, Katherine Anne Porter and Grace Paley.
gabriella.ferruggia@unige.it

Roberta Gefter Wondrich is a Lecturer in English Literature at the University of Trieste. She has specialized on contemporary Irish fiction and published *Romanzi Contemporanei d'Irlanda: Nazione e narrazioni da McGahern a McCabe* (2000) as well as several articles. She has also written on Shakespeare and XXth century British writers. Her current field of research is the Irish novel, British postmodernist Fiction, Postcolonialism and Shakespeare.
gefter@units.it

Michael Hollington is British and Australian, is currently Professor of English at the University of Toulouse (doing most of his teaching at the new University of Albi), but was for many years Professor at the University of New South Wales in Sydney. He is best known as a Dickensian, having published three books on Dickens and edited the largest collection of critical writings in print. But he also works extensively in the Modernist period, with a particular interest in Lawrence and writers connected with him, in particular Katherine Mansfield.
mhollington@wanadoo.fr

Mario Maffi is Professor at the State University of Milan, where he teaches American Studies. His most recent publications include: *Nel mosaico della città. Differenze etniche e nuove culture in un quartiere di New York* (1992, 2nd ed. 2006);

Mississippi. Il Grande Fiume. Un viaggio alle fonti dell'America (2004; French translation in progress); *New York City. An Outsider's Inside View* (2004). Professor Maffi is on the editorial board of *Ácoma. Rivista internazionale di studi nord-americani*.
mario.maffi@unimi.it

Paola Marchetti teaches English Literature and language at the Catholic University of Milan and collaborates with the State University of Milan. She is the author of *Landscapes of Meaning: Form Auden to Hughes* (2001); of *Fiction and Reality: Percorsi del Racconto Contemporaneo in Lingua Inglese* (1999), and articles on Auden, Dylan Thomas and Thom Gunn. She is currently completing a study of Auden's *libretti*.
pamarchetti@hotmail.com

Brett Neilson is Senior Lecturer in the School of Humanities and Languages at the University of Western Sydney, where he is also a member of the Centre for Cultural Research. He is author of *Free Trade in the Bermuda Triangle and Other Tales of Counterglobalization* (2004). He works across the fields of cultural studies and modernist studies, having published articles in journals such as *Comparative Literature, Textual Practice, Twentieth Century Literature* and *Studi Culturali*.
b.neilson@uws.edu.au

Caroline Patey is Associate Professor of English Literature and Culture at the State University of Milan. Her research has been mostly focused on two historical and cultural areas: on one hand, Renaissance studies, that have led to the volumes *Manierismo* (1996) and *Storie nella storia. Teatro e politica nell'Inghilterra rinascimentale* (2000); on the other, her interest for early and mature Modernism has nurtured *Tempi difficili. Su Joyce e Proust* (1991) as well as various essays on Ford Madox Ford, James Joyce, Ezra Pound, T. S. Eliot. Caroline Patey has just published a volume on Henry James and London (*Londra. Henry James e la capitale del moderno*, 2004) and is currently editing a volume of essays on James Joyce and the question of memory.
caroline.patey@unimi.it

Laura Pelaschiar is Lecturer in English literature at the University of Trieste. She is the author of *Writing the North, The Contemporary Novel in Northern Ireland* (1998) and of several articles on contemporary Irish literature and on Joyce. She is currently working on a book on Shakespeare and Joyce.
pelaschi@units.it

Giuseppina Restivo is currently Professor of English Literature at the University of Trieste. She is the author of *La nuova scena inglese: Edward Bond* (1977), *Le soglie del postmoderno: 'Finale di partita' di S. Beckett* (1991) and co-editor, with Renzo Crivelli, of *Tradurre/Interpretare 'Amleto'*, Bologna (2002) and of *Inscenare/Interpretare 'Otello'*, (2006). She has collaborated with Italian and international journals, and has contributed to many volumes. Her research interests include Shakespeare's theatre, contemporary English theatre, cultural theory.
grestivo@units.it

Laura Scuriatti is Assistant Professor of Literature at the European College of Liberal Arts, Berlin. She received a degree in Modern Languages and Literatures at the State University of Milan, and a Master in Literature and the Visual Arts from the University of Reading. At Reading she also obtained her PhD in English Literature. She has published articles on H.G. Wells, Ford Madox Ford, Mina Loy, and on contemporary art. She also edited an anthology of contemporary German literature.
l.scuriatti@t-online.de

Luisa Villa is Professor of English at the University of Genoa. She has published widely on Victorian and early modernist literature. Her publications include books on Henry James (*Esperienza e memoria*, 1989), George Eliot (*Riscrivendo il conflitto*, 1994) and resentment in late Victorian fiction (*Figure del risentimento*, 1997). Together with Marco Pustianaz she recently edited *Maschilità decadenti* (2004). She is currently engaged in a long-term research project on the representation of colonial warfare.
luisa.villa@unige.it

QUADERNI DI ACME

13. Istituto di Archeologia, *Calvatone romana. Studi e ricerhe preliminari* (a cura di Giuliana Facchini)
1991, pp. 218 ill.

14. Istituto di Geografia Umana, *Varietà delle geografie. Limiti e forza della disciplina* (a cura di Giacomo Corna Pellegrini e Elisa Bianchi)
1992, pp. 216

15. Istituto di Psicologia, Luigi Anolli, Rita Ciceri, Federico Denti, *L'incrocio fra università e lavoro. Analisi di recenti percorsi occupazionali dei laureati in lettere e filosofia dell'Università degli Studi di Milano*
1992, pp. 136

16. Istituto di Lingue e Letteratura francese, *La scoperta dell'America e le lettere francesi* (a cura di Enea Balmas)
1992, pp. 292

17. Istituto di Anglistica, *«To Make you See». Saggi su Joseph Conrad* (a cura di Marialuisa Bignami)
1992, pp. 120

18. Istituto di Germanistica, *Vincenzo Errante. La traduzione di poesia ieri e oggi* (a cura di Fausto Cercignani e Emilio Mariano)
1993, pp. 224

19. Istituto di Archeologia, Federica Chiesa, *Aspetti dell'Orientalizzante recente in Campania. La Tomba 1 di Cales*
1993, pp. 176 ill.

20. Istituto di Geografia Umana, *Eventi naturali oggi. La geografia e le altre discipline* (a cura di Giorgio Botta)
1993, pp. 288

21. Istituto di Anglistica, *L'ebreo errante. Metamorfosi di un mito* (a cura di Esther Fintz Menascé)
1993, pp. 416

22. Istituto di Archeologia, *Augusto in Cisalpina. Ritratti augustei e giulio-claudi in Italia settentrionale* (a cura di Gemma Sena Chiesa)
1995, pp. 282 ill.

23. Istituto di Filologia Moderna, *Carte Romanze. Serie I* (a cura di Alfonso D'Agostino)
    1995, pp. 253

24. Istituto di Lingua e Letteratura Francese e dei Paesi francofoni, *Don Giovanni a più voci* (a cura di Anna Maria Finoli)
    1996, pp. 382

25. Dipartimento di Filosofia, *Fondo Giuseppe Rensi. Inventario con una scelta di lettere inedite* (a cura di Lucia Ronchetti e Amedeo Vigorelli)
    1996, pp. 294

26. Dipartimento di Filosofia, *Per una storia critica della scienza* (a cura di Marco Beretta, Felice Mondella e Maria Teresa Monti)
    1996, pp. 518

27. Istituto di Filologia moderna, *Per Giovanni Della Casa* (a cura di Gennaro Barbarisi e Claudia Berra)
    1997, pp. 504

28. Istituto di Lingue e Letterature Iberiche e Iberoamericane, *La scena e la storia. Studi sul teatro spagnolo* (a cura di Maria Teresa Cattaneo)
    1997, pp. 256

29. Istituto di Archeologia, *Calvatone romana. Un pozzo e il suo contesto* (a cura di Gemma Sena Chiesa)
    1997, pp. 282 ill.

30. Istituto di Anglistica, *Wrestling with Defoe: Approaches from a Workshop on Defoe's Prose* (edited by Marialuisa Bignami)
    1997, pp. 200

31. Università degli Studi di Milano, Gioacchino Volpe, *Lezioni milanesi di Storia del Risorgimento* (a cura di Barbara Bracco)
    1998, pp. 190

32. Istituto di Geografia Umana, *Turismo sostenibile in ambienti fragili. Problemi e prospettive degli spazi rurali, della alte terre e delle aree estreme* (a cura di Maria Chiara Zerbi)
    1998, pp. 600 ill.

33. Istituto di Filologia moderna, *Interpretazioni e letture del Giorno* (a cura di Gennaro Barbarisi e Edoardo Esposito)
    1998, pp. 702 ill.

34. Dipartimento di Filosofia, *L'Archivio Giovanni Vailati* (a cura di Lucia Ronchetti)
    1998, pp. 520

35. Università degli Studi di Milano, *Pietro Verri e il suo tempo* (a cura di Carlo Capra)
    1999, pp. VII+1152

36. Dipartimento di Scienze dell'Antichità – Sezione di Filologia classica, *Ricordando Raffaele Cantarella* (a cura di Fabrizio Conca)
    1999, pp. 302

37. Istituto di Filologia Moderna, *Carte Romanze. Serie II* (a cura di Alfonso D'Agostino)
    1999, pp. 318

38. Istituto di Lingua e Letteratura Francese e dei Paesi Francofoni – Istituto di Anglistica, *Intersections. La narrativa canadese tra storia e geografia* (a cura di Liana Nissim e Carlo Pagetti)
    1999, pp. 264

39. Dipartimento di Scienze dell'Antichità – Sezione di Storia antica, *Storiografia ed erudizione. Scritti in onore di Ida Calabi Limentani* (a cura di Daniele Foraboschi)
    1999, pp. 384

40. Istituto di Filologia Moderna, *I* Triumphi *di Francesco Petrarca* (a cura di Claudia Berra)
    1999, pp. 544 ill.

41. Dipartimento di Filologia Moderna, *Studi vari di Lingua e Letteratura italiana in onore di Giuseppe Velli*
    2000, pp. 902

42. Dipartimento di Scienze del Linguaggio e Letterature straniere comparate – Sezione di Francesistica, *Le letture di Flaubert – La lettura di Flaubert* (a cura di Liana Nissim)
    2000, pp. 452 ill.

43. Dipartimento di Filologia moderna, *Fra satire e rime ariostesche* (a cura di Claudia Berra)
2000, pp. 592

44. Dipartimento di Filosofia, *Terra e storia. Itinerari del pensiero contemporaneo* (a cura di Carlo Sini)
2000, pp. 288

45. Università degli Studi di Milano, *L'amabil rito. Società e cultura nella Milano di Parini* (a cura di G. Barbarisi, C. Capra, F. Degrada, F. Mazzocca)
2001, pp. 1225 ill.

46. Dipartimento di Filologia moderna, *Prose della volgar lingua di Pietro Bembo* (a cura di Silvia Morgana, Mario Piotti, Massimo Prada)
2001, pp. 728

47. Università degli Studi di Milano, *Milano e l'Accademia scientifico-letteraria. Studi in onore di Maurizio Vitale* (a cura di Gennaro Barbarisi, Enrico Decleva, Silvia Morgana)
2001, pp. 1272 ill.

48. Dipartimento di Filosofia, *Ortega y Gasset pensatore e narratore dell'Europa* (a cura di Francesco Moiso, Marco Cipolloni, Jean-Claude Lévêque)
2001, pp. 348

49. Dipartimento di Scienze dell'Antichità – Sezione di Archeologia, Giuseppina Pavesi, Elisabetta Gagetti, *Arte e materia. Studi su soggetti di ornamento di età romana* (a cura di Gemma Sena Chiesa)
2001, pp. XII + 512 ill.

50. Dipartimento di Scienze dell'Antichità – Sezione di Filologia classica, *Tra IV e V secolo. Studi sulla cultura latina tardoantica* (a cura di Isabella Gualandri)
2002, pp. 332

51. Dipartimento di Filologia moderna. Letteratura italiana, *Aspetti dell'opera e della fortuna di Melchiorre Cesarotti* (a cura di Gennaro Barbarisi e Giulio Carnazzi)
2002, pp. XXII + 912 ill.

52. Dipartimento di Scienze dell'Antichità – Sezione di Archeologia, *Cerveteri. Importazioni e contesti delle necropoli* (a cura di Giovanna Bagnasco Gianni)
2002, pp. XXII + 678 ill.

53. Dipartimento di Scienze del Linguaggio e Letterature straniere comparate – Sezione di Francesistica, *"La cruelle douceur d'Artémis". Il mito di Artemide-Diana nelle lettere francesi* (a cura di Liana Nissim)
2002, pp. 488 ill.

54. Dipartimento di Scienze dell'Antichità, *Sviluppi recenti nella ricerca antichistica* (a cura di Violetta De Angelis)
2002, pp. 448 ill.

55. Dipartimento di Scienze dell'Antichità, λόγιος ἀνήρ. *Studi di antichità in memoria di Mario Attilio Levi* (a cura di Pier Giuseppe Michelotto)
2002, pp. XII + 516

56. Istituto di Geografia Umana, *Città Regione Territorio. Studi in memoria di Roberto Mainardi* (a cura di Guglielmo Scaramellini)
2003, pp. XI + 582 ill.

57. Dipartimento di Filologia Moderna – Letteratura italiana, *Motivi e forme delle* Familiari *di Francesco Petrarca* (a cura di Claudia Berra)
2003, pp. XII + 820 ill.

58. Dipartimento di Filosofia, *Platone e la tradizione platonica. Studi di filosofia antica* (a cura di Mauro Bonazzi e Franco Trabattoni)
2003, pp. XII + 337.

59. Università degli Studi di Milano, *Achille Vogliano cinquant'anni dopo* I (a cura di Claudio Gallazzi e Luigi Lehnus)
2003, pp. XLII + 359.

60. Dipartimento di Filosofia, *Semiotica ed ermeneutica* (a cura di Carlo Sini)
2003, pp. 375.

61. Dipartimento di Scienze dell'Antichità. Sezione di Archeologia, *Antichi Liguri sulle vie appeniniche tra Tirreno e Po. Nuovi contributi* (a cura di Cristina Chiaramonte Treré)
2003, pp. 278 ill.

62. Dipartimento di Scienze della storia e della documentazione storica, *Contado e città in dialogo. Comuni urbani e comunità rurali nella Lombardia medievale* (a cura di Luisa Chiappa Mauri)
2003, pp. 574 ill.

63. Dipartimento di Filologia moderna, Letteratura italiana, *Idee e figure del "Conciliatore"* (a cura di Gennaro Barbarisi e Alberto Cadioli)
G. Barbarisi e A. Cadioli, *Premessa* – M. Meriggi, *La società lombarda tra il 1814 e il 1821* – G. Albergoni, *I letterati e il potere politico all'epoca del "Conciliatore". Alcune linee interpretative* – C. Annoni, *Gli "Annali di Scienze e Lettere". Appunti per la storia di una rivista milanese (1810-1813)* – W. Spaggiari, *Il programma del "Conciliatore"* – A. Cadioli, *Il testo del "Conciliatore"* – A. Cottignoli, *Il Pellico "conciliatore" e la questione romantica* – M.G. Melli, *«L'atmosphère lumineuse du trône». Il regno, la corte, la nazione nel Grand Commentaire di Ludovico di Breme* – A.M. Morace, *Dal "Bardo" al " Conciliatore". Itinerario di Berchet* – G. Gaspari, *Tra letteratura e nuove scienze. La parte di Giuseppe Pecchio* – D. Tongiorgi, *Rasori, la "Biblioteca" e "Il Conciliatore" (o dell'integrazione impossibile)* – I. Becherucci, *La presenza di J.C.L. Simonde de Sismondi* – G. Turchetta, *Mescidanza di generi e pluri-stilismo nella critica del "Conciliatore"* – L. Bottoni, *Il teatro nel "Conciliatore". Antecedenti e retroscena* – A. Di Benedetto, *Apprezzare Alfieri rendendo giustizia ai suoi rivali. Un tema critico del "Conciliatore"* – A. Terzoli, *Lettere dall'Inghilterra. Foscolo e il gruppo del "Conciliatore"* – A. Colombo, *Riflessioni attorno a una collaborazione mancata: Vincenzo Monti* – L. Sozzi, *La cultura francese nel "Conciliatore"* – G.A. Camerino, *"Il Conciliatore" e la cultura letteraria tedesca* – U. Carpi, *Appunti sul caso Schiller nel romanticismo italiano* – M. Isabella, *"Il Conciliatore" e l'Inghilterra* – G. Carnazzi, *"L'Accattabrighe", satira e parti serie* – G. Nicoletti, *"Il Conciliatore" e la Toscana* – C. Varotti, *Dal "Conciliatore" all'"Antologia". La storia e la tragedia tra Giuseppe Montani e Pagani Cesa* – L. Danzi, *La fortuna del "Conciliatore"* – Indice dei nomi
2004, pp. XIV + 623 ill.

64. Dipartimento di Scienze dell'Antichità, *Il Peloponneso di Senofonte* (a cura di Giovanna Daverio Rocchi e Marina Cavalli)
L. Lehnus, *Premessa* – G. Daverio Rocchi e M. Cavalli, *Introduzione* – G. Schepens, Ἀρετή *e* ἡγεμονία. *I profili storici di Lisandro e di Agesilao nelle* Elleniche *di Teopompo* – G. Daverio Rocchi, *La città di Fliunte nelle* Elleniche. *Caso politico e modello letterario* – J. Taita, *Aspetti di geografia e di topografia dell'Elide nelle* Elleniche – C. Carità, *Un episodio di* asylia *all'Heraion della Perachora* – D.C. Gillone, *I Lacedemoni e l'autonomia degli alleati peloponnesiaci nelle* Elleniche. *Il caso di Mantinea* – G. Zanetto, *Il Peloponneso nella tradizione epica* – F. Puricelli, *Antichità sicione tra storiografia locale e poesia ellenistica* – M. Ornaghi, *Il contributo dei papiri per la ricostruzione e per la storia del testo delle* Elleniche – A. Sgobbi, *Lingua e stile di Senofonte nel giudizio degli antichi* – M. Cavalli, *Esempi di tecnica digressiva nelle* Elleniche – D. Canavero, *Scene di ambasceria nelle* Elleniche – S. Villani, *Presagi, prodigi e sacrifici nelle* Elleniche – A. Pizzone, *Storiografia e socratismo. Il ritratto di Teleutias tra* πρόνοια *e* τόλμη – G. Tentorio e S. Pozzi, *Esempi di similitudini nelle* Elleniche – Indici dei nomi
2004, pp. XII + 386 ill.

65. Dipartimento di Scienze del Linguaggio e Letterature straniere comparate, Sezione di Francesistica, *Sauver Byzance de la barbarie du monde* (a cura di Liana Nissim e Silvia Riva)
A.M. Finoli, *Presentazione* – M. Cacouros, *Vie et survie de Byzance devant les barbares avant et après 1453. Essai sur la culture et l'enseignement à Byzance et dans l'après-Byzance* – G. Lozza, *Bisanzio e il mondo slavo fra ostilità e integrazione* – F. Conca, *Bisanzio e i barbari nel* Digenis Akritas – F. Suard, *Constantinople dans la littérature épique française, jusqu'au* XIVe *siècle* – M. Colombo Timelli, *Cherchez la ville. Constantinople à la cour de Philippe le Bon (1419-1467)* – M. Barsi, *Constantinople à la cour de Philippe le Bon (1419-1467). Compte rendus et documents historiques. Avec l'édition du manuscrit B.n.F. fonds français 2691 du récit de Jacopo Tedaldi* – J. Balsamo, *Byzance à Paris: Chalcondyle, Vigenère, L'Angelier* – J. Morgante, *"Passer la mer et aller à Bisance" (1, 2). Bouleversements politiques dans l'* Astrée *d'Honoré d'Urfé* – C. Biet, *Les miroirs de la Sublime Porte. Reflets et distorsions de l'image de Constantinople dans le théâtre du* XVIIe *siècle* – M. Ferrarini, *Dalla "Constantinople" di Voltaire alla "Byzance" di Casanova* – M.G. Longhi, *"À la façon d'un conte des* Mille et Une Nuits*". Maupassant lecteur de Gustave Schlumberger* – M.-F. David de Palacio, *La tyrannie de l'Idée. Irène et la Décadence* – G. Ducrey, *Gismonda de Victorien Sardou (1894). Décadence de l'empire byzantin et féerie fin-de-siècle* – E. Sparvoli, *Screziature bizantine nella Venezia di Proust* – F. Luoni, *Le byzantinisme ou la littérature. Julien Benda face à ses cauchemars* – S. Riva, *Byzance ou la rêverie d'arrière-pays chez Yves Bonnefoy* – M. Modenesi, *Splendeurs et misères de la petite courtisane du troisième canton. Une trilogie byzantine à la fin du* XXe *siècle* – L. Nissim, *"Mais que reste-t-il de Byzance?". Quelques notes sur un poème de Paul Mathieu* – L. Nissim, *Veleggiare verso Bisanzio...* – Illustrazioni – Tavole – Indice dei nomi
2004, 502 ill.

66. Dipartimento di Filosofia e Dipartimento di Scienze della Storia e della Documentazione Storica, *Auguste Comte e la cultura francese dell'Ottocento. In ricordo di Mirella Larizza* (a cura di Marco Geuna)
M. Genua, *Premessa* – PARTE PRIMA – A. Petit, *La sociologie positiviste: entre histoire, politique et religion* – J. Lalouette, *Auguste Comte et le catholicisme* – R. Pozzi, *I fondamenti etico-religiosi della cittadinanza nel pensiero francese dell'Ottocento* – G. Lanaro, *Comte e Littré* – V. Collina, *Renouvier critico di Comte* – M.L. Cicalese, *Villari e Comte: un incontro a metà strada* – PARTE SECONDA – M. Larizza, *Tra virtù e libertà: percorsi dell'idea repubblicana nella Francia ottocentesca (1830-1875)* – M. Larizza, *Auguste Comte: la Repubblica, la scienza e le passioni* – M. Larizza, *Il pacifismo laico nella Francia ottocentesca (1820-1852)* – Mirella Larizza (1942-1998). Bibliografia (1967-2004) – Indice dei nomi
2004, pp. X + 234.

67. Dipartimento di Scienze dell'Antichità, *Momenti della ricezione omerica. Poesia arcaica e teatro* (a cura di Giuseppe Zanetto, Daniela Canavero, Andrea Capra, Alessandro Sgobbi)

L. Lehnus, *Premessa* – G. Zanetto, D. Canavero, A. Capra, A. Sgobbi, *Introduzione* – B. Graziosi, *La definizione dell'opera omerica nel periodo arcaico e classico* – J. Haubold, *Serse, Onomacrito e la ricezione di Omero* – G. Zanetto, *Omero e l'elegia arcaica* – G. Burzacchini, *Omero nella giambografia arcaica* – A. Capra, *Simonide e le corone di Omero (Simon. 47k Campbell = 10, 2 Lanata e fr. 11 West²)* – N. Stanchi, *La sede di Menelao e il destino di Agamennone in Omero ed Eschilo* – A. Sgobbi, *Tiresia. L'evoluzione di un personaggio tra rispetto e contestazione (Omero, Stesicoro, Sofocle)* – D. Canavero, *Ripresa ed evoluzione: Andromaca ed Ecuba nelle* Troiane *di Euripide* – D. Del Corno, *Odisseo fra i satiri* – M. Ornaghi, *Omero sulla scena. Spunti per una ricostruzione degli* Odissei *e degli* Archilochi *di Cratino* – M. Cavalli, *La comicità del sacrale in Aristofane* – Indici dei nomi – Indice delle fonti
2004, pp. XVIII + 286.

68. Dipartimento di Scienze dell'Antichità, *Sviluppi recenti nell'antichistica. Nuovi contributi* (a cura di Violetta de Angelis)

V. de Angelis, *Premessa* – G. Bejor, *Riscavo di uno scavo: la riscoperta di Nora tardoantica* – S. Cappelletti, *Note sulla presenza ebraica in Italia settentrionale* – C. Chiaramonte Treré, *Contesti e corredi ellenistici nell'Abruzzo settentrionale* – R. Arena, *Divagazioni su alcuni termini greci* – A. Scala, *Una lettera di Atanasio Alessandrino sull'eresia elchasaita conservata in traduzione armena* – M. Gioseffi, *Pseudo-Probo ad Virg. buc.2.48: Narciso e i suoi pittori* – P. Piacentini, *La Biblioteca e gli Archivi di Egittologia. Nuove acquisizioni e attività in corso* – D. Foraboschi, *Momenti della storiografia sulla produzione "industriale" antica* – V. de Angelis, *Marsilio Ficino al dottorato di Angelo Battista Golfo: 12 maggio 1467* – S. Bussi, *Le élites dell'Egitto romano tra I e III secolo e la loro conservazione* – L. Biondi, *Per uno studio di testi di ortografia latina nel Medioevo* – N. Negroni Catacchio, *Tra protostoria e storia. Il contributo degli scavi di Sorgenti della Nova e Sovana al processo di formazione della nazione etrusca* – A. Bonini, *Indagini Archeologiche nell'area del Capitolium di Brescia. I risultati delle nuove ricerche* – F. Cordano, *La musica e la politica, ovvero gli auloí ad Atene*
2004, pp. VIII + 356 ill.

69. *L'Archivio Storico dell'Università degli Studi di Milano. Inventario* (a cura di Stefano Twardzik)

M. Bologna, *Per conoscere un archivio – Introduzione* – ARCHIVIO PROPRIO – *Profilo storico-istituzionale (1924-1960) – Inventario* – ARCHIVI AGGREGATI – *R. Accademia scientifico-letteraria di Milano e istituti annessi – R. Scuola superiore di Medicina veterinaria di Milano poi R. Istituto superiore di Medicina veterinaria di Milano – R. Scuola superiore di Agricoltura di Milano poi R. Istituto superiore Agrario – Campi universitari d'internamento per militari italiani in Svizzera – R. Scuola di Ostetricia di Milano – RR. Istituti clinici di perfezionamento (1906-1924) – Consorzio per l'assetto degli Istituti d'istruzione superiore in Milano – Eredità Eugenio Diviani – Società cooperativa edilizia fra il personale amministrativo dell'"Università Statale' di Milano a.r.l.* – Cronotassi dei provvedimenti normativi citati – Indici degli enti – Indice dei nomi di persona
2005, pp. XXII + 282.

70. Dipartimento di Filosofia, *Fondo Mario Dal Pra* (a cura di Giuseppe Barreca e Piero Giordanetti)
R. Pettoello, *Premessa – Presentazione – Titolo 1. Corrispondenza – Titolo 2. Scritti di Mario Dal Pra – Titolo 3. Scritti su Mario Dal Pra – Titolo 4. Fascicoli vari – Titolo 5. Scritti di altri autori – Titolo 6. Attività accademica –* Indici dei nomi
2005, pp. 232.

71. Dipartimento di Filologia moderna, Letteratura italiana, *Il teatro di Machiavelli* (a cura di Gennaro Barbarisi e Anna Maria Cabrini)
G. Barbarisi, A.M. Cabrini, *Premessa –* F. Bausi, *Machiavelli e la commedia fiorentina del primo Cinquecento –* D. Fachard, *Il teatro machiavelliano e la «qualità de' tempi» –* J.-J. Marchand, *Teatralità nel primo Machiavelli. Il dispaccio ai Dieci di Balìa del 28 agosto 1506 –* F. Grazzini, *Teatralità indiretta di Machiavelli: le Lettere e la novella di Belfagor –* J.-C. Zancarini, *«Ridere delli errori delli huomini». Politique et comique chez Machiavel –* E. Fumagalli, *Machiavelli e l'esegesi terenziana –* P. Stoppelli, *La datazione dell'*Andria – C. Varotti, *Il teatro di Machiavelli e le parole degli antichi –* M. Martelli, *La* Mandragola *e il suo* Prologo – G.M. Anselmi, *Partitura della* Mandragola – C. Vela, *La doppia malizia della* Mandragola – A.M. Cabrini, *Fra' Timoteo –* G. Coluccia, *Ligurio o dell'intelligenza –* G.M. Barbuto, *Mandragola: commedia della politica –* F. Fedi, *«El premio che si spera»: il* Prologo *della* Mandragola *e il motivo dell'ingratitudine nell'opera di Machiavelli –* A. Bruni, *Gli intermedi della* Mandragola – F. Tomasi, *Letteratura e nuove tecnologie. La* Mandragola *di Niccolò Machiavelli: qualche esempio di elaborazione elettronica –* R. Rinaldi, *«Se voi non ridete». Allegorie teatrali di Machiavelli –* D. Perocco, *Il testo della* Clizia – G. Inglese, *«Le stesse cose ritornano». Considerazioni sulla* Clizia – C. Figorilli, *La presenza del teatro di Machiavelli in alcune commedie fiorentine degli anni Trenta e Cinquanta del Cinquecento –* N. Ordine, *Teatro e conoscenza: su alcuni luoghi della* Mandragola *e del* Candelaio – E. Cutinelli-Rèndina, *Sulla costituzione del* corpus *teatrale di Niccolò Machiavelli –* P. Bosisio, *Letture sceniche novecentesche del teatro di Machiavelli –* Indice dei nomi – Indice dei manoscritti
2005, pp. XII + 620 ill.

72. Dipartimento di Scienze dell'Antichità, Sezione di Storia antica, *L'opera e l'importanza di Friedrich Stählin* (a cura di Floriana Cantarelli)
F. Cantarelli, *Prefazione –* E. Olshausen, *La vita e le opere di Friedrich Stählin –* G. Daverio Rocchi, *I confini di Melitea –* F. Cantarelli, *Friedrich Stählin: motivi per la riscoperta di un grande studioso della grecità –* B. Adrymi-Sismani, *Il contributo di Friedrich Stählin alla ricerca archeologica nella regione tessala –* Z. Malakassioti, *La piana di Almyros. Dalla descrizione di Friedrich Stählin alle ricerche topografico-archeologiche in corso –* M.-F. Papakonstantinou, *I risultati della ricerca archeologica attuale da parte della XIV Eforia delle Antichità Preistoriche e Classiche in rapporto ai dati di Stählin –* P. Bouyia, *Meliteia e Narthakion nell'opera di Friedrich Stählin e l'attualità archeologica –* A. Stamoudi, *La recente scoperta di una città classica in Acaia Ftiotide –* P. Puppo, *L'interesse di Friedrich Stählin per le fasi bizantine –* G. Rovagnati, *"Pathos" ed esattezza. La lingua di Friedrich Stählin fra emozione e scienza –* Indice dei nomi
2005, pp. 184 ill.

73. Dipartimento di Scienze dell'Antichità, Sezione di Filologia classica, *Nuovo e Antico nella cultura greco-latina di IV-VI secolo* (a cura di Isabella Gualandri, Fabrizio Conca, Raffaele Passarella)

I. Gualandri, *Introduzione* – TRA ORIENTE E OCCIDENTE – A. Garzya, *Il modello della formazione culturale nella tarda antichità* – U. Criscuolo, *Interferenze fra neoplatonismo e teologia cristiana nel tardoantico* – B. Moroni, *Dopo Giuliano. Lingua e cultura greca nella famiglia imperiale fino a Teodosio* – CULTURA SECOLARE – L. Ceccarelli, *L'esametro di Ausonio tra classico e tardoantico* – N. Brocca, *Il* proditor Stilicho *e la distruzione dei* Libri Sibyllini – I. Gualandri, *Proserpina e le tre dee* (De raptu Pros. *1.214-2.54)* – L. Pernot, *L'uomo biblioteca. Intorno a una formula di Eunapio* (Vit. phil. *4.1.3:* bibliothêkê tis... empsukhos) *e alla sua fortuna* – A. Luceri, *Il carro di Venere: tradizione e innovazione in Draconzio,* Romuleon 6.72-79 – I. Canetta, Quod fecit Homerus*: i rimandi omerici nel commento di Servio all'*Eneide – M. Gioseffi, *Un libro per molte morali. Osservazioni a margine di Tiberio Claudio Donato lettore di Virgilio* – G. Moretti, *Ennodio all'incrocio fra allegoria morale e allegoria dottrinale* – CULTURA RELIGIOSA – E. Giannarelli, *Dal vescovo Cipriano al vescovo Martino: modelli "doppi" di santità e scritture anomale* – F. Gori, *L'oratoria cristiana antica: dall'improvvisazione alla ripetizione. Il ruolo della memoria* – C. Castelli, *Gregorio di Nazianzo nell'*Epitafio per Basilio il grande – C. Lo Cicero, *I* munera *del cristiano: Rufino di Aquileia lettore di Cipriano* – M.T. Messina, *Nuove tracce di Origene nel* Commento ad Osea *di Girolamo?* – F.E. Consolino, *Il senso del passato: generi letterari e rapporti con la tradizione nella 'parafrasi biblica' latina* – G. Lozza, *Postille al canto XI della* Parafrasi *di Nonno di Panopoli* – AMBROSIANA – M. Caltabiano, *Ambrogio e la comunicazione* – M. Cutino, *La tripartizione del sapere in Ambrogio* – P.F. Moretti, *Cane d'un filosofo, cane d'un eretico. Appunti sulla fortuna cristiana del cane "sillogistico"* – E. Meligrana, *L'esortazione al digiuno: rielaborazione del modello basiliano nel* De Helia *di Ambrogio* – L.F. Coraluppi, *Uso retorico del lessico giuridico nel* De Tobia *di Ambrogio: considerazioni preliminari* – R. Passarella, *Le spalle di Sichem* (Ambr. Interpell. *4.4.16)* – C. Somenzi, *Affinità di formazione scolastica tra Ambrogio e lo ps. Egesippo?* – Indice dei nomi
2005, pp. XX + 820 ill.

74. *Vincenzo Monti nella cultura italiana* (a cura di Gennaro Barbarisi). Volume I

G. Barbarisi, *Presentazione* – TOMO PRIMO – SESSANT'ANNI DI STUDI MONTIANI – A. Romano, *Gli studi biografici* – A. Colombo, *Il carteggio negli studi dell'ultimo sessantennio (considerazioni e inediti)* – S. Casini, *Le raccolte poetiche e le traduzioni* – S. Contarini, *Le tragedie e le cantate* – L. Frassineti, *La poesia epica e politica del Monti nella critica del secondo Novecento* – MONTI E LA CULTURA EMILIANO-ROMAGNOLA – F. Tarozzi, *La situazione socio-economica delle legazioni a metà Settecento* – W. Moretti, *Monti a Ferrara* – L. Pepe, *Vincenzo Monti e la cultura scientifica* – F. Sani, *Il seminario di Faenza* – A. Di Ricco, *Monti e Frugoni* – W. Spaggiari, *Monti, Minzoni, Varano: gli esordi poetici* – I. Magnani Campanacci, *Monti e i neoclassici emiliani: Savioli, Cassiani, Mazza* – OMAGGIO A VINCENZO MONTI – P. Rota, *Monti e la Bibbia* – M. Veglia, *Monti esegeta della "Commedia"* – G. Ruozzi, *Relazioni epigrammatiche: Monti e Bettinelli* – E. Ghidetti, *Di Leopardi su Monti* – D. Vanden Berghe, *Leopardi lettore di Monti poeta* – A. Cottignoli,

Carducci editore e critico del Monti (con documenti inediti) – TOMO SECONDO – MONTI NELLA CULTURA DEL SUO TEMPO – A. Bruni, *La funzione Monti* – E. Mattioda, *Monti, Ossian e l'epicizzazione del presente* – I. Becherucci, *Il primo "maestro" di Alessandro Manzoni* – A. Cristiani, *La ricezione della produzione letteraria di Vincenzo Monti nelle riviste del Settecento* – E. Parrini Cantini, *Il* Saggio di Poesie *tra citazione e autocitazione* – R. Bertazzoli, *Modelli stranieri per la «Bellezza dell'Universo» di Vincenzo Monti* – T. Matarrese, *La traduzione come poesia: sull'*Iliade *di Vincenzo Monti* – A. Dardi, *Il dialogo «Matteo Giornalista» del Monti ai primordi del dibattito sul romanticismo* – M. Fanfani, *L'Accademia della Crusca dopo la "Proposta"* – G. Lavezzi, *Il poeta ha buon orecchio: appunti sugli endecasillabi sciolti di Vincenzo Monti* – L. Frassineti, *In margine all'«Epistolario» del Monti: note sul poeta esordiente* – P. Palmieri, *Il "Giornale Arcadico" e Vincenzo Monti* – MONTI LINGUISTA, FILOLOGO, CLASSICISTA – M.M. Lombardi, *Gli scritti lessicografici di Vincenzo Monti per l'allestimento della* Proposta – P. Italia, *Monti e Leopardi: la* Proposta*, le* Annotazioni *e l'*Apologia *di Annibal Caro* – D. Martinelli, *Tommaseo e la* Proposta *del Monti* – A. Colombo, *Lo studioso del* Convivio *di Dante* – G. Lucchini, *Note e appunti sulla collaborazione tra Monti e Perticari* – L. Lehnus, *Il cavallo alato d'Arsinoe* – G. Benedetto, *Le versioni latine dell'*Iliade – P. Vecchi Galli, *Monti e Virgilio* – R. Pestarino, *Spunti di (auto)critica letteraria nel* Persio – Indice dei nomi 2005, pp. XII + 1154 ill.

75. Dipartimento di Studi Linguistici, Letterari e Filologici, Sezione di Germanistica, *Rappresentare la Shoah* (a cura di Alessandro Costazza)
A. Costazza, *La Shoah: memoria e rappresentazione* – MEMORIA – A. Appelfeld, *Arte e Shoah* – R. Marcinkowski, *La sorte dei bambini ebrei alla luce degli archivi clandestini del ghetto di Varsavia* – C. Segre, *Racconti dal lager* – G. Scaramuzza, *L'inenarrabile e la testimonianza* – M. Ciaravolo, *La voce di Cordelia Edvardson, "bambina bruciata"* – F. Melzi d'Eril, *Sarah Kofman tra parola e silenzio* – RAPPRESENTAZIONE – E. Franzini, *Rappresentazione e catarsi* – POESIA – C. Rosenzweig, *Scrivere per sopravvivere. La letteratura yiddish e lo sterminio del popolo ebraico* – M. Mayer Modena, *La poesia israeliana sulla Shoah* – M. Paleari, *I percorsi del ricordo nella lirica femminile di lingua tedesca* – S. Raimondi, *Edmond Jabès e l'interrogazione dell'assenza* – TEATRO – M. Castellari, *Documento e allegoria. Strategie di rappresentazione ne* L'istruttoria *di Peter Weiss* – E. Fintz Menascé, *L'esodo degli innocenti:* Kindertransport*, dramma di Diane Samuels* – C. Patey, *I silenzi del palcoscenico: il teatro britannico di fronte alla Shoah* – M. Mazzocchi Doglio, *Presa di coscienza e funzione taumaturgica della rappresentazione della Shoah nella drammaturgia di Jean-Claude Grumberg* – F. Haas, *La gaia Carinzia ieri e oggi.* Tanzcafé Treblinka *di Werner Kofler* – A. Bentoglio, *Teatro per non dimenticare: il* Diario di Anna Frank *sulle scene italiane* – PROSA – A.L. Callow, *Né in prosa né in poesia. I racconti di Avraham Sutzkever* – A. Zieliński, *Viaggio ai confini di una certa morale. I racconti dal lager di Tadeusz Borowski* – G. Brogi Bercoff, *La tragedia di Babyn Jar* – M.L. Roli, *Documento storico e finzione narrativa in* Jakob Littners Aufzeichnungen aus einem Erdloch *di Wolfgang Koeppen* – G. Rovagnati, *Non si emigra da se stessi. La persecuzione del ricordo in* Die Ausgewanderten *di G.W. Sebald* – A. Costazza, *La "memoria ereditaria": la Shoah nel romanzo e nel film* Gebürtig, *di Robert Schindel* – C. Pagetti, *"Making a 'story' out of extremity": la rappresentazione della Shoah*

nella cultura anglo-americana – O. Palusci, *Le tre Anne: l'Olocausto e la* children's literature – Altri media – L. Bernardini, *Una verità non artistica su Auschwitz, o la neve su Birkenau* – P. Bozzi, *Posizionalità e postmemoria. Incontri con la letteratura, la fotografia e l'architettura: Christoph Dahlhausen, Ruth Klüger, Daniel Libeskind, Paul Celan* – A. Di Liddo, *Il fardello della memoria, l'ansia della rappresentazione.* Maus: A Survivor's Tale *di Art Spiegelman* – S. Bignami, *Counter-monuments: memoria e rappresentazione tra Austria e Germania* – C. Cappelletto, *Le rotaie di Anselm Kiefer. L'immagine come rovina* – L. Bruti Liberati, *Shoah e guerra fredda attraverso lo specchio di Hollywood, 1945-1958* – E. Dagrada, *La giustezza è il fardello di chi viene dopo. A proposito di* Shoah *di Claude Lanzmann* – Immagini e documenti – Indice dei nomi 2005, pp. 576 ill.

76. Dipartimento di Storia delle Arti, della Musica e dello Spettacolo, Sezione Musica, *Musica e architettura nell'età di Giuseppe Terragni (1904-1943)* (a cura di Claudio Toscani)

C. Toscani, *Premessa* – O. Selvafolta, *Giuseppe Terragni e il razionalismo comasco: "ritmi antichi" nella modernità* – T.F. Pertusini, *Giuseppe Terragni e Alfredo Casella: architettura e musica fra tradizione e innovazione* – T. Patetta, *Ambiguità e complessità delle tradizioni nelle arti e nell'architettura del "Novecento Italiano"* – N. Cattò, *«Combattere Gabriele D'Annunzio». Il futurismo musicale e il Vate* – P. Rusconi, *«Gli occhi dei pittori di città». L'immagine di Milano nella pittura dei primi anni Trenta* – R. Valsecchi, *Musiche di scena nel teatro di Massimo Bontempelli* – M. Nahon, *I drammi "romani" di Malipiero: richiamo all'ordine o fuga dalla storia?* – Immagini e documenti – Indice dei nomi 2005, pp. X + 220 ill.

77. Dipartimento di Scienze dell'Antichità, Sezione di Archeologia, *Tarquinia e le civiltà del Mediterraneo* (a cura di Maria Bonghi Jovino)

M. Bonghi Jovino, *Introduzione* – A.M. Moretti Sgubini, *Conoscenza e valorizzazione in Etruria meridionale* – A. Giulivi, *Comune di Tarquinia e Università di Milano: risultati e prospettive* – D. Ridgway, *Riflessioni su Tarquinia. Demarato e l'"ellenizzazione dei barbari"* – G. Bartoloni, *L'inizio del processo di formazione urbana in Etruria. Analogie e differenze venute in luce nei recenti scavi* – M. Cataldi, *Tarquinia: una coppa "euboica" dalla necropoli di poggio della Sorgente* – A. Rathje, *Il sacro e il politico. Il deposito votivo di Tarquinia* – F. Prayon, *Lastre architettoniche di tipo tarquiniese da Castellina del Marangone* – N.A. Winter, *Le terrecotte architettoniche arcaiche di Tarquinia. Scambi e modelli* – D. Ciafaloni, *Nota sulle tipologie architettoniche e murarie tarquiniesi. Ulteriori corrispondenze con il Vicino Oriente antico* – G. Colonna, *Un pittore veiente del Ciclo dei Rosoni: Velthur Ancinies* – F.R. Serra Ridgway, *La ceramica del "complesso" sulla Civita di Tarquinia* – S. Stopponi, *Tecniche edilizie di tipo misto a Orvieto* – J. Gran-Aymerich, *Les confins maritimes entre Tarquinia et Caere. Civitavecchia et les recherches a La Castellina del Marangone* – S. Steingräber, *La pittura funeraria tarquiniese del periodo tardoclassico e del primo ellenismo nel contesto mediterraneo. Iconografia, stile, tecnica pittorica* – L. Cerchiai, *A proposito degli* artifices *pliniani (*Pl., N.H., *XXXV, 152)* – M. Gnade, *Tarquinia e Satricum: raffronti fra le prassi rituali* – B. d'Agostino, *I primi Greci in Etruria* – M. Torelli, *Due ritratti greci, una*

villa marittima e le coste di Gravisca – G. Bagnasco Gianni, *Ritornando ai depositi votivi del "complesso monumentale" di Tarquinia* – S. Bruni, *Le analisi chimiche nello studio dei materiali ceramici di Tarquinia* – S. Piro, *Indagini integrate ad alta risoluzione nelle aree di Tarquinia antica* – M. Bonghi Jovino, *Progettualità e concettualità nel percorso storico di Tarquinia*
2006, pp. 432 ill.

78. Dipartimento di Scienze del Linguaggio e Letterature straniere comparate, Sezione di Francesistica, *Magia, gelosia, vendetta. Il mito di Medea nelle lettere francesi* (a cura di Liana Nissim e Alessandra Preda)
A.M. Finoli, *Presentazione* – G. Lozza, *Il mito di Medea* – A. Colombini Mantovani, *Un'altra Medea: la Medea fanciulla di Benoît de Sainte-Maure* – P. Caraffi, *Il mito di Medea nell'opera di Christine de Pizan* – M. Colombo Timelli, *Entre magie du savoir et magie de la parole: Médée dans l'*Histoire de Jason *de Raoul Lefèvre (vers 1460)* – A. Preda, *La rondine e la statua: Medee cinquecentesche* – C. Biet, *Médée, caméléon sanglant. Le personnage de Médée dans les récits sanglants du début du XVIIe siècle* – D. Cecchetti, *Variazioni su tema: la* Médée *di Thomas Corneille* – K. Cahill, *Médée et le sublime* – N. Clerici Balmas, *Un'eroina raciniana. La* Médée *di Longepierre* – S. Biandrati, *La parodia della* Médée *di Longepierre* – I. Panighetti, *"D'une Médée à faire": la proposta di Delisle de Sales* – F. Claudon, *Médée révolutionnaire et romantique. Cherubini et Hoffmann* – F. Naugrette, *Le 'toi' de Médée et le Code Civil* – M. Marchetti, *"Médée la barbare" di Catulle Mendès* – M. Modenesi, *"Une idée de sorcellerie et de magie": les charmes de Médée à la fin du XIXe siècle* – V. Léonard-Roques, *Les enfants de Médée dans la littérature moderne* – E. Sparvoli, *Una mostruosa apoteosi: la* Médée *di Milhaud e Masson* – S. Genetti, *'Médée commence'. In merito a* Jason *di Elisabeth Porquerol* – F. Bruera, *Incantesimi senza magia:* Médée *di Jean Anouilh* – T. Karsenti, *Eclairer l'obscur. La* Médée *de Thomas Corneille mise en scène par Jean-Marie Villégier* – A.P. Mossetto, *La sfida africana di Medea* – L. Nissim, *Le mythe de Médée, ou de l'altérité barbare. Quelques propos en guise de conclusion* – Illustrazioni – Tavole – APPENDICE, E. Mosele, *Presentazione del volume di Enea Balmas,* Studi sul Cinquecento – Indice dei nomi
2006, pp. 464 ill.

79. Dipartimento di Scienze del Linguaggio e Letterature straniere comparate, Sezione di Iberistica, *Luoghi per il Don Chisciotte* (a cura di Mariarosa Scaramuzza Vidoni)
M. Scaramuzza Vidoni, *Presentazione* – J. Mesanza Martínez, *Ricompensa de la caballeria (poesia)* – F. Cercignani, *Caro Chisciotte (poesia)* – M. Modenesi, *Di quel che accade a Don Chisciotte andando in Québec* – C. Pagetti, *Don Chisciotte sul Mississippi* – E. Perassi, *Salvador Novo: Don Chisciotte al Messico (e altre spigolature americane)* – N. Vallorani, *La follia necessaria. Don Chisciotte e la salvezza del mondo in Kathy Acker* – G. Rovagnati, *Memoria e metamorfosi ovvero da Miguel de Cervantes a Elias Canetti* – G. Scaramuzza, *Liberarsi di Don Chisciotte? Divagazioni kafkiane* – C. Montaleone, *E quando la botte non viene vuotata?* – C. Segre, *Due Don Chisciotte in conflitto* – M.T. Cattaneo, *Don Chisciotte e la scena* – A. Bentoglio, *Teatro, cinema e televisione: il* Don Chisciotte

multimediale di Maurizio Scaparro – E. Dagrada, *Don Chisciotte secondo Orson Welles* – C. Janés, *Divertimento sobre el* Quijote *y la luna* – R. Puglise, *«D'amore furente e cieco»: la vicenda di Cardenio tra Cervantes e Donizetti* – M. Scaramuzza Vidoni, *Sui simboli alimentari nel* Quijote – G.C. Torre, *Sulla presenza del* Don Chisciotte *nelle arti e nel mondo degli ex libris* – Tavole – Indice dei nomi
2006, pp. 308 ill.

80. Dipartimento di Filologia moderna. Letteratura italiana, Dei Sepolcri *di Ugo Foscolo* (a cura di Gennaro Barbarisi e William Spaggiari)
G. Barbarisi e W. Spaggiari, *Premessa* – TOMO PRIMO – F. Gavazzeni, *Per* Dei Sepolcri *di Ugo Foscolo* – R. Bertazzoli, *La tradizione della poesia sepolcrale e i versi di Ugo Foscolo* – E. Neppi, *Ontologia dei* Sepolcri – F. Fedi, *Foscolo e i riti funebri degli antichi* – L. Sozzi, *Ancora sui* Sepolcri *e sul culto francese delle tombe* – D. Martinelli, *Alberi e fiori sui sepolcri (e altri motivi della polemica foscoliana sull'editto di Saint-Cloud)* – G. Pizzamiglio, *Pindemonte e Foscolo tra* Cimiteri *e* Sepolcri – M.A. Terzoli, *Lettura dei* Sepolcri – P. Frare, *Note sull'endecasillabo dei* Sepolcri – G. Benedetto, *I* Sepolcri *nella storia della fortuna di Pindaro* – F. Longoni, Dei Sepolcri *e* Omero – E. Farina, *La presenza di Ossian nei* Sepolcri – S. Casini, *Dei sepolcri sallustiani. A proposito di implicazioni alfieriane nei* Sepolcri – G. Venturi, *Canova e Firenze: echi canoviani nei* Sepolcri – A. Bruni, Monti nei *Sepolcri* – S. Longhi, *«Tortuose ambagi»: lo stile difficile del verso sciolto* – L. Chines, *I veli fra Petrarca e Foscolo* – E. Pasquini, *Il «ghibellin fuggiasco»* – C. Del Vento, *I* Sepolcri *e la «nuova poetica» foscoliana* – M.M. Lombardi, *La lingua dei* Sepolcri *e il* Vocabolario della Crusca – P. Vecchi Galli, *Foscolo, i* Sepolcri *e l'Inghilterra* – TOMO SECONDO – A. Cadioli, *Le prime edizioni dei* Sepolcri – R. Pestarino, *I «*Sepolcri *rifatti» di Girolamo Federico Borgno* – L. Melosi, *«... e serbi un sasso il nome»: Foscolo e l'epigrafia ottocentesca* – G. Lucchini, *La polemica tra il Guillon e il Foscolo* – W. Spaggiari, *Foscolo, Giordani e il «fumoso enigma»* – Paola Italia, *Foscolo in Leopardi: i* Sepolcri *e le canzoni «patriottiche»* – E. Ghidetti, Dei Sepolcri *o del «patriottismo disperato»* – G. Nicoletti, *Appunti sulla ricezione dei* Sepolcri *nella Toscana granducale* – A. Cottignoli, *Carducci lettore inedito dei* Sepolcri – A. Colombo, *Fortune transalpine dei* Sepolcri – J. Lindon, *Varia fortuna di Ugo Foscolo e dei* Sepolcri *in Inghilterra fino alla Seconda Guerra Mondiale* – D. Tongiorgi, *«Terribile e distruggitrice filosofia»: i* Sepolcri *nelle antologie scolastiche ottocentesche* – Indice dei nomi
2006, pp. XIV + 966 ill.

81. Istituto di Geografia umana, *Un geografo per il mondo. Studi in onore di Giacomo Corna Pellegrini* (a cura di Elisa Bianchi)
E. Bianchi, *Introduzione* – Giacomo Corna Pellegrini Spandre - Biografia - Bibliografia *dei libri principali* – M. Bergaglio, *La gran morìa. L'epidemia di "spagnola" a Milano nell'inverno del 1918* – E. Bianchi, *Il problema immigrazione: guardare indietro, guardare altrove per comprendere meglio* – L. Bonardi e F. Cavallo, *Segni di lago, segni di monte. Elementi per una geografia culturale di Montisola* – G. Botta, *Rappresentare le tradizioni* – C. Brusa, *La ricerca geografica italiana e i problemi delle migrazioni e della formazione di una società multiculturale* – M. Casari, *Ospitalità, accoglienza e ricevimento. Per una geografia di un* mondo (più) umano *del XXI secolo* – E. dell'Agnese, *Paesaggi ed eroi*

nell'Australia di Peter Weir (Gallipoli) – F. Eva, *I confini flessibili delle culture* – M. Fumagalli, *La pluralità di culture: benefici e costi* – D. Gavinelli, *Nuova Caledonia: un paesaggio culturale e socio-economico in ricomposizione e una nuova realtà di condivisione politica* – P. Inghilleri, *L'approccio bio-psico-sociale e le scienze geografiche: una nota* – F. Lucchesi, *La Geografia nella Scuola di Specializzazione per l'Insegnamento Secondario dell'Università degli Studi di Milano* – M. Morazzoni, *L'elogio della créolité nelle Antille francesi* – A. Pagani, *Il turismo ungherese tra continuità e trasformazione* – F. Ranghieri, *Politiche del villaggio globale. Clima e* partnership building *nei Paesi in via di sviluppo* – S. Rinauro, *L'immagine della Repubblica federale tedesca tra gli emigranti italiani negli anni Settanta* – G. Rocca, *Il turismo nell'isola di Wight: una verifica geostorica del modello di Miossec* – G. Roditi, *La cultura come motore di sviluppo della città contemporanea* – G. Scaramellini, *La geografia culturale tra mondo materiale e costrutti della mente. Alla ricerca di una realtà complessa e profonda* – M. Schmidt di Friedberg, *Paesaggi giapponesi: da* meisho *a* šatoyama – A. Treves, *Immagini e politiche della popolazione nell'Italia del secondo dopoguerra* – A. Turco, *Sokun. Il villaggio reticolare come struttura territoriale in Africa subsahariana* – M. Vercesi, *L'uso degli spazi extra-domestici da parte dei bambini. Tre generazioni a confronto in un quartiere di Milano* – A. Violante, *I vecchi ponti della Bosnia ed Erzegovina, simboli di identità e di nuove separazioni* – M.C. Zerbi, *Turismo sostenibile e geo-turismo: approcci della geografia*
2006, pp. XXIV + 624 ill.

82. *Vincenzo Monti nella cultura italiana.* Volume II. *Monti nella Roma di Pio VI* (a cura di Gennaro Barbarisi)
G. Barbarisi, *Presentazione* – VINCENZO MONTI NELLA ROMA DI PIO VI (Roma 27-29 ottobre 2005) a cura di Vincenzo De Caprio – P. Palmieri, *Vincenzo Monti e la Scuola erudita romagnola* – E. Graziosi, *Recitare in Arcadia: ragioni di un successo* – A. Nacinovich, *Monti e le poetiche arcadiche. Gli esordi di Autonide Saturniano* – F. Fedi, *«Il midollo dell'immagine»: Monti e le prospettive teoriche del neoclassicismo "romano"* – E. Parrini Cantini, *Monti e Vannetti* – L. Martellini, *Testi religiosi nel* Saggio di poesie *del 1779* – M. Sarnelli, *La Prosopopea di Pericle in Arcadia e oltre* – F. Grazzini, *Intorno alla* Feroniade: *Monti (con altri) e il tema delle paludi pontine* – E. Guagnini, *Monti e la poesia scientifica* – W. Spaggiari, *Monti e Metastasio* – L. Strappini, Aristodemo. *Varianti di autori* – M. Giammarco, Galeotto Manfredi. *Suggestioni shakespeariane* – A. Romano, *In margine alle polemiche romane del Monti (1778-1797)* – M. Caffiero, *Vincenzo Monti e la* Bassvilliana – M. Formica, *Misogallismi montiani* – U. Carpi, *Il cantore di Bassville* – L. Danzi, *La rima nella* Bassvilliana – L. Frassineti, *Il poeta e il* philosophe: *la fuga da Roma nel rapporto biografico-culturale Monti-de Azara* – Indice dei nomi
2006, pp. XII + 440

83. Dipartimento di Scienze dell'Antichità, *L'ufficio e il documento. I luoghi, i modi, gli strumenti dell'amministrazione in Egitto e nel Vicino Oriente antico* (a cura di Clelia Mora e Patrizia Piacentini)
O. Carruba, C. Mora, P. Piacentini, *Premessa* – BILANCIO E PROSPETTIVE DI UNA RICERCA. PER UN'INTRODUZIONE AL CONVEGNO – P. Piacentini, *Attualità delle ricerche sull'amministrazione antico-egiziana. Riflessioni metodologiche e bibliografiche* –

Produzione e conservazione della polvere da sparo nel XVI secolo: il caso veneziano – C. Zwierlein, *Fonti per una storia delle percezioni: i diari di guerra nel XVI secolo (il caso dei partecipanti alle guerre di religione in Francia)* – L. Porto, *La partecipazione dei veronesi alla difesa dello Stato veneziano nel Seicento* – D. Maffi, *Nobiltà e carriera delle armi nella Milano di Carlo II (1665-1700)* – G. Candiani, *La gestione degli equipaggi nei vascelli veneziani tra Sei e Settecento* – Indice dei nomi – Indice dei personaggi
2006, pp. 228

85. *Vincenzo Monti nella cultura italiana.* Volume III. *Monti nella Milano napoleonica e post-napoleonica* (a cura di Gennaro Barbarisi e William Spaggiari)
G. Barbarisi, *Presentazione* – A. Colombo, *Il cittadino Vincenzo Monti nella Milano cisalpina fra utopie politiche e radicalismi ideologici* – N. Mineo, *La carriera di Vincenzo Monti nella testimonianza delle lettere: tra Cispadana e Cisalpina* – L. Sozzi, *Monti traduce Voltaire* – U. Carpi, *Milano 1802: il Caio Gracco di Vincenzo Monti e il Caio Gracco di Ermolao Federigo* – C. Capra, *Intellettuali e potere nell'età napoleonica* – D. Tongiorgi, *«Né io amo d'essere il Cherilo d'Alessandro». Monti poeta del Governo Italiano* – M.A. Terzoli, *Monti e l'iconografia celebrativa napoleonica: riflessioni sulla* Visione per Napoleone Re d'Italia – B. Capaci, *In soccorso al vincitore. L'epidittico nel* Prometeo *di Vincenzo Monti* – A. De Francesco, *I* Pittagorici – P. Bosisio, *Un poeta al servizio di un nuovo modello di spettacolo: le cantate di Vincenzo Monti* – S. Garau, *Dediche di Vincenzo Monti* – A. Scardicchio, *Tumulti e insurrezioni nel Principato di Vincenzo Monti. La polemica con Francesco Gianni (con documenti inediti)* – G. Melli, *L'elogio della civiltà ne* Le nozze di Cadmo e d'Ermione *di Vincenzo Monti* – M. Palumbo, *Punti di vista sulla mitologia: il* Sermone *di Monti e le riflessioni di Foscolo* – A. Bruni, *Un nuovo manoscritto dell'*Iliade *di Vincenzo Monti* – F. Gorreri, *Il testo della* Mascheroniana – L. Frassineti, *Per il testo della* Feroniade *(con documenti inediti)* – R. Zucco, *Per un saggio sull'ordine delle parole nella* Feroniade – F. Longoni, *Testo e note della* Musogonia – A. Bentoglio, *Vincenzo Monti e il linguaggio del melodramma romantico: l'esempio di Felice Romani* – C. Chiancone, *Vincenzo Monti e la cultura veneta (con documenti inediti)* – J. Usher, *La risposta tardiva di W.S. Landor al sonetto anti-inglese «Luce ti nieghi il sole»* – Indice dei nomi
2006, pp. XII + 704 ill.

86. Dipartimento di Scienze del Linguaggio e Letterature Straniere Comparate, Sezione di Comparatistica, *Anglo-American Modernity and the Mediterranean* (edited by Caroline Patey, Giovanni Cianci and Francesca Cuojati)
Acknowledgements – C. Patey, *Foreword* – PART I, RECOLLECTIONS AND REVISITATIONS – J.B. Bullen, *W.B. Yeats, Byzantium and the Mediterranean* – M. Bacigalupo, *"Forth on the godly sea": The Mediterranean in Pound, Yeats and Stevens* – L. Colombino, *Negotiating with Gauguin's 'Solar Myth': Art, Economy and Ideology in Ford Madox Ford's Provence* – G. Ferruggia, *Edith Wharton's* Motor-Flight through France – C. Patey, *In the Mood for Provence, in the Heart of the Modern: Bloomsbury and Southern France* – M. Maffi, *Untender is the Night in the Garden of Eden: Fitzgerald, Hemingway,*

Finito di stampare nel mese di dicembre 2006 da Grafiche Speed 2000 snc,
Peschiera Borromeo (MI)